Semimodular Lattices

Lattice theory evolved as part of algebra in the nineteenth century through the work of Boole, Peirce, and Schröder, and in the first half of the twentieth century through the work of Dedekind, Birkhoff, Ore, von Neumann, Mac Lane, Wilcox, Dilworth, and others. In *Semimodular Lattices*, the author uses successive generalizations of distributive and modular lattices to outline the development of semimodular lattices from Boolean algebras. He focuses on the theory of semimodularity, its many ramifications, and its applications in discrete mathematics, combinatorics, and algebra.

The author surveys and analyzes Garrett Birkhoff's concept of semimodularity and the various related concepts in lattice theory, and he presents theoretical results as well as applications in discrete mathematics, group theory, and universal algebra. Special emphasis is given to the combinatorial aspects of finite semimodular lattices. The book also deals with lattices that are "close" to semimodularity or can be combined with semimodularity, for example, supersolvable, admissible, consistent, strong, and balanced lattices.

Researchers in lattice theory, discrete mathematics, combinatorics, and algebra will find this book valuable.

Manfred Stern, who lives in Halle, Germany, is a freelance lecturer, author, and translator.

Semimodular Lattices

Theory and Applications

MANFRED STERN

CAMBRIDGE UNIVERSITY PRESS
Cambridge, New York, Melbourne, Madrid, Cape Town, Singapore, São Paulo, Delhi

Cambridge University Press
The Edinburgh Building, Cambridge CB2 8RU, UK

Published in the United States of America by Cambridge University Press, New York

www.cambridge.org
Information on this title: www.cambridge.org/9780521118842

First published 1999
This digitally printed version 2009

A catalogue record for this publication is available from the British Library

Library of Congress Cataloguing in Publication data
Stern, Manfred.
Semimodular lattices: theory and applications. / Manfred Stern.
p.   cm. – (Encyclopedia of mathematics and its applications ;
v. 73)
Includes bibliographical references and index.
I. Title.   II. Series.
QA171.5.S743   1999
511.3′3 – dc21   98-44873
                                                                 CIP

ISBN 978-0-521-46105-4 hardback
ISBN 978-0-521-11884-2 paperback

Dedicated to the memory of
Garrett Birkhoff (1911–1996) and Robert P. Dilworth (1914–1993)

Contents

Preface

This book aims at giving a survey of semimodularity and related concepts in lattice theory as well as presenting a number of applications. The book may be regarded as a supplement to certain aspects of vol. 26 (*Theory of Matroids*), vol. 29 (*Combinatorial Geometries*), and vol. 40 (*Matroid Applications*) of this encyclopedia.

Classically semimodular lattices arose out of certain closure operators satisfying what is now usually called the Steinitz–Mac Lane exchange property. Inspired by the matroid concept introduced in 1935 by Hassler Whitney in a paper entitled "On the abstract properties of linear dependence," Garrett Birkhoff isolated the concept of semimodularity in lattice theory. Matroids are related to geometric lattices, that is, to semimodular atomistic lattices of finite length. The theory of geometric lattices was not foreshadowed in Dedekind's work on modular lattices. Geometric lattices were the first class of semimodular lattices to be systematically investigated.

The theory was developed in the thirties by Birkhoff, Wilcox, Mac Lane, and others. In the early forties Dilworth discovered locally distributive lattices, which turned out to be important new examples of semimodular lattices. These examples are the first cryptomorphic versions of what became later known in combinatorics as antimatroids. The name *antimatroid* hints at the fact that this combinatorial structure has properties that are very different from certain matroid properties. In particular, antimatroids have the so-called antiexchange property. While matroids abstract the notion of linear independence, antimatroids abstract the notion of Euclidean convexity. The theory was further developed by Dilworth, Crawley, Avann, and others in the fifties and sixties.

During the years 1935–55 and later many ramifications and applications of semimodularity were discovered. It was my intention to survey these results and to give at least some indication of the many places where semimodularity and/or related conditions occur. By a condition related to semimodularity I mean a condition equivalent to semimodularity in lattices of finite length. If the hypothesis of finite length is dropped, the concept of semimodularity "ramifies in a most unpleasant way," as Birkhoff states in the 1967 edition of his *Lattice Theory*. For example, in 1953 Croisot discussed the interdependence of 23 such ramifications. Some of these conditions (in particular the conditions found by Wilcox, Mac Lane, and Dilworth) turned out to be extremely useful: they could be considered adequate substitutes for semimodularity in lattices having continuous chains.

It was also my aim to point out the interrelationships between these and other conditions (cf. e.g. Sections 3.2, 3.3, 4.6) and to show what classes of lattices there are "close" to semimodularity or can be combined with semimodularity (see Chapters 4 and 5). For example, Stanley's supersolvable lattices may or may not be semimodular. Other classes of lattices that may or may not be semimodular are Kung's consistent lattices (whose origins go back to Crawley's work of 1961), Faigle's strong lattices of 1980, and the related balanced lattices introduced by Reuter in 1989.

Many proofs have had to be omitted or were only briefly sketched. In particular, results that have already appeared in other books have not been given proofs in general.[1] In some places this has led to a disproportion in the length of the presentation. For example, the characterization of local distributivity given by Dilworth and Crawley for strongly atomic algebraic lattices is touched upon only very briefly and without proof in Section 7.1 [a detailed proof can be found in the book *Algebraic Theory of Lattices* by Crawley and Dilworth (1973)]. On the other hand, the most important steps of Avann's characterization for the finite case are given in some detail in Section 7.2.

My approach also results in some repetitions as well as in a certain overlap between chapters. The presentation is not deductive. For example, Chapter 6 gives a number of major results for the finite-length case and, in particular, for the finite case. However, specific results on the finite-length case and on the finite case are also given in the other chapters.

Let me now briefly review the chapters of this book.

Chapter 1 traces the way leading from Boolean lattices to upper and lower semimodular lattices by successive generalizations via distributive and modular lattices. Apart from modularity, semidistributivity is the most significant generalization of distributivity. Free lattices are the most important examples of semidistributive lattices, but semidistributivity also appears in the congruence lattices of meet semilattices (these congruence lattices are semimodular) and it plays an important role in the investigation of lattice varieties. We have a look at various complementedness conditions and at the existence of meet decompositions and irredundant meet decompositions in certain lattices of infinite length. We mention the facts with later applications elsewhere in mind. In the section on upper and

---

[1] When finishing this book, the author did not yet possess a copy of the second, completely revised edition of the *General Lattice Theory* by Grätzer (1998).

lower semimodularity we give a number of fundamental properties and examples. Important examples of semimodular lattices are Birkhoff's geometric lattices and Dilworth's locally distributive lattices. We close the chapter with results on the Jordan–Dedekind chain condition.

Chapter 2 deals with Wilcox's criticism of Birkhoff's concept of semimodularity. Wilcox's aim was to find a suitable condition equivalent to semimodularity in lattices of finite length but substituting semimodularity in arbitrary lattices. Such a condition is Wilcox's concept of the symmetry of modular pairs, M-symmetry, which he discovered in 1938–39. We also mention related concepts of symmetry and their interdependence. In particular we consider AC lattices, that is, atomistic lattices satisfying the covering property. We close the chapter by discussing some specific properties of atomistic Wilcox lattices, M-symmetric AC lattices, orthomodular M-symmetric lattices, and complements in finite-modular AC lattices.

Chapter 3 deals with Mac Lane's criticism of Birkhoff's concept of semimodularity. In 1936 this criticism led Mac Lane to a condition that is related to semimodularity but goes in another direction than Wilcox's M-symmetry. Mac Lane's condition was refined by Dilworth in 1941. Still other conditions related to semimodularity were found by Croisot in 1951. We describe the interdependence of these conditions. Some of them coincide in lattices satisfying the ascending chain condition, while others coincide in lattices satisfying the descending chain condition. As already indicated, for lattices of finite length all these conditions turn out to be equivalent, that is, they characterize semimodularity in these lattices. We then have a look at local versions of modularity at 0 (0-modularity), Mac Lane's condition at 0, the notion of $\perp$-symmetry (M-symmetry at 0), and semimodularity at 0 (which is simply the covering property). Further extensions of these "0-properties" are the so-called disjointness properties. We investigate the implications between these concepts as well as the reversibility of implications for lattices carrying some kind of complementation.

In Chapter 4 we recall some facts concerning the Möbius function on partially ordered sets. We briefly indicate the connection between complements and fixed points as well as some results of Stanley, Björner, and others dealing with supersolvable lattices and admissible lattices. Supersolvable lattices generalize modularity in another direction than upper semimodularity. Both finite supersolvable lattices and finite upper semimodular lattices are admissible lattices. Admissible lattices are instances of shellable posets, which in turn are Cohen–Macaulay posets. In other words, Cohen–Macaulay posets may be viewed as a far-reaching generalization of finite semimodular lattices. Still other generalizations of modularity are consistency (which can be traced back to Crawley) and strongness (which has its roots in Faigle's work). These concepts go in another direction than upper semimodularity or supersolvability. Consistency and strongness (which will also be considered in Sections 6.5 and 8.2) prove to be particularly interesting when combined with upper or lower semimodularity. For example, we present Reuter's method of obtaining all finite consistent semimodular lattices by suitably gluing finite geometric lattices. This approach generalizes Herrmann's S-glued sums of lattices, which is itself an extension of Dilworth's gluing. The results of Faigle and Reuter show, in particular, that consistency and strongness have the same meaning for semimodular lattices of finite length. Consistency and strongness will also be considered in Sections 6.5 and 8.2.

In Chapter 5 we investigate the covering graph of posets and, in particular, graph isomorphisms of graded balanced lattices (a balanced lattice is a lattice that is – together with its dual – strong in the sense of Faigle). The point of departure is Problem 8 from the 1948 edition of Birkhoff's *Lattice Theory*. More generally, Oystein Ore formulated in 1962 the problem of characterizing those finite unoriented graphs G for which there exists a poset P such that the covering graph of P, $G(P)$, is graph isomorphic with G. This problem is still one of the major unsolved problems of order theory. For lattices answers have been given for example in the distributive, in the modular, and in the geometric cases. Within the class of graded and balanced lattices we give a common framework for the results obtained by Jakubík (in the modular case) and Duffus and Rival (in the geometric case). We briefly mention the results of Ratanaprasert and Davey on semimodular lattices with isomorphic covering graphs (who thus gave an affirmative answer to a problem posed by Jakubík) and deal with centrally symmetric graphs and lattices. Ward proved in 1939 that a modular lattice of finite length is distributive if and only if its covering graph contains no subgraph isomorphic to the covering graph of the diamond. We close the chapter by indicating an analogue of Ward's theorem given by Duffus and Rival for finite dismantlable semimodular lattices.

Chapter 6 deals with a selected number of important results on semimodular lattices of finite length and, in particular, with the finite case. First we look at Kung's approach to the matching problem for finite lattices, which leads to a number of new rank and covering inequalities. For example, Kung generalized the Dowling–Wilson inequalities from finite geometric lattices to finite semimodular lattices. The approach he developed enabled him both to prove a strengthening of Dilworth's covering theorem for finite modular lattices and to give an affirmative answer to Rival's matching conjecture for finite modular lattices. We will then give some of the results on lattice embeddings going back to Dilworth and Finkbeiner and describe the main steps of the Grätzer–Kiss approach to isometric embeddings. We have a look at Faigle's geometric closure operators and at Faigle's geometries on partially ordered sets. Certain strings of elements in a Faigle geometry form a selector as introduced by Crapo. Selectors (which are special greedoids in the sense of Korte and Lovász) were also used by Crapo to obtain a representation for finite semimodular lattices. We close the chapter by giving some more results on consistent semimodular lattices as well as results on pseudomodular lattices by Björner and Lovász.

Chapter 7 deals with local distributivity in the sense of Dilworth. We briefly indicate the results obtained by Crawley and Dilworth for strongly atomic algebraic lattices and consider in more detail Avann's approach to the finite case, which leads to a very broad characterization theorem including earlier results of Dilworth. We have a look at applications to abstract convexity as given by Edelman and Jamison. Finally we give a number of further characterizations of local distributivity in the finite case, including Avann's upper splitting property and Crapo's representation via locally free selectors.

Chapter 8 gives further results on local modularity by Dilworth and Crawley. First we indicate that for strongly atomic algebraic lattices there is a characterization of the Kurosh–Ore replacement property that is analogous to the finite case. We then give some results of Gragg and Kung, as well as applications to lattices of subnormal subgroups, and close the chapter with a combinatorial characterization of modularity among finite consistent lower semimodular lattices.

Chapter 9 reviews some results on congruence semimodularity, a concept which has gained in importance since the discovery that partition lattices and congruence lattices of semilattices are semimodular. Some results concerning congruence lattices of semilattices are given in more detail. A systematic investigation of the role of congruence semimodularity in semigroups was carried out by Jones. Agliano and Kearnes extended these investigations to algebras. These results – which could only very briefly be touched upon – rely heavily on methods of semigroup theory and universal algebra developed during the last years.

The book is based on my Teubner-Text *Semimodular Lattices* (Teubner-Verlag 1991, Stuttgart). This text was substantially revised, enlarged, and updated.

I would like to thank several persons and institutions for their generous support. First of all, I am grateful to Professor Gian-Carlo Rota, the Series Editor, for suggesting and encouraging this project. Thanks are due to Lauren Cowles, Cambridge University Press, New York, and Andrew Wilson, TechBooks, Fairfax, Virginia, for their patience and answering numerous technical questions.

I would like to thank the University of Jyväskylä for giving me the opportunity of staying there on several occasions and the University of Bielefeld for one stay in 1991. I would like to express my gratitude to the Consiglio Nazionale di Ricerche of Italy for an invitation for two months to the University of Milan and to the Politecnico Milan in 1993. I am also grateful to the CNRS of France for an invitation for one month to the Laboratoire de Mathématiques Discrètes in Marseille-Luminy in 1994. During these stays I worked on parts of the manuscript and I would like to express my sincerest thanks to Mikko Saarimäki (Jyväskylä), Andreas Dress (Bielefeld), Celestina Bonzini, Alessandra Cherubini and Daniele Mundici

(Milan), Gérard Rauzy (Marseille), and Bertram Scharf (Boston and Marseille) for their support. I am grateful to Wolfgang Vogel (1940–1996).

I acknowledge to having freely borrowed from the sources cited in the references. It would not be possible to mention all those by name who helped me in one way or another answering my questions and with their remarks and comments. For numerous remarks and comments over many years I would like to thank Garrett Birkhoff, Ulrich Faigle, Shuichiro Maeda, Gerd Richter, and Lee Roy Wilcox.

I also thank Sherwin P. Avann, Mary Katherine Bennett, Gerhard Betsch, Melvin F. Janowitz, Jürgen Köhler, Bruno Leclerc, Léonce Lesieur, Leonid Libkin, Werner Loch, Rudolf Maier, Eli Maor, Heinz Mitsch, Bernard Monjardet, Nimbakrishna Thakare, and last but not least Meenakshi P. Wasadikar.

Halle an der Saale
February 1999

Manfred Stern
manfred@stern.hal.uunet.de
Kiefernweg 8
D-06120 Halle

1

From Boolean Algebras
to Semimodular Lattices

1.1 Sources of Semimodularity

Summary. We briefly indicate some milestones in the general development of lattice theory. In particular, we outline the way leading from Boolean algebras to semimodular lattices. Most of the concepts mentioned in this section will be explained in more detail later. We also give a number of general references and monographs on lattice theory and its history.

Boolean Algebras and Distributive Lattices

Lattice theory evolved in the nineteenth century through the works of George Boole, Charles Saunders Peirce, and Ernst Schröder, and later in the works of Richard Dedekind, Garrett Birkho21 , Oystein Ore, John von Neumann, and others during the first half of the twentieth century. Boole [1847] laid the foundation for the algebras named after him. Since then the more general distributive lattices have been investigated whose natural models are systems of sets. There are many monographs on Boolean algebras and their applications, such as Halmos [1963] and Sikorski [1964]. For the theory of distributive lattices we refer to the books by Grätzer [1971] and Balbes & Dwinger [1974].

Modular Lattices

Dedekind [1900] observed that the additive subgroups of a ring and the normal subgroups of a group form lattices in a natural way (which he called *Dualgruppen*) and that these lattices have a special property, which was later referred to as the *modular law*. Modularity is a consequence of distributivity, and Dedekind's observation gave rise to examples of nondistributive modular lattices.

Lattice theory became established in the 1930s due to the contributions of Garrett Birkhoff, Ore, Menger, von Neumann, Wilcox, and others.

In a series of papers Ore generalized the classical results of Dedekind and investigated decomposition theorems known from algebra in the context of modular lattices (cf. in particular Ore [1935], [1936]). Kurosch [1935] published a note in

1

which he proved a result, independently of Ore and at the same time, that became later known as the Kurosh–Ore theorem. The motivation behind this work was the conjecture that the corresponding result for commutative rings (which was proved by Noether [1921]) could also be proved in the noncommutative case. Kurosh and Ore reduced Noether's theorem to its basic ingredients. However, their work amounted to more than a mere repetition of the former proof in the more general framework of modular lattices. Modularity is the best-known generalization of distributivity. Distributive and modular lattices are dealt with in all books on lattice theory and universal algebra. There are several monographs treating modular lattices within the framework of continuous geometries (von Neumann [1960], Maeda [1958], and Skornjakov [1961]).

Semimodular Lattices

An important source of examples leading to lattices is based on the idea of considering various collections of points, lines, planes, etc., as geometrical "configurations." For example, projective incidence geometries lead to complemented modular lattices: the lattices of flats or closed subspaces of the geometry. However, if one considers affine incidence geometries, the corresponding lattices of flats are no longer modular, although they retain certain important features of complemented modular lattices. These lattices are special instances of so-called geometric lattices. Properties of lattices of this kind were studied by Birkhoff [1933], [1935a]. During the years 1928–35 Menger and his collaborators independently developed ideas that are closely related (see Menger [1936]). Birkhoff's work [1935b] was inspired by the matroid concept introduced by Whitney [1935] in a paper entitled "On the abstract properties of linear dependence." A matroid is a finite set endowed with a closure operator possessing what is now usually called the Steinitz–Mac Lane exchange property. Matroid theory has developed into a rich and flourishing subject. Crapo & Rota [1970a] present in the introduction to their book a survey of the development of matroid theory and geometric lattices. For more details see also Crapo & Rota [1970b] and Kung [1986b]. For a comprehensive account of older and more recent developments in matroid theory, including numerous contributions and historical notes on the relationship with lattice theory and other fields, we refer to the three volumes White [1986], [1987], [1992].

As Garrett Birkhoff stated, the theory of geometric lattices was not foreshadowed in Dedekind's work. Geometric lattices are atomistic lattices of finite length satisfying the *semimodular implication*

(Sm) If $a \wedge b$ is a lower cover of a, then b is a lower cover of $a \vee b$.

We shall call a lattice (of finite length or not) *upper semimodular* or simply *semimodular* if it satisfies the implication (Sm). Birkhoff originally introduced another condition, namely

(Bi) If $a \wedge b$ is a lower cover of a and b, then a and b are lower covers of $a \vee b$.

In lattices of finite length both conditions are equivalent, but in general *Birkhoff's condition* (Bi) is weaker than (Sm). This is why lattices satisfying (Bi) are sometimes called *weakly semimodular*. In our use of the word *semimodular* we have adopted the terminology used by Crawley & Dilworth [1973]. We remark, however, that the notion *semimodular* has also been used to denote other related conditions, some of which will be considered later. The name *semimodular* was coined by Wilcox [1939].

From Dedekind's isomorphism theorem for modular lattices it is immediate that any modular lattice is semimodular. On the other hand, matroids lead to semimodular lattices that are not modular in general. The implication reversed to (Sm) will be denoted by (Sm*). Lattices satisfying (Sm*) are called *lower semimodular* or *dually semimodular*. For lattices of finite length, (Sm) together with (Sm*) yields modularity. In this sense semimodularity is indeed just "one half" of modularity. However, an upper and lower semimodular lattice of infinite length need not be modular. An example is provided by the orthomodular lattice of closed subspaces of an infinite-dimensional Hilbert space.

Conditions Related to Semimodularity

The semimodular implication (Sm) and Birkhoff's condition (Bi) are stated in terms of the covering relation. Hence they only trivially apply to infinite lattices with continuous chains. For example, the lattice of projection operators of a von Neumann algebra has no atoms; it therefore trivially satisfies (Sm) and could formally be cited as an example of a semimodular lattice. However, not much insight is gained from this observation.

Wilcox and Mac Lane were the first to introduce conditions that do not involve coverings and that may be considered as substitutes for the semimodular implication (Sm) in arbitrary lattices.

Wilcox [1938], [1939] showed that affine geometry as developed algebraically by Menger can be axiomatized without the use of points (now usually called atoms). Wilcox's central concept is the symmetry of modular pairs, called *M-symmetry* and briefly denoted by (Ms). This notion came to play a decisive role in nonmodular lattices. Looking back Wilcox (1988, personal communication) wrote:

> The affine geometries seemed a natural place to start. Karl Menger had done some work here, but not in a way that would lend itself to generalization. As I recall my approach which led naturally to the idea of modular pairs, I noted the obvious fact that the failure of an affine geometry (as a lattice) to be modular stems from the presence of parallel pairs, i.e. pairs whose meets

are "too small".... Parallel pairs may be thus viewed as non-modular pairs. As to the symmetry of modularity, I noted that in the affine case parallelism is symmetric, i.e. non-modularity of pairs is symmetric, so that modularity of pairs is also symmetric. Hence I began to look for generalizations of affine geometries in which modularity was symmetric.

We have already stated that the lattice of projection operators of a von Neumann algebra is trivially upper semimodular. Topping [1967] proved the highly nontrivial result that this lattice is M-symmetric. We shall return to M-symmetry on several occasions, especially in Chapter 2.

The condition introduced by Mac Lane [1938] and briefly denoted by (Mac) is somewhat more complicated (for details see Section 3.1). In his investigations on "exchange lattices" Mac Lane was led to this condition when looking for "point-free" substitutes for an axiom due to Menger. For more details on the background of these investigations see Mac Lane [1976].

Wilcox's condition of M-symmetry (Ms) and Mac Lane's condition (Mac) are both consequences of modularity. On the other hand, neither (Ms) nor (Mac) implies modularity. Also, both (Ms) and (Mac) imply upper semimodularity, but not conversely. Moreover, (Ms) and (Mac) are independent of each other, that is, neither of these conditions implies the other. However, for lattices of finite length, the conditions (Sm), (Bi), (Ms), and (Mac) are all equivalent.

By a *condition related to semimodularity* I mean a condition that is equivalent to upper semimodularity for lattices of finite length. In this sense Birkhoff's condition (Bi), Wilcox's condition (Ms) of M-symmetry, and Mac Lane's condition (Mac) are conditions related to semimodularity.

Figure 1.1 visualizes the interrelationships between the classes of lattices mentioned before. An arrow indicates proper inclusion, that is, if X and Y are classes of lattices, then X → Y means X ⊂ Y.

Other conditions related to semimodularity were discovered by Dilworth and Croisot; these conditions will be considered in Section 3.2.

We shall insert inclusion charts in many places. In particular, we shall give inclusion charts refining Figure 1.1. In these inclusion charts arrowheads will sometimes be omitted with the understanding that, where two concepts are connected by an ascending line, the "lower" concept implies the "upper" one.

Local Distributivity and Local Modularity

The early papers by Dilworth [1940], [1941a] were further milestones and important sources of semimodularity. Many of the decomposition theorems in algebra had already been extended to the more general domains of distributive lattices and modular lattices in the 1930s, for example in the above-mentioned works of Ore and Kurosh. Dilworth observed that there are lattices with very simple arithmetical

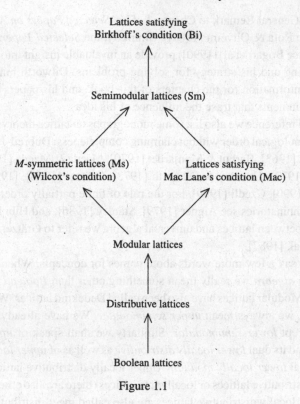

Lattices satisfying
Birkhoff's condition (Bi)

Semimodular lattices (Sm)

M-symmetric lattices (Ms)   Lattices satisfying
(Wilcox's condition)   Mac Lane's condition (Mac)

Modular lattices

Distributive lattices

Boolean lattices

Figure 1.1

properties that come under neither of these classifications. For example, the lattices
with unique irreducible meet decompositions that were considered by Dilworth
[1940] are upper semimodular. This paper led to the concept of *local distributivity*
and marked the origin of a combinatorial structure that is nowadays also referred to
as an *antimatroid*. More generally, Dilworth [1941a] investigated *local modularity*; this was the first paper dealing with an extension of the Kurosh–Ore theorem
from modular lattices to semimodular lattices.

Notes

Let us first mention some further sources of lattice theory and its history. Mehrtens
[1979] gives a detailed account of the development of lattice theory from the very
beginnings until about 1940 (for a comprehensive review of Mehrtens's book see
Dauben [1986]). Much information on logics as a source of lattice theory can be
found in Chapter 2 of Mangione & Bozzi [1993]. An excellent source for tracing
the historical development of lattice theory in general and of semimodular lattices
in particular is the three editions of Birkhoff's treatise *Lattice Theory* (Birkhoff
[1940a], [1948], [1967]). Details on the early history of lattice theory can be found

in Birkhoff's General Remark to Chapter I of his *Selected Papers on Algebra and Topology* (see Rota & Oliveira [1987]). Similarly, the *Selected Papers of Robert P. Dilworth* (see Bogart et al. [1990]) provide an invaluable insight into Dilworth's way of thinking and his strategy for solving problems. Dilworth himself wrote background information for the chapters of this book, and his papers are supplemented by comments that trace the influence of his ideas.

For general reference we also list some monographs on lattice theory and related topics in chronological order without claiming completeness: Dubreil-Jacotin et al. [1953], Szász [1963], Barbut & Monjardet [1970], Maeda & Maeda [1970], Blyth & Janowitz [1972], Crawley & Dilworth [1973], Grätzer [1978], [1998], Davey & Priestley [1990], Czédli [1999]. For the role of finite partially ordered sets and lattices in combinatorics see Aigner [1979], Stanley [1986], and Hibi [1992]. For the interplay between lattices and universal algebra we refer to Grätzer [1979] and McKenzie et al. [1987].

Let us also say a few more words about names for concepts. When we use the term *Boolean algebra* we really mean something other than *Boolean lattice* (see Section 1.2). Modular lattices have also been called Dedekind lattices. When we say *semimodular*, we always mean *upper semimodular*. We have already mentioned the dual concept *lower semimodular*. Similarly we shall speak of *upper locally distributive* and its dual *lower locally distributive* as well as of *upper locally modular* and its dual *lower locally modular*. Upper locally distributive lattices are also called join-distributive lattices or locally free lattices (there are also other names for them). Lower locally distributive lattices are also called meet-distributive lattices.

References

Aigner, M. [1979] *Combinatorial Theory*, Springer-Verlag, Berlin.

Balbes, R. and P. Dwinger [1974] *Distributive Lattices*, Univ. of Missouri Press, Columbia, Mo.

Barbut, M. and B. Monjardet [1970] *Ordre et classification*, Vols. 1, 2, Hachette, Paris.

Birkhoff, G. [1933] On the combination of subalgebras, Proc. Camb. Phil. Soc. 29, 441–464.

Birkhoff, G. [1935a] Abstract linear dependence and lattices, Amer. J. Math. 57, 800–804.

Birkhoff, G. [1935b]

Birkhoff, G. [1940a] *Lattice Theory*, Amer. Math. Soc. Colloquium Publications, Vol. 25, New York.

Birkhoff, G. [1948] *Lattice Theory* (2nd edition), Amer. Math. Soc. Colloquium Publications, Vol. 25, New York.

Birkhoff, G. [1967] *Lattice Theory* (3rd edition), Amer. Math. Soc. Colloquium Publications, Vol. 25, Providence, R.I.; reprinted 1984.

Birkhoff, G. [1987] General remark to Chapter I, in: Rota & Oliveira [1987], pp. 1–8.

Blyth, T. S. and M. F. Janowitz [1972] *Residuation Theory*, Pergamon Press, Oxford.

Bogart, K. P., Freese, R., and Kung, J. P. S. (eds.) [1990] *The Dilworth Theorems. Selected Papers of Robert P. Dilworth*, Birkhäuser, Boston.

Boole, G. [1847] *The Mathematical Analysis of Logic. Being an Essay Towards a Calculus of Deductive Reasoning*, Cambridge, reprinted, Oxford, 1951.

Crapo, H. H. and G.-C. Rota [1970a] *On the Foundations of Combinatorial Theory: Combinatorial Geometries*, MIT Press, Cambridge, Mass.

Crapo, H. H. and G.-C. Rota [1970b] Geometric lattices, in: *Trends in Lattice Theory*, (ed. J. C. Abbott), van Nostrand-Reinhold, New York, pp. 127–172.

Crawley, P. and R. P. Dilworth [1973] *Algebraic Theory of Lattices*, Prentice-Hall, Englewood Cliffs, N.J.

Czédli, G. [1999] *Lattice Theory* (Hungarian), JATEPress, Szeged.

Dauben, J. W. [1986] Review of Mehrtens [1979], Order 3, 89–102.

Davey, B. A. and H. A. Priestley [1990] *Introduction to Lattices and Order*, Cambridge Univ. Press, Cambridge.

Dedekind, R. [1900] Über die von drei Moduln erzeugte Dualgruppe, Math. Ann. 53, 371–403, Gesammelte Werke, Vol. 2, pp. 236–271.

Dilworth, R. P. [1940] Lattices with unique irreducible decompositions, Ann. of Math. 41, 771–777.

Dilworth, R. P. [1941a] The arithmetical theory of Birkhoff lattices, Duke Math. J. 8, 286–299.

Dubreil-Jacotin, M. L., L. Lesieur, and R. Croisot [1953] *Leçons sur la théorie des treillis, des structures algébriques ordonnées et des treillis géométriques*, Gauthier-Villards, Paris.

Grätzer, G. [1971] *Lattice Theory: First Concepts and Distributive Lattices*, Freeman, San Francisco.

Grätzer, G. [1978] *General Lattice Theory*, Birkhäuser, Basel.

Grätzer, G. [1979] *Universal Algebra* (2nd edition), Springer-Verlag, New York.

Grätzer, G. [1998] *General Lattice Theory*, Birkhäuser, Basel. Second, completely revised edition.

Halmos, P. R. [1963] *Lectures on Boolean Algebras*, Van Nostrand, Princeton, N.J.

Hibi [1992] *Algebraic Combinatorics on Convex Polytopes*, Carslaw, Glebe, Australia.

Kung, J. P. S. [1986a] *A Source Book in Matroid Theory*, Birkhäuser, Boston.

Kurosch, A. G. [1935] Durchschnittsdarstellungen mit irreduziblen Komponenten in Ringen und in sogenannten Dualgruppen, Mat. Sbornik 42, 613–616.

Mac Lane, S. [1938] A lattice formulation for transcendence degrees and p-bases, Duke Math. J. 4, 455–468.

Mac Lane, S. [1976] Topology and logic as a source of algebra, Bull. Amer. Math. Soc. 82, 1–40.

Maeda, F. [1958] *Kontinuierliche Geometrien*, Springer-Verlag, Berlin.

Maeda, F. and S. Maeda [1970] *Theory of Symmetric Lattices*, Springer-Verlag, Berlin.

Mangione, C. and S. Bozzi [1993] *Storia della Logica da Boole ai Nostri Giorni*, Garzanti Editore, Milano.

McKenzie, R., G. F. McNulty, and W. Taylor [1987] *Algebras, Lattices, Varieties*, Vol. 1, Wadsworth & Brooks/Cole, Monterey, Calif.

Mehrtens, H. [1979] *Die Entstehung der Verbandstheorie*, Gerstenberg, Hildesheim.

Menger, K. [1936] New foundations of projective and affine geometry, Ann. of Math. 37, 456–482.

Noether, E. [1921] Idealtheorie in Ringbereichen, Math. Ann. 83, 24–66.

Ore, O. [1935] On the foundations of abstract algebra I, Ann. of Math. 36, 406–437.

Ore, O. [1936] On the foundations of abstract algebra II, Ann. of Math. 37, 265–292.

Rota, G.-C. and J. S. Oliveira (eds.) [1987] *Selected Papers on Algebra and Topology by Garrett Birkhoff*, Birkhäuser, Boston.

Sikorski, R. [1964] *Boolean Algebras* (2nd edition), Academic Press, New York.

Skornjakov, L. A. [1961] *Complemented Modular Lattices and Regular Rings* (Russian), Gosudarst. Izdat. Fiz.-Mat. Lit., Moscow.

Stanley, R. [1986] *Enumerative Combinatorics I*, Wadsworth & Brooks/Cole, Monterey, Calif.

Szász, G. [1963] *Introduction to Lattice Theory*, Akadémiai Kiadó, Budapest, and Academic Press, New York.

Topping, D. M. [1967] Asymptoticity and semimodularity in projection lattices, Pacific J. Math. 20, 317–325.

von Neumann, J. [1960] *Continuous Geometry* (ed. I. Halperin), Princeton Mathematical Ser. 25, Princeton Univ. Press, Princeton, N.J.

White, N. L. (ed.) [1986] *Theory of Matroids*, Encyclopedia of Mathematics and Its Applications, Vol. 26, Cambridge Univ. Press, Cambridge.

White, N. L. (ed.) [1987] *Combinatorial Geometries*, Encyclopedia of Mathematics and Its Applications, Vol. 29, Cambridge Univ. Press, Cambridge.

White, N. L. (ed.) [1992] *Matroid Applications*, Encyclopedia of Mathematics and Its Applications, Vol. 40, Cambridge Univ. Press, Cambridge.

Whitney, H. [1935] On the abstract properties of linear dependence, Amer. J. Math. 57, 509–533.

Wilcox, L. R. [1938] Modularity in the theory of lattices, Bull. Amer. Math. Soc. 44, 50.

Wilcox, L. R. [1939] Modularity in the theory of lattices, Ann. of Math. 40, 490–505.

1.2 Boolean Lattices, Ortholattices, and Orthomodular Lattices

Summary.   We recall the definitions of Boolean lattices, distributive lattices, modular lattices, and orthocomplemented and orthomodular lattices. The notion of modular pair is introduced. Some important results and examples are given.

A lattice is called *distributive* if

(D) $c \vee (a \wedge b) = (c \vee a) \wedge (c \vee b)$

holds for all triples (a, b, c) of lattice elements. A lattice is distributive if and only if $c \wedge (a \vee b) = (c \wedge a) \vee (c \wedge b)$ holds for all triples (a, b, c) of lattice elements.

For the visual representation of posets and lattices we frequently use Hasse diagrams. The lattices in Figure 1.2(a), (d) are distributive, whereas the lattices in Figure 1.2(b), (c) are not. The lattices in Figure 1.2(b) and (c) will be denoted by M_3 and N_5, respectively.

Let us explain some more concepts. We say that x is a *lower cover* of y and we write $x \prec y$ if $x < y$ and $x \leq t < y$ implies $t = x$. Equivalently we say in this case that y is an *upper cover* of x and write $y \succ x$. If a lattice has a *least element*, denoted by 0, we also say that the lattice is *bounded below*. If a lattice has a *greatest element*, denoted by 1, we also say that the lattice is *bounded above*. A *bounded lattice* is a lattice having both a least element and a greatest element. In a lattice bounded below, an upper cover of the least element is called an *atom*. In Figure 1.2(c) the elements a and c are atoms, but b is not. A lattice bounded below is said to be *atomistic* if every of its elements ($\neq 0$) is a join of atoms. The lattice in Figure 1.2(b) is atomistic, but the lattices in Figure 1.2(a), (c), and (d) are not. In a lattice bounded below an element z ($\neq 0$) is called a *cycle* if the interval $[0, z]$ is a chain. Every atom is a cycle; in Figure 1.2(c) the element b is a

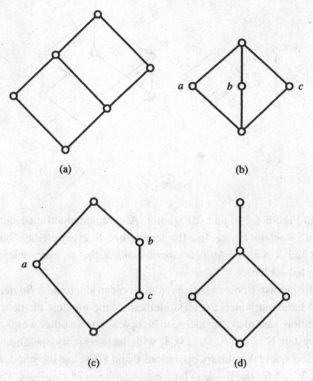

Figure 1.2

cycle that is not an atom. A lattice bounded below is called *cyclically generated* if every element ($\neq 0$) is a join of cycles. The lattices in Figure 1.2(a), (b), (c) are cyclically generated, but the lattice in Figure 1.2(d) is not. Any atomistic lattice is cyclically generated. The name *cycle* has its origin in the theory of abelian groups. Figure 1.2(a) is isomorphic to the lattice of all subgroups of the cyclic group Z_{18}. Figure 1.2(b) is isomorphic to the the subgroup lattice of the Klein 4-group.

A sublattice K of a lattice L is called a *diamond* or a *pentagon* if K is isomorphic to M_3 or N_5, respectively. Distributivity is characterized by the absence of diamonds and pentagons:

Theorem 1.2.1 *A lattice is distributive if and only if it does not contain a diamond or a pentagon.*

This is Birkhoff's distributivity criterion (Birkhoff [1934]). For more on distributive lattices see Section 1.3.

In a bounded lattice, an element $\bar{a}$ is a complement of a if $a \wedge \bar{a} = 0$ and $a \vee \bar{a} = 1$. A *complemented lattice* is a bounded lattice in which every element has a complement. The lattices of Figure 1.2(b) and (c) are complemented, whereas

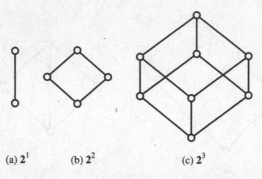

(a) 2^1 (b) 2^2 (c) 2^3

Figure 1.3

the lattices in Figure 1.2(a) and (d) are not. A complemented distributive lattice will be called a *Boolean lattice*. In a Boolean lattice B, every element has a unique complement and B is also *relatively complemented*, that is, every interval of B is a complemented sublattice.

We shall distinguish Boolean lattices from Boolean algebras. A *Boolean algebra* is a Boolean lattice in which the least element 0, the greatest element 1 and the complementation $^-$ are also considered to be operations. In other words, a Boolean algebra is a system $\mathbf{B} = \langle B, \wedge, \vee, ^-, 0, 1 \rangle$ with the two binary operations $\wedge$, $\vee$, the unary operation $^-$, and the nullary operations 0 and 1. We use the standard notation 2^n ($n = 1, 2, 3, \ldots$) for the Boolean lattice consisting of 2^n elements. The Boolean lattices 2^1, 2^2, and 2^3 are shown in Figure 1.3(a), (b), and (c), respectively.

Any distributive lattice obviously satisfies the implication

(M) $c \leq b \Rightarrow c \vee (a \wedge b) = (c \vee a) \wedge b$

for all elements a, b, c. A lattice is called *modular* if it satisfies (M) for all a, b, c. Modularity is the most important generalization of distributivity. The diamond M_3 [Fig. 1.2(b)] is modular but not distributive. The pentagon N_5 [Fig. 1.2(c)] is not modular. Modularity can be characterized by the absence of pentagons:

Theorem 1.2.2 *A lattice is modular if and only if it does not contain a pentagon.*

This characterization is due to Dedekind [1900]. Some more properties of modular lattices will be given in Section 1.6. However, let us note here that for some special classes of lattices, the forbidden-sublattice characterizations for modular and distributive lattices can be sharpened by showing the existence of very large or very small pentagons or diamonds.

For instance, a bounded relatively complemented nonmodular lattice always contains a pentagon as $\{0, 1\}$-sublattice. The same is true of the diamond in certain complemented modular lattices (von Neumann [1936–7]). If a lattice is finite and nonmodular, then the pentagon it contains can be required to satisfy $b \succ c$ [the notation referring to Fig. 1.2(c)]. The modularity criterion (Theorem 1.2.2) simplifies as follows in the case of complemented atomic lattices (a lattice L with 0 is called *atomic* if for every $x \in L$, $x \neq 0$, there exists an atom $p \in L$ such that $x \geq p$).

Theorem 1.2.3   *If a complemented atomic lattice contains no pentagon including both the least and greatest elements, then the lattice is modular.*

This theorem is due to McLaughlin [1956]. For a proof we also refer to Salii [1988], pp. 27–30, and to Dilworth [1982], pp. 333–353.

In nonmodular lattices we shall be interested in modular pairs and dual modular pairs, which were introduced by Wilcox [1938], [1939]. We say that an ordered pair (a, b) of elements of a lattice L is a *modular pair* and we write $a \, M \, b$ if, for all $c \in L$,

$$c \leq b \quad \text{implies} \quad c \vee (a \wedge b) = (c \vee a) \wedge b.$$

We say that (a, b) is a *dual modular pair* and we write $a \, M^* \, b$ if, for all $c \in L$,

$$c \geq b \quad \text{implies} \quad c \wedge (a \vee b) = (c \wedge a) \vee b.$$

If (a, b) is not a modular pair, then we write $a \, \bar{M} \, b$. It is clear that a lattice is modular if and only if every ordered pair of elements is modular. In the nonmodular lattice of Figure 1.2(c) we have $b \, M \, a$ but $a \, \bar{M} \, b$, which shows that the relation of being a modular pair is not symmetric in this lattice. Similarly, this example also shows that the relation of being a dual modular pair is not symmetric in general.

Let L be a lattice with 0 and 1. An *orthocomplementation* on L is a unary operation $a \to a^{\perp}$ on L satisfying the following three conditions:

(i) $a \wedge a^{\perp} = 0$, $a \vee a^{\perp} = 1$, that is, $a^{\perp}$ is a complement of a;

(ii) $a \leq b$ implies $b^{\perp} \leq a^{\perp}$;

(iii) $a^{\perp\perp} = a$ for every $a \in L$.

Note that $a^{\perp\perp}$ stands for $(a^{\perp})^{\perp}$. We call $a^{\perp}$ the *orthocomplement* of a. An *orthocomplemented lattice* (briefly: *ortholattice*, *OC lattice*) is a lattice with 0 and 1 carrying an orthocomplementation.

Any Boolean lattice is an ortholattice (the Boolean complement of an element being its orthocomplement). Two other examples of ortholattices are shown in Figure 1.4. The "benzene ring" of Figure 1.4(a) is also be called the *hexagon*. The lattice in Figure 1.4(b) is the *horizontal sum* of the *blocks* 2^2 and 2^3, that is, $2^2 \cap 2^3 = \{0, 1\}$.

For elements a, b of an ortholattice the *De Morgan laws*

$$(a \vee b)^{\perp} = a^{\perp} \wedge b^{\perp} \quad \text{and} \quad (a \wedge b)^{\perp} = a^{\perp} \vee b^{\perp}$$

hold, since the orthocomplementation $a \to a^{\perp}$ is a dual isomorphism of the lattice onto itself. Conversely, either of the two De Morgan laws implies (i) in the

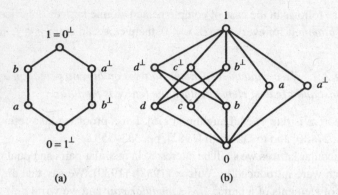

Figure 1.4

preceding definition of ortholattices. Hence we may replace (i) by $(a \vee b)^{\perp} = a^{\perp} \wedge b^{\perp}$ or by its dual. For elements a, b of an ortholattice we define a binary relation $\perp$ ("orthogonality") by

$$a \perp b \quad \text{if and only if} \quad a \leq b^{\perp}.$$

If $a \perp b$ holds, we also say that a is *orthogonal* to b.

Using the above-introduced orthogonality relation and the notion of modular pairs, we now define orthomodular lattices. An orthocomplemented lattice L is called *orthomodular* if, for all $a, b \in L$, $a \perp b$ implies $a\,M\,b$. In other words, every orthogonal pair is a modular pair. This explains the expression *orthomodular*.

Orthomodularity can be characterized in several ways within the class of ortho-complemented lattices. Some of these characterizations are gathered in

Theorem 1.2.4 *In an orthocomplemented lattice L the following five statements are equivalent:*

(i) *L is orthomodular;*
(ii) *$a\,M\,a^{\perp}$ holds for all $a \in L$;*
(iii) *$a\,M^*\,a^{\perp}$ holds for all $a \in L$;*
(iv) *$a \leq b$ implies $a \vee (a^{\perp} \wedge b) = b$;*
(v) *$a \leq b$ implies the existence of $c \in L$ such that $a \perp c$ and $a \vee c = b$.*

For a proof see Maeda & Maeda [1970], Theorem 29.13, p. 132. Any Boolean lattice is orthomodular. The lattice in Figure 1.4(a) is the simplest example of an orthocomplemented lattice that is not orthomodular. The lattice of all subspaces of the three-dimensional real Euclidean space is modular and orthocomplemented and hence orthomodular. The lattice in Figure 1.4(b) is orthomodular, but not modular.

The lattices in Figures 1.2–1.4 are finite and hence of finite length. Let us now give examples of orthomodular lattices of infinite length. (We refer to Section 1.9 for a formal definition of the notion of length.)

Our first example is the lattice of closed subspaces of a Hilbert space. Let H denote a Hilbert space. As a metric space, H is complete, which means that every Cauchy sequence in H converges. A subspace M of H is *closed* if for every Cauchy sequence γ_m in M with $\gamma_m \to x \in H$ implies $x \in M$. A closed subspace is also called a *flat*. By $L_c(H)$ we denote the set of all closed subspaces of H. The lattice $\langle L_c(H); \subseteq \rangle$ is a complete lattice $\langle L_c(H); \wedge, \vee \rangle$, the meet of two closed subspaces being their set-theoretic intersection and the join being the closure of their sum. This lattice will also briefly be denoted by $L_c(H)$. For a subspace M of H we define $M^{\perp} = \{x \in H : \langle x, y \rangle = 0$ for all $y \in M\}$ with $\langle x, y \rangle$ denoting the inner product of x and y defined on H. The operation $M \to M^{\perp}$ is an orthocomplementation, and $L_c(H)$ is a complete orthomodular lattice (Sasaki [1954]). We can interpret the orthomodular identity as a basic fact about the geometry of Hilbert space: If the closed subspace M is contained in the closed subspace N, then N is the *orthogonal direct sum* of M and $N - M$ (the orthocomplement of M in N), that is, $M \leq N$ implies $N = M \oplus (N - M)$. For more details see for example Halmos [1957]. The lattice of all closed subspaces of an infinite-dimensional Hilbert space H is atomistic but not modular. Indeed, one can show that the lattice $L_c(H)$ is modular if and only if H is finite-dimensional.

Without going into details, we mention as a second example the projection lattice of a von Neumann algebra. This lattice is orthomodular, but neither modular nor atomistic. For some more information see Section 2.6. There are several monographs dealing with the theory and applications of orthomodular lattices, such as Maeda & Maeda [1970], Kalmbach [1983], [1986], Beran [1984]. Other references will be given in the Notes below and in Section 2.6.

We close this section with a brief look at varieties of lattices.[1] Let $p_i = q_i$ be identities for $i \in I$. The class K of all lattices satisfying all identities $p_i = q_i$, $i \in I$, is called a *variety* (or *equational class*) of lattices. A variety is *trivial* if and only if it contains one-element lattices only. Let K be a class of lattices. We use the following notation:

H (K) denotes the class of all homomorphic images of members of K.
S (K) denotes the class of all sublattices of members of K.
P (K) denotes the class of all isomorphic images of direct products of members of K.

We say that K is *closed under the formation of homomorphic images, under the formation of sublattices*, and *under the formation of direct products* if $H(\mathsf{K}) \subseteq \mathsf{K}$, $S(\mathsf{K}) \subseteq \mathsf{K}$, and $P(\mathsf{K}) \subseteq \mathsf{K}$, respectively. The following result is known as Birkhoff's *HSP* theorem (Birkhoff [1935b]).

---

[1]For this topic see the monograph *Varieties of Lattices* by P. Jipsen and H. Rose, Springer-Verlag, Berlin (1992).

Theorem 1.2.5   *A class* K *of lattices is a variety if and only if* K *is closed under the formation of homomorphic images, under the formation of sublattices, and under the formation of direct products.*

Corollary 1.2.6   *Let* K *be a class of lattices. Then* **HSP** (K) *is the smallest variety containing* K.

The corollary is due to Tarski [1946]. The *smallest variety containing* K will be denoted by $V(\mathsf{K})$. We shall also say that $V(\mathsf{K})$ is *the variety generated by* K. Several important classes of lattices form a variety. We have already mentioned the class of one-element lattices (the trivial variety); the class L of all lattices is also a variety. It is immediate from the definition that distributive lattices form a variety. A Boolean algebra $\mathbf{B} = \langle B, \wedge, \vee, \bar{\ }, 0, 1 \rangle$ can be defined by equations. Hence the class of Boolean algebras is a variety. An orthomodular lattice considered as an algebra $\mathbf{L} = \langle L, \wedge, \vee, \perp, 0, 1 \rangle$ can be defined by equations. Hence the class of orthomodular lattices is a variety.

There are many statements equivalent to modularity. It can be shown, for example, that a lattice $\mathbf{L} = \langle L, \wedge, \vee \rangle$ is modular if and only if it satisfies $((b \wedge c) \vee a) \wedge c = (b \wedge c) \vee (a \wedge c)$ for all $a, b, c \in L$. Hence the class of modular lattices is a variety. In contrast to this, the class of semimodular lattices does not form a variety (see Section 1.7).

Notes

The theory of orthomodular lattices has its roots in functional analysis, and its origins go back to the theory of von Neumann algebras. For a thorough study of this background and an analysis of the historical sequence von Neumann algebras → continuous geometries → orthomodular lattices we refer to Holland [1970]. The theory of continuous geometries was also invented by von Neumann and developed in the period 1935–7 (for von Neumann's contribution to lattice theory see Birkhoff [1958]).

The investigation of the lattice-theoretical foundations of quantum-mechanical systems was initiated by Birkhoff & von Neumann [1936], who set up a model of what they called the *logic of quantum mechanics*. However, their lattices were modular, and this turned out to be too restrictive a condition. The interpretation of observables as operators in Hilbert space and, in particular, the investigation of lattices of projection operators led to nonmodular orthomodular lattices.

The relationship between certain orthomodular lattices (or more general structures) and quantum mechanics belongs to the vast field vaguely described as *quantum logics*. A popular account of some problems arising in this field is McGrath [1991]. For details we refer to the monographs Beltrametti & Cassinelli [1981], Cohen [1989], and Pták & Pulmannová [1991].

References

Beltrametti, E. G. and G. Cassinelli [1981] *The Logic of Quantum Mechanics*,
    Encyclopedia of Mathematics and its Applications, Vol. 15, Addison-Wesley,
    Reading, Mass.

Beran, L. [1984] *Orthomodular Lattices. Algebraic Approach*, Reidel, Dordrecht.

Birkhoff, G. [1934] Applications of lattice algebra, Proc. Camb. Phil. Soc. 30, 115–122.

Birkhoff, G. [1935b] On the structure of abstract algebras, Proc. Camb. Phil. Soc. 31,
    433–454.

Birkhoff, G. [1958] von Neumann and lattice theory, in: *John von Neumann, 1903–1957*,
    Bull. Amer. Math. Soc. 64, 50–56.

Birkhoff, G. and J. von Neumann [1936] The logic of quantum mechanics, Ann. of Math.
    37, 823–843.

Cohen, D.W. [1989] *An Introduction to Hilbert Space and Quantum Logic*,
    Springer-Verlag, New York.

Dedekind, R. [1900] Über die von drei Moduln erzeugte Dualgruppe, Math. Ann. 53,
    371–403; *Gesammelte Werke*, Vol. 2, pp. 236–271.

Dilworth, R. P. [1982] The role of order in lattice theory, in: Rival [1982a], pp. 333–353.

Grätzer, G. [1978] *General Lattice Theory*, Birkhäuser, Basel.

Halmos, P. R. [1957] *Introduction to Hilbert Space and the Theory of Spectral
    Multiplicity*, Chelsea, New York.

Holland, S. S., Jr. [1970] The current interest in orthomodular lattices, in: *Trends in
    Lattice Theory* (ed. J.C. Abbott), van Nostrand Reinhold, New York, pp. 41–126.

Kalmbach, G. [1983] *Orthomodular Lattices*, Academic Press, London.

Kalmbach, G. [1986] *Measures and Hilbert Lattices*, World Scientific, Singapore.

Maeda, F. and S. Maeda [1970] *Theory of Symmetric Lattices*, Springer-Verlag, Berlin.

McGrath, J. H. [1991] Quantum theory and the lattice join, Math. Intelligencer 13, 72–79.

McLaughlin, J. E. [1956] Atomic lattices with unique comparable complements, Proc.
    Amer. Math. Soc. 7, 864–866.

Pták, P. and S. Pulmannová [1991] *Orthomodular Structures as Quantum Logics*, Kluwer
    Academic, Dordrecht.

Salii, V. N. [1988] *Lattices with Unique Complements*, Translations of Mathematical
    Monographs, Vol. 69, Amer. Math. Soc., Providence, R.I. (translation of the 1984
    Russian edition, Nauka, Moscow).

Sasaki, U. [1954] Orthocomplemented lattices satisfying the exchange axiom, J. Sci.
    Hiroshima Univ. A 17, 293–302.

Tarski, A. [1946], A remark on functionally free algebras, Ann. of Math. 47, 163–165.

von Neumann, J. [1936–7] *Lectures on Continuous Geometries*, Inst. for Advanced Study,
    Princeton, N.J.

Wilcox, L. R. [1938] Modularity in the theory of lattices, Bull. Amer. Math. Soc. 44, 50.

Wilcox, L. R. [1939] Modularity in the theory of lattices, Ann. of Math. 40, 490–505.

1.3 Distributive and Semidistributive Lattices

Summary. The early work on distributive lattices included representation theorems, embedding theorems, and structure theorems. Later it was discovered that the congruence
lattice of a lattice and the lattice of lattice varieties are distributive. We mention some of
these results with a view to later applications and generalizations. We also give some facts
concerning semidistributivity, which is next to modularity the most important generalization
of distributivity.

An element j (m) of a lattice L is called *join-irreducible* (*meet-irreducible*) if, for all $x, y \in L$,

$$j = x \vee y \text{ implies } j = x \text{ or } j = y \qquad (m = x \wedge y \text{ implies } m = x \text{ or } m = y).$$

For a lattice L of finite length let $J(L)$ denote the set of all nonzero join-irreducible elements, regarded as a poset under the partial ordering of L. Sometimes we emphasize this by writing more precisely $(J(L), \leq)$ instead of $J(L)$. By j' we denote the uniquely determined lower cover of $j \in J(L)$. For a poset P, we call a subset $I \subseteq P$ an *order ideal* (or *hereditary subset*) if $x \in I$ and $y \leq x$ imply $y \in I$. Let ord(P) denote the set of all order ideals of the poset P, and regard ord(P) as partially ordered by set inclusion. With respect to this partial order, ord(P) forms a lattice in which meet and join are the set-theoretic intersection and union, respectively. Thus ord(P) is a distributive lattice. For finite distributive lattices we have the following structural result

Theorem 1.3.1   *Let L be a finite distributive lattice. Then the map*

$$\varphi : a \rightarrow \{j : j \leq a, \ j \in J(L)\} = (a] \cap J(L)$$

is an isomorphism between L and ord$(J(L))$.

For a proof see Grätzer [1978], pp. 61–62, or Stanley [1986], p. 106.

Corollary 1.3.2   *The correspondence $L \rightarrow J(L)$ makes the class of all finite distributive lattices (with more than one element) correspond to the class of all finite posets. Isomorphic lattices correspond to isomorphic posets, and vice versa.*

The preceding corollary is called the *fundamental theorem on finite distributive lattices*. It has several important consequences. Recall that a subset S of the power set of a set is called a *ring of sets* if $X, Y \in S$ implies both $X \cap Y \in S$ and $X \cup Y \in S$. Since ord$(J(L))$ is a ring of sets, we have

Corollary 1.3.3   *A finite lattice is distributive if and only if it is isomorphic to a ring of sets.*

In particular, we have

Corollary 1.3.4   *A finite lattice is Boolean if and only if it is isomorphic to the lattice of all subsets of a finite set.*

If a is an element of a lattice L, then a representation $a = j_1 \vee \cdots \vee j_n$ of a as a join of finitely many join-irreducible elements $j_1, \ldots, j_n \in J(L)$ is called a *finite join decomposition* of a. This join decomposition is said to be *irredundant* if, for each $i = 1, \ldots, n$, one has $a \neq j_1 \vee \cdots \vee j_{i-1} \vee j_{i+1} \vee \cdots \vee j_n$. Dually one defines *finite meet decompositions* and *irredundant finite meet decompositions* (see Crawley & Dilworth [1973], p. 38). Finite join decompositions or finite meet decompositions do not always exist. However, if a lattice element has a finite join decomposition, then it clearly has an irredundant join decomposition, which is obtained by omitting superfluous elements from the given join decomposition. A similar statement holds for finite meet decompositions. It is easy to see that in a distributive lattice an element has at most one irredundant join decomposition and at most one irredundant meet decomposition. In the finite case we have

Corollary 1.3.5 *Every element of a finite distributive lattice has a unique irredundant join decomposition and a unique irredundant meet decomposition.*

Similar results have also been proved for distributive lattices that are not finite but satisfy certain relaxations of finiteness. For example, if a distributive lattice satisfies the *ascending chain condition* (ACC for short), then each of its elements has a unique irredundant meet decomposition (see Birkhoff [1948], p. 142), which is necessarily finite. In Section 1.8 we shall have a look at the existence of possibly infinite meet decompositions and the question of irredundant meet decompositions. The uniqueness property for irredundant meet decompositions will be discussed in more detail in Chapter 7.

Let $M(L)$ denote *the set of meet irreducible elements* ($\neq 1$) of a lattice of finite length L. By m^* we denote the uniquely determined upper cover of $m \in M(L)$. In a finite distributive lattice there is a natural one-to-one correspondence between the meet irreducibles ($\neq 1$) and the join irreducibles ($\neq 0$): For every $m \in M(L)$, there exists a unique minimal join irreducible $j \in J(L)$ such that $j \nleq m$. In turn, m is the unique maximal meet irreducible not containing j. This correspondence implies

Corollary 1.3.6 *If L is a finite distributive lattice, then $|J(L)| = |M(L)|$.*

For a proof see Grätzer [1978], pp. 62–63. For a finite distributive lattice L it is even true that $J(L) \cong M(L)$. More precisely, we have $(J(L), \leq) \cong (M(L), \leq)$, where $\leq$ denotes the partial ordering induced by L (see Pezzoli [1984] for a proof). This property of finite distributive lattices is illustrated in Figure 1.5.

Consider now Con(L), the set of congruence relations on a lattice L, and let Σ be a subset of Con(L). Define the relation π in L by $a \ \pi \ b$ if $a \ \theta \ b$ holds for all $\theta \in \Sigma$, and define the relation σ by the rule $a \ \sigma \ b$ if there exist a sequence $a = a_0, a_1, \ldots, a_n = b$ in L and congruence relations $\theta_1, \ldots, \theta_n \in \Sigma$ such that $a_{i-1}\theta_i a_i$ for each $i = 1, \ldots, n$. It is easy to see that π and σ are congruence relations, that π is the meet in Con(L) of the subset Σ, and that σ is the join in Con(L) of Σ. Hence Con(L) is a complete lattice. Funayama & Nakayama [1942] proved that the lattice Con(L) of congruence relations on a lattice L is distributive and algebraic. For a proof see also Crawley & Dilworth [1973], p. 75. The fact that, for an arbitrary lattice L, the congruence lattice Con(L) is distributive can be reformulated by saying that the variety of all lattices is *congruence-distributive*.

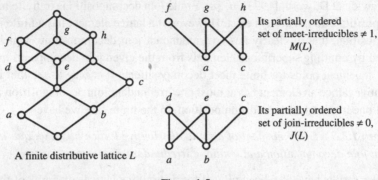

A finite distributive lattice L

Its partially ordered set of meet-irreducibles $\neq 1$, $M(L)$

Its partially ordered set of join-irreducibles $\neq 0$, $J(L)$

Figure 1.5

We have already remarked that modularity is the most important generalization of distributivity. There are weakenings of distributivity going in other directions than modularity. Among these, semidistributivity turned out to be particularly fruitful. A lattice L is called *meet-semidistributive* if

(SD∧) $a \wedge b = a \wedge c$ implies $a \wedge b = a \wedge (b \vee c)$ for all $a, b, c \in L$.

A lattice L is called *join-semidistributive* if

(SD∨) $a \vee b = a \vee c$ implies $a \vee b = a \vee (b \wedge c)$ for all $a, b, c \in L$,

and a lattice is called *semidistributive* – (SD) for short – if it satisfies both (SD∧) and (SD∨). In what follows we briefly recall some results on (SD), (SD∧), and (SD∨) with an eye to later applications (e.g. in Section 9.3).

Semidistributivity was introduced by Jónsson [1961] in his investigations of free lattices.[2] Jónsson proved that (SD∧) and (SD∨) hold in a free lattice. Hence any sublattice of a free lattice is semidistributive. Let us note that sublattices of free lattices also satisfy the following condition due to Whitman [1941]:

(W) $x \wedge y \leq u \vee v$ implies $[x \wedge y, u \vee v] \cap \{x, y, u, v\} \neq \emptyset$.

Nation [1982] proved Jónsson's longstanding conjecture that a finite lattice is isomorphic to a sublattice of a free lattice if and only if it is a semidistributive lattice satisfying *Whitman's condition* (W). Free lattices provide the most important examples of semidistributive lattices. However, semidistributivity has also shown up in other areas of lattice theory. In particular, meet semidistributivity appears in the congruence lattice of meet semilattices, and it plays an important role in the study of lattice varieties. Let us give some more examples. We begin with lattices that are *not* semidistributive.

The lattice M_3 is not semidistributive; in fact, it satisfies neither (SD∧) nor (SD∨). Next we consider the lattices $S_7, S_7^*, L_3, L_4, L_5$ shown in Figure 1.6. From

---

[2]For this topic see the monograph *Free Lattices* by R. Freese, J. Ježek, and J. B. Nation Amer. Math. Soc., Providence, R.I. (1995).

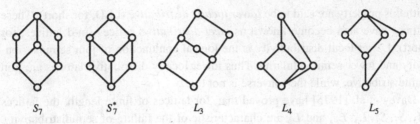

S_7 S_7^* L_3 L_4 L_5

Figure 1.6

L_3 on we adopt in Figure 1.6 and in Figure 1.8 the numbering system of Jónsson & Rival [1979]. However, instead of their L_1 and L_2 we use S_7^* and S_7, respectively. Let us remark here that S_7^* and S_7 are also denoted by some authors as D_1 and D_2, respectively – see e.g. Hobby & McKenzie [1988].

The lattices S_7 and S_7^*, together with M_3 and N_5, play a significant role in the classification scheme for locally finite varieties (cf. Hobby & McKenzie [1988]), to which we briefly return in Section 9.3. The lattices S_7 and S_7^* are the smallest lattices satisfying one, but not both, of (SD∧) and (SD∨). In fact, S_7 satisfies (SD∧) but not (SD∨), while S_7^* satisfies (SD∨) but not (SD∧).

S_7^* is isomorphic to the lattice of convex subsets of a three-element linearly ordered set. This relationship will be considered in more detail in Section 7.3. S_7 is isomorphic to the congruence lattice of E_2^2, where E_2 is the two-element meet semilattice. This example reflects the fact that, for arbitrary semilattices S, Con(S) always satisfies (SD∧) (Papert [1964]). We also say that the variety of semilattices, denoted by Semilattices, is *congruence meet-semidistributive*. We shall return at several points to this example, especially in Section 9.1.

Following a suggestion of Garrett Birkhoff, we shall use the name *centered hexagon* for S_7, the name *hexagon* being reserved for the lattice of Figure 1.4(a) (Section 1.2). If a lattice L (perhaps with no least element) has the property that the interval $[x, y]$ contains an atom whenever $x < y$ in L, we say that L is *strongly atomic*.

Let us note here that S_7 has the property that for every $x \neq 1$ the interval $[x, x^+]$ is a Boolean sublattice (x^+ denoting *the join of all elements covering x*). Finite (more generally: strongly atomic algebraic) lattices with this property are called *upper locally distributive* or just *locally distributive* (ULD, for short). They are also called *join-distributive* lattices (j-d lattices, for short) or *locally free* lattices. Locally distributive lattices are meet-semidistributive, while the converse is not true. In fact, for finite (more generally: strongly atomic algebraic) lattices, upper local distributivity is the logical conjunction of meet semidistributivity and upper semimodularity. This result (cf. Theorem 7.1.2) is due to Dilworth & Crawley [1960] (see also Crawley & Dilworth [1973], p. 51).

Dually, S_7^* has the property that for every $x \neq 0$ the interval $[x_+, x]$ is a Boolean sublattice (here x_+ denotes the meet of all elements covered by x). Finite lattices with this property are said to be *lower locally distributive* (LLD, for short). These lattices have also become known as *meet-distributive* lattices (m-d lattices, for short). Lower local distributivity is the logical conjunction of join semidistributivity and lower semimodularity. Thus lower locally distributive lattices are join semidistributive, while the converse is not true.

Davey et al. [1975] have proved that, for lattices of finite length, the lattices M_3, S_7, S_7^*, L_3, L_4, and L_5 are characteristic of the failure of semidistributivity. More precisely, they proved

Theorem 1.3.7   *A lattice L of finite length is (join + meet) semidistributive if and only if it contains no sublattice isomorphic to one of M_3, S_7, S_7^*, L_3, L_4, and L_5. In fact, L satisfies (SD$\vee$) if and only if it has no sublattice isomorphic to M_3, S_7, L_3, L_4, and L_5.*

An example due to Wille [1974] shows that this result fails for lattices of infinite length. A variety K of lattices is called *join-semidistributive* (*meet-semidistributive, semidistributive*) if every member of K is join-semidistributive (meet-semidistributive, semidistributive, respectively). The above cited result of Davey et al. [1975] was extended to varieties of lattices by Jónsson & Rival [1979], who proved that a variety of lattices is semidistributive if and only if it contains none of the lattices M_3, S_7, S_7^*, L_3, L_4, and L_5. In fact, they proved

Theorem 1.3.8   *A variety of lattices is join-semidistributive if and only if it contains none of the lattices M_3, S_7, L_3, L_4, and L_5.*

Since the intersection of varieties is again a variety, it follows that the varieties of similar algebras of a given type form a lattice. In particular, the lattice varieties form a lattice. We collect some facts about this lattice. For more details see Crawley & Dilworth [1973], Grätzer [1978], and the surveys given by Dilworth [1984] and Nation [1994], upon which we have drawn here.

From the congruence distributivity of the variety of lattices it follows that *the lattice of lattice varieties is distributive*. Next we give some results concerning the elements of this lattice up to rank 3 (see Section 1.9 for a formal definition of rank). The least element of this lattice is the variety of all one-element lattices (the trivial variety). Denoting by D the class of distributive lattices, we state that the variety $V(D) = D$ of all distributive lattices is the only atom in the lattice of lattice varieties.

A class K of lattices may consist of a finite set of finite lattices. Let us consider now, in particular, the variety $V(N_5)$ generated by the pentagon N_5 and the variety $V(M_3)$ generated by the diamond M_3. From the forbidden-sublattice characterization of modularity (Theorem 1.2.2) and from the forbidden-sublattice characterization of distributivity (Theorem 1.2.1) it follows that the varieties $V(N_5)$ and $V(M_3)$ are the only upper covers of $V(D)$. For going further upwards and finding the covers of $V(N_5)$ and $V(M_3)$ in the lattice of lattice varieties, one needs Jónsson's improvement of Birkhoff's *HSP* theorem (Theorem 1.2.5). To state Jónsson's result (cf. Jónsson [1967]), we have to consider two more operators acting on classes of lattices. For a class K of lattices we denote by

$P_s(K)$ the class of all lattices that are isomorphic with subdirect products of members of K;

$P_u(K)$ the class of all lattices that are isomorphic with ultraproducts of members of K.

Theorem 1.3.9 *Let* K *be a class of lattices. Then* $V(K) = P_s HSP_u(K)$.

Equivalently, Jónsson's theorem may be formulated as follows: Let K be a class of lattices. Then $P_s(K) \subseteq HS\,P_u(K)$, that is, the subdirectly irreducible members of $V(K)$ belong to $HSP_u(K)$.

Jónsson's theorem (which holds more generally for varieties of congruence distributive algebras) has many important consequences. For example, McKenzie [1972] uses this theorem in his theory of splitting varieties. Also, the investigation of Mal'cev's conditions (cf. Section 9.3) can be traced back to Jónsson [1967]. The theory of congruence modular varieties was partly motivated by Jónsson's results on congruence distributive varieties. The theory of congruence semimodular varieties in turn (cf. Section 9.3) seeks to extend results obtained for congruence modularity. One of the consequences of Jónsson's aforementioned result is

Theorem 1.3.10 *If* K *is a finite set of finite lattices, then* $P_u(K) \subseteq K$.

Corollary 1.3.11 *If* K *is a finite set of finite lattices, then* $V(K) = P_s HS(K)$. *In particular, every subdirectly irreducible lattice in* $V(K)$ *belongs to* $HS(K)$.

This means that a variety $V(K)$ generated by a finite set of finite lattices has only finitely many subvarieties, and these subvarieties can be found by looking at the subdirectly irreducible lattices in $HS(K)$. It is therefore easy to find the subvarieties of a given finitely generated lattice variety. It is, however, difficult to find the upper covers of a lattice variety. On the other hand, if one has a candidate for an upper cover, then it is easy to check whether this candidate is really an upper cover.

We list some further results concerning the determination of covers in the lattice of lattice varieties. Grätzer showed that the only finitely generated varieties covering $V(M_3)$ are the varieties generated by M_4, $M_{3,3}$ (see Figure 1.7) and by $M_3 \times N_5$. Jónsson proved this result for general lattice varieties and showed that $V(M_4)$, $V(M_{3,3})$, and $V(M_3 \times N_5)$ are all the covers of the variety generated by the diamond.

McKenzie [1972] constructed sixteen varieties covering $V(N_5)$, namely $V(M_3 \times N_5)$, the five join-irreducible varieties $V(S_7)$, $V(S_7^*)$, $V(L_3)$, $V(L_4)$, $V(L_5)$, plus the ten join-irreducible varieties $V(L_i)(i = 6, \ldots, 15)$ generated by the

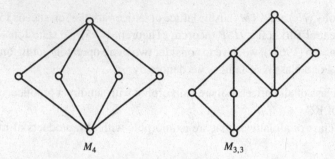

M_4 $M_{3,3}$

Figure 1.7

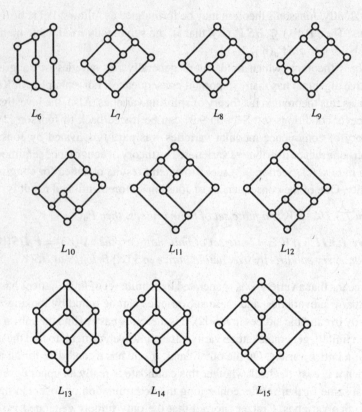

L_6 L_7 L_8 L_9

L_{10} L_{11} L_{12}

L_{13} L_{14} L_{15}

Figure 1.8

subdirectly irreducible lattices $L_i (i = 6, \ldots, 15)$ shown in Figure 1.8. McKenzie [1972] conjectured that these were all varieties covering the variety generated by the pentagon. Rival proved that these are in fact all the finitely generated covers of the variety generated by the pentagon. For general lattice varieties this was again proved by Jónsson, yielding an affirmative answer to McKenzie's conjecture.

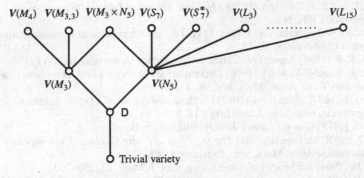

Figure 1.9

Thus there are 18 varieties having rank 3 in the lattice of lattice varieties (cf. Figure 1.9). With the exception of the covers of $V(M_3 \times N_5)$, all covers of these 18 varieties are also known.[3]

Let us note that a lattice is called *p-modular* if it does not contain a sublattice isomorphic with the lattice L_{15} of Figure 1.8. McKenzie [1972] proved that a lattice is p-modular if and only if it satisfies the identity

$$(x \vee (y \wedge z)) \wedge (z \vee (x \wedge y)) = (z \wedge (x \vee (y \wedge z))) \vee (x \wedge (z \vee (x \wedge y))).$$

Every modular lattice is p-modular but not conversely (the pentagon is p-modular).

Notes

For a survey of the role of distributivity in lattice theory, including decompositions, congruence distributivity, semidistributivity, and results on the lattice of lattice varieties, see Dilworth [1984]. We also refer to Baker [1994] for a survey containing a section on algebras whose congruence lattices are distributive. Rose [1984] deals, among other things, with nonmodular lattice varieties whose members are semidistributive. Nation [1985] gives a survey on varieties of semidistributive lattices. Mitsch [1983] surveys semigroups (in particular, semilattices) and their lattices of congruences.

References

Baker, K. [1994] Bjarni Jónsson's contributions in algebra, Algebra Universalis 31, 306–336.
Birkhoff, G. [1948] *Lattice Theory* (2nd edition), Amer. Math. Soc. Colloquium Publications, Vol. 25, New York.

[3] For the variety $V(M_3 \times N_3)$ only the finitely generated covers are known at this point, cf. P. Jipsen and H. Rose, Varieties of Lattices, Appendix F in the second edition of the *General Lattice Theory* by Grätzer (1998).

Crawley, P. and R. P. Dilworth [1973] *Algebraic Theory of Lattices*, Prentice-Hall, Englewood Cliffs, N.J.

Davey, B. A., W. Poguntke, and I. Rival [1975] A characterization of semidistributivity, Algebra Universalis 5, 72–75.

Dilworth, R. P. [1984] Aspects of distributivity, Algebra Universalis 18, 4–17.

Dilworth, R. P. and P. Crawley [1960] Decomposition theory for lattices without chain conditions Trans. Amer. Math. Soc. 96, 1–22.

Funayama, N. and T. Nakayama [1942] On the distributivity of a lattice of lattice congruences, Proc. Imp. Acad. Tokyo 18, 553–554.

Grätzer, G. [1978] *General Lattice Theory*, Birkhäuser, Basel.

Hobby, D. and R. McKenzie [1988] *The Structure of Finite Algebras*, Contemporary Mathematics, Amer. Math. Soc., Providence, R.I.

Jónsson, B. [1961] Sublattices of a free lattice, Can. J. Math. 13, 256–264.

Jónsson, B. [1967] Algebras whose congruence lattices are distributive, Math. Scand. 21, 110–121.

Jónsson, B. and I. Rival [1979] Lattice varieties covering the smallest non-modular variety, Pacific J. Math. 82, 463–478.

McKenzie, R. [1972] Equational bases and nonmodular lattice varieties, Trans. Amer. Math. Soc. 74, 1–43.

Mitsch, H. [1983] Semigroups and their lattices of congruences, Semigroup Forum 26, 1–63.

Nation, J. B. [1982] Finite sublattices of a free lattice, Trans. Amer. Math. Soc. 269, 311–337.

Nation, J. B. [1985] Some varieties of semidistributive lattices, in: Universal Algebra and Lattice Theory (ed. by S. D. Comer), Proceedings of a Conference held at Charleston, July 11–14, 1984, Lecture Notes in Mathematics 1149, Springer-Verlag, Berlin, pp. 198–223.

Nation, J. B. [1994] Jónsson's contributions to lattice theory, Algebra Universalis 31, 430–445.

Papert, D. [1964] Congruence relations in semi-lattices, J. London Math. Soc. 39, 723–729.

Pezzoli, L. [1984] On D-complementation, Adv. in Math. 51, 226–239.

Rose, H. [1984] *Nonmodular Lattice Varieties*, Memoirs Amer. Math. Soc., Vol. 47, No. 292.

Stanley, R. [1986] *Enumerative Combinatorics I*, Wadsworth & Brooks/Cole, Monterey, Calif.

Whitman, Ph. M. [1941] Free lattices, Ann. of Math. 42, 325–329.

Wille, R. [1974] Jeder endlich erzeugte modulare Verband endlicher Weite ist endlich, Mat. Čas. 24, 77–80.

1.4 Pseudocomplemented Lattices

Summary. We recall the notion of a pseudocomplemented lattice and cast a glance at the special subclasses of relatively pseudocomplemented lattices and pseudocomplemented distributive lattices. Then we have a closer look at a special property of the centered hexagon S_7: each of its intervals is a pseudocomplemented sublattice. For later use we clarify the connection of this property to meet semidistributivity and to other related conditions.

Let L be a lattice with least element 0. An element $t \in L$ is said to have a *meet pseudocomplement* $g(t)$ if $g(t) \wedge t = 0$ and $x \wedge t = 0$ implies $x \leq g(t)$. When we use the expression *pseudocomplement* we always mean meet pseudocomplement. The element $g(t)$ is obviously the greatest element in the set of all $x \in L$ for which

$x \wedge t = 0$. In other words, the subset of all elements disjoint from t is required to form a principal ideal. Hence an element can have at most one pseudocomplement. A lattice is called *meet-pseudocomplemented* if each of its elements has a meet pseudocomplement. When we say *pseudocomplemented lattice* we always mean a meet-pseudocomplemented lattice. Dually, a lattice L with greatest element 1 is said to be *join-pseudocomplemented* if, for every $t \in L$, the set of all $x \in L$ for which $x \vee t = 1$ has a least element which is called the *join pseudocomplement* of t. A lattice that is both meet-pseudocomplemented and join-pseudocomplemented is often called a *double p-lattice*. The algebra $(L; \vee, \wedge, g, 0, 1)$ where L is pseudocomplemented is also called a *p-algebra*.

Two classes of meet-pseudocomplemented lattices have been studied extensively: the class of relatively pseudocomplemented lattices and the class of pseudocomplemented distributive lattices.

Let a and b be elements of a lattice L. The *pseudocomplement of a relative to b* is an element $a * b$ of L satisfying $a \wedge x \leq b$ if and only if $x \leq a * b$. The element $a * b$ is unique if it exists and $a * a$ exists if and only if the lattice has a greatest element. If $a * b$ exists for all elements a and b, then the lattice is called *relatively pseudocomplemented*.

In any Boolean lattice, the complement $\bar{a}$ of an element a is the greatest element x such that $a \wedge x = 0$. More generally, $a \wedge x \leq b$ holds if and only if $x \leq \bar{a} \vee b$, and thus any Boolean lattice is relatively pseudocomplemented. Every chain is relatively pseudocomplemented, and so is the complete distributive lattice of all open subsets of any topological space. Relatively pseudocomplemented lattices are distributive, but not conversely: the complete distributive lattice of all closed subsets of the real line is not relatively pseudocomplemented (see Birkhoff [1967], p. 46).

The investigation of relatively pseudocomplemented lattices was started in connection with the foundations of logic by Glivenko, Brouwer, and Heyting around 1930. A comprehensive source on the history of logic, including many hints about lattices, is Mangione & Bozzi [1993]. Relatively pseudocomplemented lattices are also called Brouwerian, implicative, Heyting, or pseudoboolean lattices.

The property of being relatively pseudocomplemented is very strong, since it implies distributivity. If a lattice is only pseudocomplemented, it need not even be modular, as the centered hexagon S_7 shows (see Figure 1.11). On the other hand, pseudocomplemented lattices that are in addition distributive have been studied by many authors. The following result shows that these lattices are ubiquitous.

Theorem 1.4.1 *Any complete lattice that satisfies the join-infinite distributive identity*

(JID) $x \wedge \bigvee(x_i : i \in I) = \bigvee(x \wedge x_i : i \in I)$

is a pseudocomplemented distributive lattice.

For a proof see Grätzer [1978], Theorem 1, p. 111. Since every algebraic distributive lattice satisfies the join-infinite distributive law (JID), these lattices are pseudocomplemented. In particular, any finite distributive lattice is pseudocomplemented. Boolean lattices are pseudocomplemented distributive lattices.

A particularly interesting class of pseudocomplemented distributive lattices is the class of Stone lattices: a pseudocomplemented distributive lattice is a *Stone lattice* if $g(t) \vee g(g(t)) = 1$. The investigation of these lattices was initiated by Stone [1937]. Stone lattices were the first class of pseudocomplemented distributive lattices different from the class of Boolean lattices. Detailed investigations of pseudocomplemented lattices and, in particular, of Stone lattices can be found in Varlet [1963], [1974–5] and Grätzer [1978]. For a thorough study of finite pseudocomplemented lattices we refer to Chameni-Nembua & Monjardet [1992]. We shall be interested here in some results on the class of finite pseudocomplemented lattices. Let us first note several examples. The lattice L_{14} in Figure 1.8 (Section 1.3) is meet-pseudocomplemented, join-pseudocomplemented, and complemented. The lattice in Figure 1.10 is meet-pseudocomplemented and join-pseudocomplemented but not complemented.

The centered hexagon S_7 is pseudocomplemented. In Figure 1.11 we indicate the elements x of S_7 together with their pseudocomplements $g(x)$. The lattice S_7 even has the stronger property that each of its intervals is pseudocomplemented.

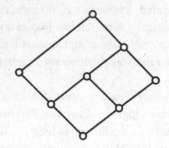

Figure 1.10

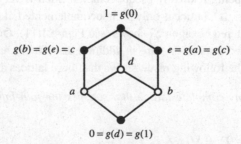

Figure 1.11

This property is also referred to as *interval pseudocomplementedness*. However, S_7 is clearly not relatively pseudocomplemented (this would imply distributivity).

For finite lattices there is a simple criterion for pseudocomplementedness (Chameni-Nembua & Monjardet [1992], Theorem 3.3). The same result can also be proved more generally for strongly atomic algebraic lattices.

Theorem 1.4.2   *In a finite lattice L the following two conditions are equivalent:*

(i) *L is pseudocomplemented;*
(ii) *Each atom of L has a pseudocomplement.*

It is not difficult to see that finite pseudocomplemented lattices L can be characterized as those lattices in which every atom p induces a partition of L into a dual ideal $[p)$ and an ideal $(g(p)]$, that is, $L = [p) + (g(p)]$, where $+$ denotes the *disjoint union*.

Another class of finite lattices sharing the property of interval pseudocomplementedness with the centered hexagon are the *Tamari lattices* T_n. Tamari [1951] defined a partial order in the set of all possible binary bracketings of n letters under a semiassociativity law. We shall not go into details here, but rather refer to Grätzer [1978], Exercises 26–36, pp. 14–15. Friedman & Tamari [1967] showed that this partial order is a lattice T_n. The Tamari lattice T_3 is isomorphic to N_5, and the Tamari lattice T_4 is shown in Figure 1.12. It can be shown that T_n is complemented and that each interval of T_n is pseudocomplemented (see Grätzer [1978], Exercises 30 and 31, p. 51).

In what follows we clarify the connection between interval pseudocomplementedness [condition (ii) below] and two related conditions [conditions (i) and (iii) below]. (Cf. R. Freese and J. B. Nation, Clarification to "Congruence lattices of semilattices," unpublished remarks of December 8, 1993, and of July 24, 1995; available on the web at http://www.math.hawaii.edu/~ralph/papers.html.)

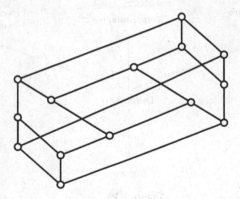

Figure 1.12

The requirement that each principal dual ideal of a given lattice with least element be pseudocomplemented is expressed by the following condition:

(i) For each $x \geq z$ there is a y such that $y \wedge x = z$ and $u \wedge x = z$ implies $u \leq y$.

In a strongly atomic algebraic lattice this condition is already fulfilled if it holds for each $x = p \succ z$ (cf. Theorem 1.4.2 and the remarks before that theorem). If the lattice also has a greatest element, condition (i) is equivalent to the following property shared by the lattices S_7 and T_n:

(ii) Each interval is a pseudocomplemented sublattice.

Moreover, in algebraic lattices each of the conditions (i) and (ii) is equivalent to meet semidistributivity [condition (SD∧)], and for strongly atomic algebraic lattices L each of the conditions (i), (ii), and (SD∧) is equivalent to the following condition:

(iii) If $a, p, x, y \in L, x, y \geq a$ and $p \succ a$ then $p \wedge (x \vee y) = (p \wedge x) \vee (p \wedge y)$,

which means that any atom p of the principal dual ideal $[a)$ is dually distributive in this dual ideal (for the notion of a distributive element see Section 2.2). The

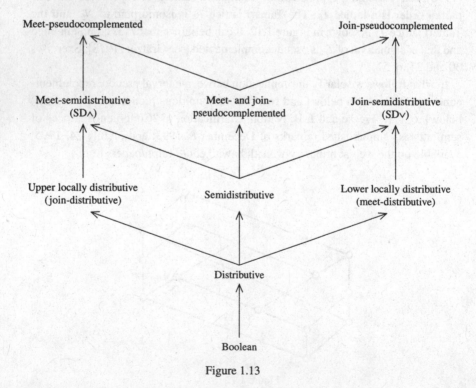

Figure 1.13

preceding condition (iii) is just condition (2) of Crawley & Dilworth [1973], p. 51 (cf. also Theorem 7.1.2).

For finite lattices the interrelationships between some of the concepts mentioned above are shown in Figure 1.13.

References

Birkhoff, G. [1967] *Lattice Theory*, American Math. Soc. Colloquium Publications, Vol. 25, Providence, R.I. (reprinted 1984).

Chameni-Nembua, C. and B. Monjardet [1992] Les treillis pseudocomplémentés finis, Eur. J. Combin. 13, 89–107.

Crawley, P. and R. P. Dilworth [1973] *Algebraic Theory of Lattices*, Prentice-Hall, Englewood Cliffs, N.J.

Freese, R. and J. B. Nation [1973] Congruence lattices of semilattices, Pacific J. Math. 49, 51–58.

Friedman, H. and D. Tamari [1967] Une structure de treillis fini induite par une loi demi-associative, J. Combin. Theory Ser. A 2, 215–242.

Grätzer, G. [1978] *General Lattice Theory*, Birkhäuser, Basel.

Mangione, C. and S. Bozzi [1993] *Storia della Logica da Boole ai Nostri Giorni*, Garzanti Editore, Milano.

Stone, M. H. [1937] Topological representation of distributive lattices and Brouwerian logics, Časopis. Pešt. Mat. 67, 1–25.

Tamari, D. [1951] Monoides préordonnés et chaînes de Malcev.Thése, Université de Paris.

Varlet, J. C. [1963] Contribution à l'étude des treillis pseudo-complémentés et des treillis de Stone, Mém. Soc. Roy. Sci. Liège 8, 5–71.

Varlet, J. C. [1974–5] *Structures algébriques ordonnées*, Univ. de Liège.

1.5 Complementation

Summary. We have introduced complementation and pseudocomplementation. In this section we survey some other conditions related to complementation in lattices. Several of these conditions will be needed later, especially in Section 3.5. We first collect the definitions, then indicate their interrelationships and give some examples. For more details we refer to Grillet & Varlet [1967] and Blyth & Janowitz [1972].

Recall that a bounded lattice is complemented if, for each element x, there exists at least one element y such that $x \wedge y = 0$ and $x \vee y = 1$. We also recall that if each interval of a lattice is complemented as a sublattice, then the lattice is called *relatively complemented* (RC). If each element of a bounded lattice has exactly one complement, the lattice is said to be *uniquely complemented*, or briefly *unicomplemented* (UC). If each interval of a lattice is unicomplemented, we say that the lattice is *relatively unicomplemented* (RUC). It is well known that every relatively unicomplemented lattice is distributive. Moreover, if a relatively unicomplemented lattice has a least element, it is a generalized Boolean lattice, and if it is bounded, it is a Boolean lattice. In a lattice L bounded below an element y is called a *semicomplement* of x if $x \wedge y = 0$; and L is said to be *semicomplemented* (SC) if each $x \in L$ (with $x \neq 1$ if 1 exists in L) admits at least one nonzero semicomplement. Hence the (meet) pseudocomplement of an element in a lattice with 0 is a special semicomplement. We have already indicated in Section 1.4 that a pseudo-complemented lattice is a lattice with 0 in which the semicomplements of each element form a principal ideal. If, in a lattice bounded below, every interval of the form $[0, a]$ is complemented, the lattice is called *section-complemented* (SeC). A lattice with 0 is *section-semicomplemented* (SeSC) if every interval $[0, a]$ is semi-complemented. For the definition of orthocomplemented (OC) lattices we refer to Section 1.2.

Most of the above defined notions require the existence of a least element and sometimes the existence of a greatest element is also required. In what follows we shall assume the existence of these two elements. In any bounded lattice, the logical implications connecting the previous concepts are given by the arrows in Figure 1.14. All implications are well known and easy to verify.

The reverse implications do not hold in a general bounded lattice. Without attempting at a detailed study of the logical interdependence of all the conditions involved in the diagram, we merely state some examples and refer to Grillet & Varlet [1967] for a comprehensive treatment, in particular concerning the falseness of the implications (UC) $\Rightarrow$ (OC) and (UC) $\Rightarrow$ (SeC).

The pentagon N_5 is complemented but not section-semicomplemented. The diamond M_3 is relatively complemented (and thus section-complemented) but neither orthocomplemented nor uniquely complemented. The lattice L_4 of Figure 1.6 (Section 1.3) is section-complemented but not relatively complemented. The lattice S_7^* (cf. Figure 1.6, Section 1.3) is section-semicomplemented but not section-

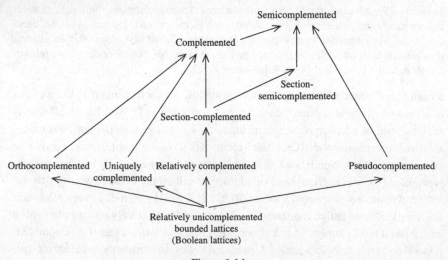

Figure 1.14

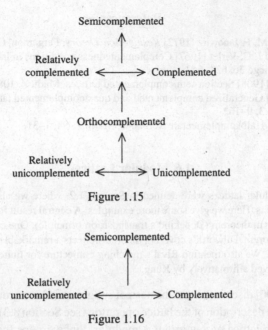

Figure 1.15

Figure 1.16

complemented; it is also semicomplemented but not complemented. A semicomplemented lattice that is not section semicomplemented can be found in Blyth & Janowitz [1972], Example 1.4.

If the lattices are modular or distributive, some of the above concepts turn out to be equivalent. For example, every complemented modular lattice is relatively complemented, and hence a bounded modular lattice is complemented if and only if it is relatively complemented. We indicate the interrelationships in Figures 1.15 and 1.16 (with the symbol ↔ denoting logical equivalence) and refer to Grillet & Varlet [1967] for more details.

Figure 1.15 shows the interrelationships in a bounded modular lattice. Figure 1.16 shows the interrelationships for bounded distributive lattices:

Notes

We have restricted ourselves to some types of complementation to which we return occasionally, in particular in Section 3.5. A number of other concepts related to complementation have been introduced, for example, generalized complements, quasicomplements (Szász [1953]) and "almost complemented" lattices (Wille [1966]). Some special types of complements in certain atomistic lattices will be considered in Section 2.5. The relationship between complements and fixed points in finite lattices is briefly indicated in Section 4.2.

References

Blyth, T. S. and M. F. Janowitz [1972] *Residuation Theory*, Pergamon, Oxford.
Grillet, P. A. and J. C. Varlet [1967] Complementedness conditions in lattices, Bull. Soc. Roy. Sci. Liège 36, 628–642.
Janowitz, M. F. [1968] Section semicomplemented lattices, Math. Z. 108, 83–76.
Szász, G. [1953] Generalized complemented and quasicomplemented lattices, Publ. Math. (Debrecen) 3, 9–16.
Wille, R. [1966] Halbkomplementäre Verbände, Math. Z. 94, 1–31.

1.6 Modular Lattices

Summary.   Modular lattices were defined in Section 1.2, where we also provided some examples and facts. Here we give some more examples. A central result for modular lattices is the isomorphism theorem (Dedekind's transposition principle). One consequence is the Kurosh–Ore theorem. Dilworth's covering theorem reflects a remarkable property of finite modular lattices. We also mention Rival's matching conjecture for finite modular lattices, which was resolved affirmatively by Kung.

Dedekind [1900] discovered that the normal subgroups of any group form a modular lattice (for a description of the lattice operations see Section 8.3). In other words, the congruence lattice of any group is modular. Since groups form a variety, we may also say that the variety of groups is *congruence-modular*. For example, the lattice of normal subgroups (coinciding with the lattice of all subgroups) of the Klein 4-group is isomorphic to M_3.

Congruence lattices of (universal) algebras are algebraic lattices, but they do not seem to possess other properties in general. However, in addition to groups, also other important kinds of algebras, such as rings, modules, and lattices, have modular congruence lattices. This underlines the importance of congruence modularity. In fact, we had already mentioned that the congruence lattices of lattices are distributive (cf. Section 1.3). On the other hand, congruence lattices of semilattices are not modular in general (S_7 is isomorphic to the congruence lattice of E_2^2, where E_2 is the two-element meet semilattice – cf. Section 1.3).

Figure 1.2(a) (Section 1.2) shows the lattice of congruences of Z_{18}, the cyclic group of order 18. We had already noted that this lattice is cyclically generated and distributive. It is a consequence of the fundamental theorem for finitely generated abelian groups that the subgroup lattice of any finite abelian group is cyclically generated.

Examples of modular atomistic lattices arise as *lattices of closed subspaces* (or *flats*) of projective incidence geometries. For example, the *Fano plane* (cf. Figure 1.17) has the lattice of flats shown in Figure 1.18. For more details on the interplay between directly indecomposable atomistic modular lattices and projective incidence geometries see Crapo [1986].

We have indicated above that both projective incidence geometries and finite abelian groups give rise to certain cyclically generated modular lattices. The

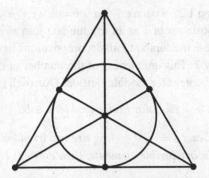

Figure 1.17

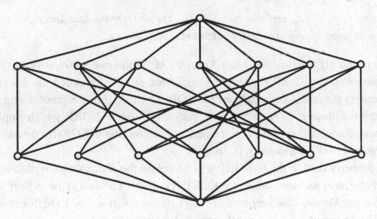

Figure 1.18

similarities in the behavior of these two structures led Baer [1942] to develop a unified theory for both, initiating thereby an important line of research.

Two intervals of the form $[a \wedge b, b]$ and $[a, a \vee b]$ are said to be *transposed*. The following result is referred to as the *isomorphism theorem for modular lattices* or *Dedekind's transposition principle*.

Theorem 1.6.1 *Transposed intervals of a modular lattice are isomorphic.*

We shall prove this result in a somewhat broader context in Section 2.1 (Theorem 2.1.4). For the special case of congruence lattices of groups, the transposition principle yields one of the group-theoretic isomorphism theorems (sometimes called the second isomorphism theorem; see e.g. Fraleigh [1989], Theorem 3.2, p. 182). Dedekind's transposition principle is essential for Ore's decomposition theory (see Ore [1936]).

The uniqueness property for finite distributive lattices (cf. Corollary 1.3.5) does not hold for modular lattices. as the diamond M_3 shows: referring to the notation of Figure 1.2(b) (Section 1.2), we have $1 = a \vee b = a \vee c = b \vee c$, that is, there are three different decompositions of 1 as an irredundant join of join-irreducible elements. However, all these irredundant join decompositions have the same number of components, namely 2. This uniqueness of the number of components follows from the Kurosh–Ore theorem for modular lattices (Kurosch [1935], Ore [1936]):

Theorem 1.6.2 *Let L be a modular lattice and let $a \in L$.*

(i) *If $a = x_1 \vee \cdots \vee x_m$ and $a = y_1 \vee \cdots \vee y_n$ are two irredundant decompositions of a as joins of join-irreducible elements, then every x_i can be replaced by a y_j such that*

$$a = x_1 \vee \cdots \vee x_{i-1} \vee y_j \vee x_{i+1} \vee \cdots \vee x_m.$$

(ii) *If $a = x_1 \vee \cdots \vee x_m$ and $a = y_1 \vee \cdots \vee y_n$ are two irredundant decompositions of a as joins of join-irreducible elements, then $m = n$.*

Conclusion (i) is also called the *Kurosh–Ore replacement property for join decompositions* ($\vee$-KORP, for short). Conclusion (ii) is also called the *Kurosh–Ore property for join decompositions* ($\vee$-KOP, for short). For the proof, using the isomorphism theorem (Theorem 1.6.1), one shows the $\vee$-KORP, which implies (by repeated application of the replacement procedure) the $\vee$-KOP (for details see e.g. Grätzer [1978], Theorem 5, p. 163).

The property dual to the $\vee$-KORP will be called the *Kurosh–Ore replacement property for meet decompositions* ($\wedge$-KORP, for short). The dual to the $\vee$-KOP will be called the *Kurosh–Ore property for meet decompositions* ($\wedge$-KOP, for short). Dualizing Theorem 1.6.2, we get that for finite irredundant meet decompositions in a modular lattice the $\wedge$-KORP and hence also the $\wedge$-KOP holds.

There are also nonmodular lattices for which the statement of the Kurosh–Ore theorem is true. For example, the pentagon satisfies both the $\wedge$-KORP and the $\vee$-KORP. A part of the investigations of Dilworth and Crawley centers around the problem of characterizing lattices that have the $\wedge$-KORP (see Crawley & Dilworth [1973]). We return to these questions in Sections 4.5 and 6.5 and in Chapter 8.

We remarked in Section 1.3 that $(J(L), \leq) \cong (M(L), \leq)$ [briefly: $J(L) \cong M(L)$] holds in a finite distributive lattice L. This is no longer true in the modular case, as the lattice of Figure 1.19 shows. Nevertheless Dilworth [1954] showed that in a finite modular lattice the number of join irreducibles equals the number of meet irreducibles. This follows from Dilworth's covering theorem, for which we refer to Section 6.1 (Theorem 6.1.9). For the modular lattice of Figure 1.19 we have $|J(L)| = |M(L)| = 5$.

In 1972 Rival proposed the following conjecture (see Rival [1990]): Given a finite modular lattice, there is a one-to-one mapping ϕ of the join irreducibles to the meet irreducibles such that $j \leq \phi(j)$ for every join irreducible j. Such a mapping is called a *matching* of the join-irreducible elements into the meet-irreducible

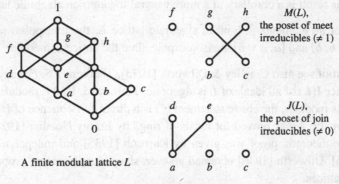

Figure 1.19

elements. The preceding conjecture has become known as *Rival's matching conjecture*. For the modular lattice of Figure 1.19 a matching is given by the mapping

$$0 \to 1, \qquad a \to f, \qquad b \to b, \qquad c \to c, \qquad d \to g, \qquad e \to h.$$

The matching conjecture was solved affirmatively by Kung [1985], who also succeeded in giving a common generalization of Dilworth's covering theorem and his own proof of Rival's matching conjecture (see also Kung [1986b], [1987]). These results of Kung will be discussed in more detail in Section 6.1.

Notes

Several results on normal subgroups of a group (and ideals of a ring) gave rise to generalizations to modular and to semimodular lattices. This underlines the importance of group theory as a source for modular and semimodular lattices. In addition to the second group-theoretic isomorphism theorem, we mention the Jordan–Hölder theorem (cf. McKenzie et al. [1987], Theorem 2.37), part of which is reflected in the Jordan–Dedekind chain condition treated in Section 1.9.

Ward [1939] proved that if a lattice satisfies the ACC or the DCC, then the property that transposed intervals are isomorphic is equivalent to modularity. If the chain conditions are dropped, however, then there are simple examples of nonmodular lattices L such that $[a \wedge b, b]$ and $[a, a \vee b]$ are isomorphic for all $a, b \in L$. Crawley [1959] showed that the isomorphism of all transposed intervals does characterize modularity, provided this condition is applied to the ideals of the lattice. More precisely, his result is the following:

(*) An arbitrary lattice L is modular if and only if for every pair of ideals A, B of L, the intervals $[A \wedge B, B]$ and $[A, A \vee B]$ in the ideal lattice of L are isomorphic.

In fact, this result is a corollary of a more general theorem on algebraic lattices:

(**) If, for all elements a, b of an algebraic lattice L, the transposed intervals $[a \wedge b, b]$ and $[a, a \vee b]$ are isomorphic, then the lattice is modular.

For a proof see also Crawley & Dilworth [1973], Theorem 3.5. For any lattice L, the lattice $\mathbf{I}(L)$ of all ideals of L is algebraic. Thus, since $\mathbf{I}(L)$ is modular if and only if L is modular, the above statement (*) is a direct consequence of (**).

Theorem 1.6.2 was proved for ideals of rings by Emmy Noether [1921]. The first lattice-theoretic proof was given by Kurosch [1935] and independently by Ore [1936]. Dilworth [1946] obtained an even sharper replacement property for modular lattices.

For modular lattices of finite length, there is a stronger decomposition theorem, the direct join decomposition theorem (cf. McKenzie et al. [1987], p. 66). This result, which has its roots likewise in group theory, is known as the Schmidt–Ore theorem or the Krull–Schmidt theorem.

References

Baer, R. [1942] A unified theory of projective spaces and finite abelian groups, Trans. Amer. Math. Soc. 52, 283–343.

Bogart, K. P., Freese, R., and Kung, J. P. S. (eds.) [1990] *The Dilworth Theorems. Selected Papers of Robert P. Dilworth*, Birkhäuser, Boston.

Crapo, H. H. [1986] Examples and basic concepts, in: White [1986], Chapter 1.

Crawley, P. [1959] The isomorphism theorem in compactly generated lattices, Bull. Amer. Math. Soc. 65, 377–379.

Crawley, P. and R. P. Dilworth [1973] *Algebraic Theory of Lattices*, Prentice Hall, Englewood Cliffs, N.J.

Dedekind, R. [1900] Über die von drei Moduln erzeugte Dualgruppe, Math. Ann. 53, 371–403; *Gesammelte Werke*, Vol. 2, pp. 236–271.

Dilworth, R. P. [1946] Note on the Kurosch–Ore theorem, Bull. Amer. Math. Soc.52, 659–663.

Dilworth, R. P. [1954] Proof of a conjecture on finite modular lattices, Ann. of Math. 60, 359–364.

Fraleigh, J. B. [1989] *A First Course in Abstract Algebra* (4th edition), Addison-Wesley, Reading, Mass.

Grätzer, G. [1978] *General Lattice Theory*, Birkhäuser, Basel.

Kung, J. P. S. [1985] Matchings and Radon transforms in lattices I. Consistent lattices, Order 2, 105–112.

Kung, J. P. S. [1986b] Radon transforms in combinatorics and lattice theory, in: *Combinatorics and Ordered Sets* (ed. I. Rival), Contemporary Mathematics 57, pp. 33–74.

Kung, J. P. S. [1987] Matchings and Radon transforms in lattices II. Concordant sets, Math. Proc. Camb. Phil. Soc. 101, 221–231.

Kurosch, A.G. [1935] Durchschnittsdarstellungen mit irreduziblen Komponenten in Ringen und in sogenannten Dualgruppen, Math. Sbornik 42, 613–616.

McKenzie, R., G.F. McNulty, and W. Taylor [1987] *Algebras, Lattices, Varieties*, Vol. 1, Wadsworth & Brooks/Cole, Monterey, Calif.

Mehrtens, H. [1979] *Die Entstehung der Verbandstheorie*, Gerstenberg, Hildesheim.

Noether, E. [1921] Idealtheorie in Ringbereichen, Math. Ann. 83, 24–66.

Ore, O. [1935] On the foundations of abstract algebra I, Ann. of Math. 36, 406–437.

Ore, O. [1936] On the foundations of abstract algebra II, Ann. of Math. 37, 265–292.

Rival, I. [1990] Dilworth's covering theorem for modular lattices, in: Bogart et al. [1990], pp. 261–264.

Ward, M. [1939] A characterization of Dedekind structures, Bull. Amer. Math. Soc. 45, 448–451.

White, N. L. (ed.) [1986] *Theory of Matroids*, Encyclopedia of Mathematics and its Applications, Vol. 26, Cambridge Univ. Press, Cambridge.

1.7 Upper and Lower Semimodularity

Summary. We recall the condition (Sm) of upper semimodularity and the more general Birkhoff condition (Bi) as well as the dual conditions (Sm*) and (Bi*). Moreover we consider the atomic covering property (C) and the Steinitz–Mac Lane exchange property (EP) and study the interrelationships between these conditions. The origins of semimodular lattices go back to the investigation of closure operators satisfying this exchange property. We give some examples of semimodular lattices. Geometric lattices and upper locally distributive (join-distributive) lattices are the most important instances. We also have a look at the independence of atoms and consider operators on the class of semimodular lattices.

An interval $[x, y]$ is said to be *prime* if y is an upper cover of x. Let $[a \wedge b, a]$ and $[b, a \vee b]$ be transposed intervals of a modular lattice. If $[a \wedge b, a]$ is a prime interval, then the isomorphism theorem (Theorem 1.6.1) implies that the transposed interval $[b, a \vee b]$ is also prime. Hence in a modular lattice

(Sm) $a \wedge b \prec a$ implies $b \prec a \vee b$

for arbitrary elements a, b. A lattice satisfying (Sm) will be called an *upper semi-modular lattice* or just *semimodular lattice*. (Sm) will also be referred to as the *semimodular implication*. Similarly, in any modular lattice,

(Sm*) $b \prec a \vee b$ implies $a \wedge b \prec a$

for arbitrary elements a, b. Implication (Sm*) is the reverse to (Sm). A lattice satisfying (Sm*) will be called *lower semimodular* or *dually semimodular*. The dual of an upper semimodular lattice is lower semimodular and vice versa.

The centered hexagon S_7 is the smallest nonmodular upper semimodular lattice, and its dual, the lattice S_7^*, is the smallest nonmodular lower semimodular lattice.

While any modular lattice is both upper and lower semimodular, the converse is not true, as the following example shows: the disjoint union of two continuous chains is not modular, but satisfies both (Sm) and (Sm*) in a trivial way, since there do not exist covers at all. This example immediately exhibits a decisive weakness of (Sm) when applied to lattices that do not possess "enough" covers, a situation that rather frequently occurs in lattices of infinite length. Such lattices are semimodular in a trivial way. To remedy this situation, other conditions were devised by Wilcox, Mac Lane, Dilworth, and Croisot, equivalent to (Sm) in lattices

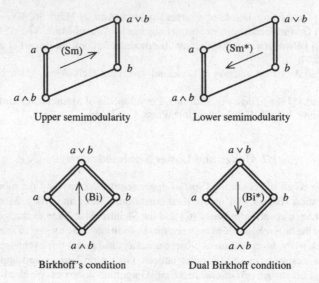

Figure 1.20

of finite length, but replacing (Sm) in a nontrival way for arbitrary lattices. These and further related conditions will be discussed in Chapters 2 and 3.

On the other hand, in lattices of finite length the conjunction of upper and lower semimodularity turns out to be equivalent to modularity, which follows, for example, from Corollary 2.3.11. This equivalence even holds for strongly atomic algebraic lattices (see Crawley & Dilworth [1973], Theorem 3.6, p. 24). Next we clarify the connection of (Sm) with *Birkhoff's condition*, a related implication involving the covering relation. This condition means that

(Bi) $a \wedge b \prec a, b$ implies $a, b \prec a \vee b$

for arbitrary elements a, b. The converse implication will be denoted by (Bi*). Upper and lower semimodularity as well as Birkhoff's condition and its dual are visualized in Figure 1.20, with double lines indicating the covering relation.

Both condition (Bi) and condition (Sm) were first considered by Birkhoff [1933]. In an arbitrary lattice, upper semimodularity obviously implies Birkhoff's condition.

The lattice in Figure 1.21 shows that (Bi) does not imply (Sm): Birkhoff's condition (Bi) is trivially satisfied. On the other hand, we have $a \wedge b \prec a$, but b is not a lower cover of $a \vee b$ and thus (Sm) does not hold. Thus Birkhoff's condition is weaker than semimodularity. This is why some authors also use the name *weakly semimodular* for lattices satisfying Birkhoff's condition (Bi). On the other hand, it is not difficult to see that the two conditions are equivalent in lattices of finite length. In fact, (Sm) and (Bi) are equivalent in a broader class of lattices, as follows from

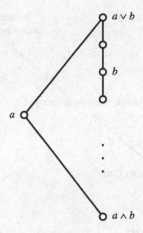

Figure 1.21

Theorem 1.7.1 *In an upper continuous strongly atomic lattice, Birkhoff's condition implies upper semimodularity.*

The proof is that of Crawley & Dilworth [1973], Theorem 3.7, pp. 25–26. (They formulate the assertion for algebraic strongly atomic lattices, but their proof uses only upper continuity and strong atomicity.)

Next we consider a special case of (Sm). In a lattice with 0, the condition

(C) If p is an atom and $b \wedge p = 0$, then $b \prec b \vee p$

is called the *atomic covering property* or briefly the *covering property*. In a lattice bounded below, (Sm) implies (C), but not conversely. If a lattice has no atoms, then (C) is trivially fulfilled. Hence it is not interesting to consider (C) in such lattices. In a lattice with atoms, condition (C) may or may not be satisfied. The lattice in Figure 1.21 satisfies Birkhoff's condition (Bi), but it does not possess the covering property (C). The pentagon with a new zero element added satisfies (C) but not (Bi). Hence neither of these conditions implies the other one. We also note the following characterization of the atomic covering property (C) (for a proof see Maeda & Maeda [1970], Theorem 7.6, p. 31):

Theorem 1.7.2 *Let L be a lattice with 0. Then the following conditions are equivalent:*

(i) *L has the atomic covering property (C);*
(ii) *if p is an atom then $p \, M \, x$ holds for all $x \in L$;*
(iii) *if p is an atom then $p \, M^* \, x$ holds for all $x \in L$.*

An element b of a lattice L is called *left modular* if bMx holds for all $x \in L$. Thus condition (ii) of the preceding theorem can be reformulated by saying that each atom is a left modular element.

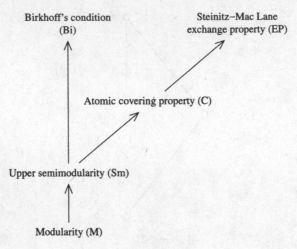

Figure 1.22

Let us now consider lattices with 0 that are atomic or atomistic. In these lattices we consider an exchange property related to upper semimodularity and to the atomic covering property:

(EP) If p and q are atoms and if $b \wedge p = 0$, then $p \leq b \vee q$ implies $q \leq b \vee p$.

(EP) is called the *Steinitz–Mac Lane exchange property*; it is the lattice-theoretic counterpart of the following well-known exchange property of a closure operator cl on a set X that is finite (or of finite character): Let $B \subseteq X$ and $p, q \in X$. Then $p, q \notin \text{cl}(B)$ and $p \in \text{cl}(B \cup q)$ imply $q \in \text{cl}(B \cup p)$. (Note that we briefly write $B \cup x$ for $B \cup \{x\}$.) In fact, the theory of semimodular lattices originated from this type of closure operators.

In lattices with 0, the atomic covering property (C) implies the Steinitz–Mac Lane exchange property (EP) (cf. Maeda & Maeda [1970], 7.9, p. 32) but not conversely (cf. Figure 1.21). On the other hand, (EP) and Birkhoff's condition (Bi) are unrelated, since (C) and (Bi) are unrelated.

The interrelationships are visualized in Figure 1.22.

In atomistic lattices several of the aforementioned concepts coincide (for a proof see Maeda & Maeda [1970], Theorem 7.10, p. 32):

Theorem 1.7.3 *In an atomistic lattice L, the following four statements are equivalent:*

 (i) *L has the atomic covering property (C);*
 (ii) *L has the Steinitz–Mac Lane exchange property (EP);*

(iii) *L is upper semimodular, that is, it satisfies* (Sm);

(iv) $a \wedge b \prec a$ *implies* $a \ M^* \ b$.

Following the terminology of Maeda & Maeda [1970], an atomistic lattice having the atomic covering property will be called an *AC lattice*. Algebraic AC lattices
are also called *matroid lattices*. In our usage, a geometric lattice is an AC lattice
of finite length.

Let us now give some examples. We already noted that modular lattices are
semimodular. In fact, we isolated upper and lower semimodularity as properties
shared by all modular lattices. Examples of modular lattices were given in Section
1.6. In particular, lattices of flats of projective incidence geometries are geometric
modular lattices, briefly *geomodular lattices*. Another example of a geomodular
lattice is shown in Figure 1.23(b). This lattice originates from the set S_1 (as a subset
of $\mathbb{R}^2$) shown in Figure 1.23(a) in the following way: take the subsets of S_1 of the
form $S_1 \cap W$ where W is an affine subspace of the affine space over $\mathbb{R}$, and order
these subsets by set inclusion (for more details see Crapo [1986]).

Nonmodular examples of geometric lattices arise from affine incidence geometries. For details we again refer to Crapo [1986]. A concrete example is the affine
incidence geometry on four points and six lines of Figure 1.24(a), whose lattice of
flats is shown in Figure 1.24(b).

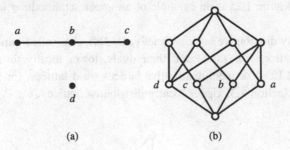

(a)   (b)

Figure 1.23

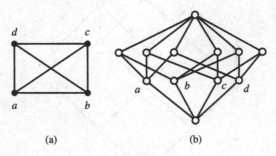

(a)   (b)

Figure 1.24

The nonmodular geometric lattice of Figure 1.24(b) has the property that for every $x > 0$ the principal dual ideal $[x)$ is a modular sublattice. Lattices with 0 having this property will be called *weakly modular*. A lattice with 0 is weakly modular if and only if $e \wedge f \neq 0$ implies $e \ M \ f$. Weakly modular lattices are not semimodular in general.

An important class of matroid lattices is given as follows. Denote by $\Pi(X)$ the set of all partitions (equivalence relations) of a set X. Give a partial order on $\Pi(X)$ by defining $\sigma \leq \tau$ if and only if $x \ \sigma \ y$ implies $x \ \tau \ y$. With this partial order $\Pi(X)$ forms a lattice, called the *partition lattice* (or *equivalence lattice*) of X. The partition lattices of a one-element, a two-element, and a three-element set are isomorphic with the one-element lattice, the two-element chain, and the diamond M_3, respectively. The partition lattice of a four-element set is shown in Figure 6.5 (Section 6.2). A thorough investigation of partition lattices was started by Ore [1942], who proved, among other things, that they are simple lattices, that is, they have no nontrivial congruence relation.

The notion of p-modularity was introduced in Section 1.3. While every modular lattice is p-modular, the pentagon shows that a p-modular lattice is in general neither upper nor lower semimodular. The lattice $AG_n(D)$ of affine subspaces of the n-dimensional vector space D^n over a field D that does not have characteristic 2 is p-modular and upper semimodular but not modular (see Gedeonová [1972]). The lattice in Figure 1.25 is an example of an upper semimodular lattice that is not p-modular.

Upper locally distributive lattices (briefly: ULDs), also called join-distributive lattices (j-d lattices, for short) and their duals, lower locally distributive lattices (briefly: LLDs) or meet-distributive lattices (m-d lattices, for short), were introduced in Section 1.3. Upper locally distributive lattices (e.g. S_7) are upper

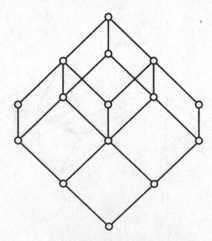

Figure 1.25

semimodular, while lower locally distributive lattices (e.g. S_7^*) are lower semi-modular. ULDs and LLDs will be dealt with in more detail in Chapter 7.

We have already stated that the lattice of convex subsets of a three-element chain is isomorphic with the lower locally distributive lattice S_7^* (see Section 1.3). This can be generalized in the following way. Assume the set E is linearly ordered as, for instance, the chain $C_n = \{1 < 2 < 3 < \cdots < n\}$. Then the set $\text{Int}(C_n)$ of all intervals of C_n, endowed with set inclusion, forms a lattice (the *lattice of subintervals*): the meet of two intervals is their set-theoretic intersection, which may be empty (the empty set is considered an interval), and their join is the convex closure of the set-theoretic union of the two intervals. This lattice is lower locally distributive. For the four-element chain of Figure 1.26(a) the lattice $\text{Int}(C_4)$ of subintervals is shown in Figure 1.26(b).

Next we turn to a class of upper semimodular lattices properly containing the class of upper locally distributive lattices. The lattice of Figure 1.27 is upper semimodular but not modular. However, the interval sublattices generated by the elements covering a given element ($\neq 1$) are all modular. A lattice having this property will be called *upper locally modular* or just *locally modular* (briefly:

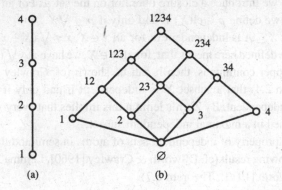

(a)   (b)

Figure 1.26

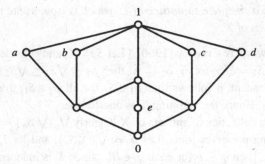

Figure 1.27

ULM). A lattice will be called *lower locally modular* (briefly: LLM) if its dual is upper locally modular.

The lattice of Figure 1.27 is a nonmodular semimodular lattice that satisfies the Kurosh–Ore replacement property for meet decompositions ($\wedge$-KORP) and hence also the Kurosh–Ore property for meet decompositions ($\wedge$-KOP) (cf. Section 1.6). For example, the $\wedge$-KOP is illustrated by the irredundant meet decompositions of 0, which are given by $0 = a \wedge c = b \wedge c = b \wedge d = b \wedge e = c \wedge e = d \wedge e$, showing that all such decompositions have the same number of components.

The lattice of subgroups of the dihedral group D_4 (the group of symmetries of the square) is the dual of the lattice shown in Figure 1.27, and hence it is lower semimodular (see also Section 8.3). For more details on the group-theoretic background of this example see for example Fraleigh [1989], pp. 70–71.

Next we consider a concept of independence. Let E be a set endowed with a closure operator cl. An element y of E is said to be *dependent* on a subset X of E if $y \in \text{cl}(X)$. Otherwise the element y is called *independent* of X. The subset X is said to be *independent* if, for all $x \in X$, $x \notin \text{cl}(X - x)$; otherwise X is said to be *dependent*.

Let now L be a complete atomic lattice, and denote by $A = A(L)$ its set of atoms. In a natural way we introduce a closure operator on the set A: For an atom p and a subset $X \subseteq A$ we define $p \in \text{cl}(X)$ if and only if $p \leq \bigvee X$.

Thus a subset $X \subseteq A$ is independent if, for all $x \in X$, $x \nleq \bigvee (X - x)$. The independence of X as defined here means that, for all $x \in X$, we have $x \wedge \bigvee (X - x) = 0$. Hence if L is upper continuous, then it can be shown (cf. Crawley & Dilworth [1973], Theorem 2.4) that a subset X is independent if and only if every finite subset of X is independent. By Zorn's lemma this implies that every independent subset is contained in a maximal independent subset.

An important property of independent sets of atoms in semimodular lattices is given in the following result (cf. Dilworth & Crawley [1960], Lemma 3.7; see also Crawley & Dilworth [1973], Theorem 6.7).

Theorem 1.7.4 *If L is a semimodular, algebraic, strongly atomic lattice and if X is an independent set of atoms of L, then the elements of L that are joins of subsets of X form a complete sublattice of L, which is isomorphic to the Boolean lattice of all subsets of X.*

Proof. (See Dilworth & Crawley [1960].) Let $S \subseteq X$, and consider the mapping $f : S \to \bigvee S$. If $\bigvee S_1 = \bigvee S_2$ for $S_1, S_2 \subseteq X$, then $p_1 \leq \bigvee S_1 \leq \bigvee S_2$ for all $p_1 \in S_1$. Since X is independent, it follows that $p_1 \in S_2$ for all $p_1 \in S_1$, that is, $S_1 \subseteq S_2$. Similarly $S_2 \subseteq S_1$. Hence the mapping f is one-to-one.

Now let S_α be a collection of subsets of X. Clearly $\bigvee_\alpha (\bigvee S_\alpha) = \bigvee (\cup_\alpha S_\alpha)$ and hence the mapping preserves joins. Let $b = \bigvee (\cap_\alpha S_\alpha)$, and let $T_\alpha = \{b \vee p : p \in S_\alpha - \cap_\alpha S_\alpha\}$. Then $q \succ b$ for each $q \in T_\alpha$, since X is independent and L is semimodular. Also $\bigvee T_\alpha = b \vee \bigvee(S_\alpha - \cap_\alpha S_\alpha) = \bigvee(\cap_\alpha S_\alpha) \vee \bigvee(S_\alpha - \cap_\alpha S_\alpha) = \bigvee(S_\alpha)$. Now $b \vee p_1 = b \vee p_2$, where $p_1, p_2 \in S_\alpha - \cap_\alpha S_\alpha$ implies that $p_1 = p_2$ since X is independent. Thus $\cap_\alpha T_\alpha = \emptyset$. Let us suppose that $\bigwedge_\alpha(\bigvee T_\alpha) > b$. Then $\bigwedge_\alpha(\bigvee T_\alpha) \geq r \succ b$ by strong atomicity, and hence $r \leq \bigvee T_\alpha$ for all α. But r is compact in $[b, b^+]$, and hence $r \leq \bigvee T_\alpha'$ for some finite subset T_α' for each α. For a fixed α there exist by semimodularity $q \in T_\alpha'$ such that $q \leq \bigvee(T_\alpha' - q) \vee r$. Since $\cap_\alpha T_\alpha = \emptyset$, there exists β such that $q \notin T_\beta$. But then $r \leq \bigvee T_\beta$ implies $q \leq \bigvee(T_\alpha' - q) \vee \bigvee(T_\beta \leq \bigvee(T_\alpha \cup (T_\beta - q))$. If $q = b \vee p$, then we have $p \leq \bigvee(S_\alpha \cup (S_\beta - p))$, contrary to the independence of X . Hence we conclude that $\bigwedge_\alpha(\bigvee S_\alpha) = \bigwedge_\alpha(\bigvee T_\alpha) = b = \bigvee(\cap_\alpha S_\alpha)$, that is, the mapping also preserves meets. ∎

Corollary 1.7.5 *If a is an element of a semimodular, algebraic, strongly atomic lattice, then $[a, a^+]$ is complemented and each element of $[a, a^+]$ is a meet of elements covered by a^+.*

Proof. (See Dilworth & Crawley [1960].) If $b \in [a, a^+]$, let S be a maximal independent set of atoms of $[a, a^+]$ contained in b. Extend S to a maximal independent set of atoms P. Then $b \vee \bigvee(P - S) = \bigvee P = a^+$ and clearly $b \wedge \bigvee(P - S) = a$, since $\bigwedge S \wedge \bigwedge(P - S) = a$. Now $p \vee b \succ b$ for each $p \in P - S$. Let $P_1 \subseteq P - S$ be such that the set $\{p \vee b : p \in P_1\}$ is maximal independent. Then if $p_1 \in P_1$ we have $a^+ \succ \bigvee\{p \vee b : p \in P_1 - p_1\}$ and $\bigwedge_p \bigvee\{p \vee b : p \in P_1 - p_1\} = b$ by the preceding theorem. ∎

A sublattice S of a lattice L is said to be *cover-preserving* in L if, for a, $b \in S$, $a \succ b$ in S implies $a \succ b$ in L. The preceding corollary implies that the set of independent atoms of a semimodular strongly atomic algebraic lattice generates a cover-preserving Boolean sublattice.

Let us now have a look at operators acting on the class of semimodular lattices. We have mentioned that Boolean algebras, distributive lattices, and modular lattices form varieties. In contrast to this, the class of semimodular lattices, that is, the lattices satisfying the semimodular implication (Sm), does not form a variety. More precisely: a nonmodular lattice always has the pentagon N_5 as a sublattice (see Theorem 1.2.2), and N_5 is not semimodular. Hence, every variety of semimodular lattices consists entirely of modular lattices. This means that a great part of the algebraic machinery that can be applied to modular lattices is not applicable to semimodular lattices.

However, the class of semimodular lattices is closed under a number of class operators. For example, it is not difficult to see that semimodular lattices form an elementary class and hence are closed under the formation of ultraproducts and elementary sublattices (for the notions of universal algebra not defined here see Grätzer [1979] or Burris & Sankappanavar [1981].)

A lattice homomorphism is *bounded* if each class of the kernel contains a least and a greatest element. It is easily seen that a homomorphism between complete lattices is bounded if and only if it respects the complete lattice operations. A class of lattices is *full* if it is closed under the formation of subdirect products, interval sublattices, and bounded homomorphic images. The following result of Kearnes [1991] shows that semimodular lattices form a full class.

Theorem 1.7.6 *The image of an upper semimodular complete lattice under a complete homomorphism is upper semimodular. Any convex sublattice of an upper semimodular lattice is upper semimodular. Any subdirect product of upper semimodular lattices is upper semimodular.*

Proof.   Suppose that $\varphi : L \to K$ is a complete lattice homomorphism from the complete lattice L onto the lattice K and that L is upper semimodular. Assume that K fails to be semimodular. Then we can find $w, x, y, z \in K$ such that $x \prec y$ and $x \vee z < w < y \vee z$. Choose $a, b, c, d \in L$ such that b is the least element satisfying $\phi(b) = y$, such that a is the greatest element of L satisfying $\varphi(a) = x$ and $a \le b$, and such that $\varphi(c) = z$, $\varphi(d) = w$. It is easily verified that the lattice of Figure 1.28(a) is a sublattice of L. Since L is upper semimodular and $a \vee (a \vee c)$ is not a lower cover of $b \vee (a \vee c)$, it follows that a is not a lower cover of b. Hence there is an element e such that $a < e < b$. Our choice of a and b implies $\varphi(a) < \varphi(e) < \varphi(b)$ or $x < \varphi(e) < y$. This contradicts $x \prec y$.

For the second assertion of the theorem we note that any failure of upper semimodularity in a convex sublattice of a lattice is a failure in the whole lattice. For the third statement, suppose that M is not an upper semimodular lattice. Then there are elements $p, q, r, s \in M$ such that $p \prec q$ and $p \vee r < s < q \vee r$ [see Fig. 1.28(b), with the double line indicating the covering relation]. The image of this pentagon under any surjective homomorphism ψ prevents the image of M from being upper semimodular unless we have at least $\psi(s) = \psi(p \vee r)$. Thus the elements s and $p \vee r$ cannot be separated by homomorphisms onto upper semimodular lattices. It

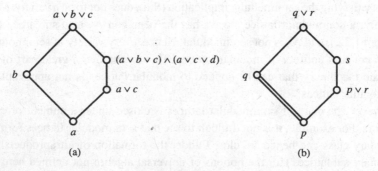

Figure 1.28

follows that M cannot be represented as a subdirect product of upper semimodular lattices. Since M was an arbitrarily chosen lattice failing upper semimodularity, every subdirect product of upper semimodular lattices is upper semimodular. ∎

In fact, semimodular lattices form a very large full class. Let Sem denote the class of semimodular lattices and L the class of all lattices. Whitman's proof that every lattice embeds in a partition lattice (Whitman [1946]) shows that $S(\mathrm{Sem}) = \mathrm{L}$. The fact that Sem is closed under subdirect products also shows that $\mathrm{L} = S(\mathrm{Sem}) \subseteq V(\mathrm{Sem}) = HP_s(\mathrm{Sem}) = H(\mathrm{Sem})$, that is, every lattice is a homomorphic image of a semimodular lattice.

Notes

Birkhoff [1933], Theorem 8.1, showed that a modular lattice satisfies what he calls condition (ξ) [= (Bi)] and condition (ξ') [= (Bi*)]. Birkhoff [1948], p. 66, uses (ξ') for (Bi). The name "Birkhoff's condition" is due to Ore [1943b], p. 564, who used it to obtain a lattice version of the classical Jordan–Hölder theorem for groups (cf. Section 1.9).

Lattices of finite length satisfying (Bi) [or equivalently (Sm)] were called *Birkhoff lattices* by Klein-Barmen [1937]. Note, however, that Dilworth [1941b] uses the term "Birkhoff lattice" in a stronger sense, namely for a lattice whose lattice of dual ideals satisfies (Bi) (see Section 3.2). In Mac Lane [1938] condition (Sm) appears as (E_4).

Matroids, that is, finite (or more general) sets endowed with a closure operator satisfying the Steinitz–Mac Lane exchange property, were introduced by Whitney [1935]. He and other authors (above all van der Waerden, Mac Lane, Teichmüller) recognized that different concepts of dependence in algebra (e.g. algebraic dependence and p-dependence of elements in a field extension) have many properties in common with the linear dependence of vectors. Matroids arise in many combinatorial and algebraic contexts. The ideas of independence and bases as in vector spaces, dependence as in algebraic dependence or p-dependence, circuits as in graphs, flats as in projective and affine geometries, geometric lattices, and the like all have the same underlying structure, that is, these concepts can be used to give cryptomorphic descriptions of matroids. That matroids show up in so many places and disguises makes them a worthwhile object of study. We have already mentioned matroids as an important source for semimodular lattices: for example, Birkhoff [1935a] was stimulated by the matroid concept of Whitney.

A closure operator on a set X (being finite or at least having finite character) and satisfying the Steinitz–Mac Lane exchange property defines a (possibly infinite) matroid on the set X. The closed subsets or flats of this matroid form with respect to set inclusion a matroid lattice, that is, an atomistic semimodular algebraic lattice.

In what follows we list a number of monographs on matroids and their applications: Crapo & Rota [1970a], Welsh [1976], White [1986], [1987], [1992], Recski [1989], Oxley [1993]. For a concise survey we refer to Mason [1977] (finite case) and Oxley [1992] (infinite matroids). For the historical development and the background of matroid theory see Kung [1986a]. Let us also note that Karzel & Kroll [1988], pp. 19–20, give a brief historical survey of the role of lattice theory in the development of geometry and deal, in particular, with the Steinitz–Mac Lane exchange property.

Upper locally distributive lattices form a class of semimodular lattices that are very different in behavior from matroid lattices. These lattices were discovered by Dilworth [1940]. Locally distributive lattices are cryptomorphic versions of what is today frequently called an antimatroid. See Chapter 7 for some more details.

It is not possible to characterize semimodular lattices by forbidden finite sublattices, since due to a result of Dilworth (cf. Crawley & Dilworth [1973], Chapter 14) every finite lattice can be isomorphically embedded in a semimodular lattice (see also Section 6.2). Quackenbush [1985] made an attempt to circumvent these difficulties by insisting that each lattice have a spanning M_3. He introduced nullary operations (constants) $a, b, c, 0, 1$ so that $a \wedge b = a \wedge c = b \wedge c = 0, a \vee b = a \vee c = b \vee c = 1$, and $0 \leq x \leq 1$ for all x and used the tensor product of semilattices to construct numerous semimodular lattices. On the other hand, we shall see that, for lattices of finite length, semimodularity can be contrasted to modularity by means of certain forbidden cover-preserving sublattices (cf. Theorem 3.1.10).

Hall [1971] showed that, for a semilattice S, the congruence lattice $\mathrm{Con}(S)$ is upper semimodular (cf. also Freese & Nation [1973]). A stronger result [viz., Con (S) is M-symmetric] follows from Theorem 9.1.1. This, together with the results of Papert [1964], shows that, for a semilattice S, $\mathrm{Con}(S)$ belongs to the class K of lattices L that have the following properties:

(i) L is (upper) semimodular;
(ii) every interval $[a, b]$ of L is pseudocomplemented;
(iii) L is coatomistic (dually atomistic);
(iv) L is algebraic.

Suzuki [1956] calls a group an LM group if its subgroup lattice is lower semimodular. For example, the above-mentioned dihedral group D_4 is an LM group. Let us mention here that the subgroups of a finite p-group form a lower semimodular lattice (cf. e.g. Hall [1959], Theorem 8.3.7, p. 123), that is, finite p-groups are LM groups. We shall return to finite p-groups in Section 8.3 in the broader context of subnormal subgroups. Upper semimodularity of finite subgroup lattices was characterized by Regonati [1996].

The set of all closure operators on a poset $P = (P, \leq)$ may be partially ordered in the following way: For two closure operators cl_1 and cl_2, set $cl_1 \leq cl_2$ if and only if $cl_1(x) \leq cl_2(x)$ for all $x \in P$. With respect to this partial order, the set of all closure operators on a poset P forms a complete lattice, which we denote by C_P. (For a totally unordered ground set this was already observed by Birkhoff [1936].) If $P = 2^S$, the set of all subsets of S, then we speak of closure operators on the set S (rather than 2^S) and denote the corresponding lattice by C_S (instead of C_{2^S}). Ore [1943b] showed that C_S is coatomistic and semimodular. The lattice C_P for a poset P was investigated by Hawrylycz & Reiner [1993]. For the sake of simplicity, they assume that the poset P is finite. However, many of their results are valid under weaker finiteness conditions on P (such as when P has no infinite chains or no infinite ascending chains). C_P is an upper semimodular lattice. In fact, Hawrylycz & Reiner [1993], Corollary 6, show that C_P is an upper locally distributive (join-distributive) lattice. For a more detailed treatment of related questions and, in particular, of topological closure operators and lattices of topologies, see Birkhoff [1967], Chapters V.4 and IX. For a comprehensive survey of properties of lattices of topologies $\Sigma(X)$ on a set X see Larson & Andima [1975].

Let Sub(L) denote the set of all subsets A of a lattice L that are closed under $\wedge$ and $\vee$ and partially ordered under set inclusion. Thus, if $A \in$ Sub(L) and $A \neq \emptyset$, then A is a sublattice of L. Clearly, Sub(L) is closed under arbitrary intersections (see Grätzer [1978], p. 24), and thus it forms a lattice. Koh [1973] proved that the lattice Sub(L) of all sublattices of a lattice L is modular if and only if L is a chain. This was extended for finite lattices L by Lakser [1973], who provided several necessary and sufficient conditions for the lattice of sublattices of L to be lower semimodular (one of these conditions is that L has no sublattice isomorphic to the direct product of a two-element chain with a three-element chain). Tan [1978] gives several characterizations of modular lattices that have a lower semimodular lattice of sublattices. For example, this is the case if and only if the lattice is a linear sum of lattices of dimension at most two.

For a lattice L we denote by Int(L) the set of all intervals of L including the empty interval. With respect to set-theoretic inclusion, Int(L) is a lattice with the empty interval as least element (Duthie [1942]). The lattice operations are given by $[a, b] \vee [c, d] = [a \wedge c, b \vee d]$ and $[a, b] \wedge [c, d] = [a \vee c, b \wedge d]$, it being understood that $[x, y] = \emptyset$ unless $x \leq y$.

By Csub(L) we denote the set of all convex sublattices of a lattice L including the empty sublattice. Csub(L) is a complete lattice with respect to set inclusion. Int(L) is a sublattice of Csub(L), but in general it is not a complete sublattice (Duthie [1942], Koh [1972]). Grätzer [1971], [1978], p. 56, poses several problems concerning Csub(L) and Sub(L) that can also be formulated for Int(L). We refer to the exercises below for some results on these problems.

Let us finally remark that Maeda [1974a] uses the concepts of local distributivity and local modularity in another sense than we do here.

Exercises

1. Show that upper semimodularity [condition (Sm)] is equivalent to each of the following two conditions (Croisot [1951], cf. also Dubreil-Jacotin et al. [1953], Lemme 2, pp. 90–91):
   (a) $y \succ x \wedge y$ and $x \leq z$ imply $x \vee (y \wedge z) = (x \vee y) \wedge z$,
   (b) $y \succ x \wedge z$ and $x \leq z$ imply $x \vee (y \wedge z) = (x \vee y) \wedge z$.

2. Upper semimodularity implies the following condition:

   (*) If $z \succ x$ and $y \succ t$ then $x \vee [y \wedge (z \vee t)] = [(x \vee y) \wedge z] \vee t$ (Croisot [1951]).

3. Give an example showing that the converse to condition (*) of the preceding exercise does not hold, that is, condition (*) does not imply upper semimodularity (Croisot [1951], 21°, p. 221; cf. also Dubreil-Jacotin et al. [1953], Lemme 3, p. 92). Show that in strongly atomic lattices, condition (*) and upper semimodularity are equivalent (Dubreil-Jacotin et al. [1953], Lemme 5, p. 97).

4. Let L be a semimodular lattice of finite length. If the greatest element of L is a join of atoms, then L is complemented. If each element of L is a join of atoms then L is relatively complemented. (Birkhoff).

5. Give an example of a weakly modular lattice that is not semimodular.

6. A nonempty subset of I of a finite lattice L is called a $2n$-ideal ($n = 1, 2, 3, \ldots$) of L if the following two conditions hold: (i) $a \leq i \in I$ implies $a \in I$ and (ii) If $a, b \in I, a \neq b$, and there exists a maximal chain of length n from a to $a \vee b$ as well as from b to $a \vee b$, then $a \vee b \in I$. Clearly every lattice ideal of a finite lattice is also a $2n$-ideal, but not conversely (example?). For $n = 1$ one obtains the notion of a 2-ideal. A finite lattice is called a 2-lattice if every 2-ideal is also an ideal.
   (a) Show that every 2-ideal of a finite semimodular lattice is an ideal.
   (b) Give an example of a nonsemimodular finite lattice in which every 2-ideal is an ideal.
   (c) Prove that a finite lattice L is semimodular if and only if for every $a \in L$ the dual ideal $[a)$ is a 2-lattice (Nieminen [1985a], [1985b]).

7. Show that an upper semimodular lattice of finite length that is complemented and (meet + join) pseudocomplemented is Boolean (cf. Reeg & Weiß [1991], Theorem 6.5.4). Remark: In the special case of upper locally distributive (join-distributive) lattices, this result is due to Dilworth [1940]. From the more general result it follows that if a lattice of finite length is complemented, (meet + join) pseudocomplemented, and non-Boolean, then it is not

semimodular. Specific examples of nonsemimodular complemented and (meet + join) pseudocomplemented lattices are the Tamari lattices T_n.

8. Prove that the lattice Int(L) of all intervals of a lattice L is upper semimodular if and only if L is the one-element or the two-element lattice (Igoshin [1980] and Birkhoff [1982], Lemma 5).

9. Prove that (Igoshin [1988])

   (a) The lattice Int(L) of all intervals of a lattice L is lower semimodular if and only if, for two arbitrary incomparable elements x and y of L, there exist elements u and v such that $x \wedge y < u < x < x \vee y$.

   (b) Let L be a lattice satisfying the ascending or descending chain condition. Then Int(L) is lower semimodular if and only if L is a chain.

Remark. Note that (a) implies (b). For lattices of finite length, assertion (b) was also noted in Igoshin [1980], Koh [1972], Chen & Koh [1972], and Birkhoff [1982]. If L is a finite chain, we have already indicated that Int(L) is lower semimodular; indeed, it is a lower locally distributive (or meet-distributive) lattice. Figure 1.26(b) shows this lattice for the four-element chain of Figure 1.26(a).

10. Give an example of a lattice L that is not a chain such that Int(L) is lower semimodular (Igoshin [1988]).

References

Birkhoff, G. [1933] On the combination of subalgebras, Proc. Camb. Phil. Soc. 29, 441–464.

Birkhoff, G. [1935a] Abstract linear dependence and lattices, Amer. J. Math. 57, 800–804.

Birkhoff, G. [1936] On the combination of topologies, Fund. Math. 29, 156–166.

Birkhoff, G. [1948] *Lattice Theory* (2nd edition), Amer. Math. Soc. Colloquium Publications, Vol. 25, New York.

Birkhoff, G. [1967] *Lattice Theory*, Amer. Math. Soc. Colloquium Publications, Vol. 25, Providence, R.I. (reprinted 1984).

Birkhoff, G. [1982] Some applications of universal algebra, in: *Colloq. Math. Soc. J. Bolyai, vol. 29: Universal Algebra, Esztergom 1977*, North-Holland, Amsterdam, pp. 107–128.

Burris, S. and H. P. Sankappanavar [1981] *A Course in Universal Algebra*, Springer-Verlag, New York.

Chen, C.C. and K.M. Koh [1972] On the lattice of convex sublattices of a finite lattice, Nanta Math. 5, 93–95.

Crapo, H. H. [1986] Examples and basic concepts, in: White [1986], Chapter 1.

Crapo, H. H. and G.-C. Rota [1970a] *On the Foundations of Combinatorial Theory: Combinatorial Geometries*, MIT Press, Cambridge, Mass.

Crawley, P. and R. P. Dilworth [1973] *Algebraic Theory of Lattices*, Prentice-Hall, Englewood Cliffs, N.J.

Croisot, R. [1951] Contribution à l'étude des treillis semi-modulaires de longueur infinie, Ann. Sci. Ecole Norm. Sup. (3) 68, 203–265.

Dilworth, R. P. [1940] Lattices with unique irreducible decompositions, Ann. of Math. 41, 771–777.

Dilworth, R. P. [1941b] Ideals in Birkhoff lattices, Trans. Amer. Math. Soc. 49, 325–353.

Dilworth, R. P. [1990c] Background to Chapter 5, in: Bogart et al. [1990], pp. 265–267.

Dilworth, R. P. and P. Crawley [1960] Decomposition theory for lattices without chain conditions, Trans. Amer. Math. Soc. 96, 1–22.

Dubreil-Jacotin, M. L., L. Lesieur, and R. Croisot [1953] *Leçons sur la théorie des treillis, des structures algébriques ordonnées et des treillis géométriques*, Gauthier-Villards, Paris.

Duthie, W. D. [1942] Segments of ordered sets, Trans. Amer. Math. Soc. 51, 1–14.

Fraleigh, J. B. [1989] *A First Course in Abstract Algebra* (4th ed.), Addison-Wesley, Reading, Mass.

Freese, R. and J. B. Nation [1973] Congruence lattices of Semilattices, Pacific J. Math. 49, 51–58.

Gedeonová, E. [1972] Jordan-Hölder theorem for lines, Mat. Časopis 22, 177–198.

Grätzer, G. [1971] *Lattice Theory: First Concepts and Distributive Lattices*, Freeman, San Francisco.

Grätzer, G. [1978] *General Lattice Theory*, Birkhäuser, Basel.

Grätzer, G. [1979] *Universal Algebra* (2nd edition), Springer-Verlag, New York.

Hall, M. [1959] *The Theory of Groups*, Macmillan, New York.

Hall, T. E. [1971] On the lattice of congruences on a semilattice, J. Austral. Math. Soc. 12, 456–460.

Hawrylycz, M. and V. Reiner [1993] The lattice of closure relations on a poset, Algebra Universalis 30, 301–310.

Igoshin, V. I. [1980] Lattices of intervals and lattices of convex sublattices of lattices (Russian), Uporjadočennyje množestva i rešotki 6, 69–76.

Igoshin, V. I. [1988] Semimodularity in interval lattices (Russian), Math. Slovaca 38, 305–308.

Karzel, H. and H.-J. Kroll [1988] *Geschichte der Geometrie seit Hilbert*, Wissenschaftliche Buchgesellschaft Darmstadt.

Kearnes, K. K. [1991] Congruence lower semimodularity and 2-finiteness imply congruence modularity, Algebra Universalis 28, 1–11.

Klein-Barmen, F. [1937] Birkhoffsche und harmonische Verbände, Math. Z. 42, 58–81.

Koh, K. M. [1972] On the lattice of convex sublattices of a lattice, Nanta Math. 5, 18–37.

Koh, K. M. [1973] On sublattices of a lattice, Nanta Math. 6, 68–79.

Kung, J. P. S. [1986a] *A Source Book in Matroid Theory*, Birkhäuser, Boston.

Lakser, H. [1973] A note on the lattice of sublattices of a finite lattice, Nanta Math. 6, 55–57.

Larson, R. E., and S. J. Andima [1975] The lattice of topologies: a survey, Rocky Mountain J. Math. 5, 177–198.

Mac Lane, S. [1938] A lattice formulation for transcendence degrees and p-bases, Duke Math. J. 4, 455–468.

Maeda, F. and S. Maeda [1970] *Theory of Symmetric Lattices*, Springer-Verlag, Berlin.

Maeda, S. [1974a] Locally modular and locally distributive lattices, Proc. Amer. Math. Soc. 44, 237–243.

Mason, J. H. [1977] Matroids as the study of geometrical configurations, in: *Higher Combinatorics* (ed. M. Aigner), Reidel, Dordrecht, pp. 133–176.

Nieminen, J. [1985a] 2-ideals of finite lattices, Tamkang J. Math. 16, 23–37.

Nieminen, J. [1985b] A characterization of the Jordan–Hölder chain condition, Bull. Inst. Math. Acad. Sinica 13, 1–4.

Ore, O. [1942] Theory of equivalence relations, Duke Math. J. 9, 573–627.

Ore, O. [1943b] Combinations of closure relations, Ann. of Math. 44, 514–533.

Oxley, J. G. [1992] Infinite matroids, in: White [1992], pp. 73–90.

Oxley, J. G. [1993] *Matroid Theory*, Oxford University Press.

Papert, D. [1964] Congruence relations in semi-lattices, J. London Math. Soc. 39, 723–729.

Quackenbush, R. W. [1985] Non-modular varieties of semimodular lattices with a spanning M_3, Discrete Math. 53, 193–205.

Recski, A. [1989] *Matroid Theory and its Applications in Electric Network Theory and in Statics*, Akadémiai Kiadó, Budapest, and Springer-Verlag, Berlin.

Reeg, S. and W. Weiß [1991] Properties of finite lattices, Diplomarbeit, Darmstadt.

Regonati, F. [1996] Upper semimodularity of finite subgroup lattices, Europ. J. Combinatorics 17, 409–420.

Suzuki, M. [1956] *Structure of a Group and the Structure of Its Lattice of Subgroups*, Ergebnisse, Vol. 10, Springer-Verlag, Berlin.

Tan, T. [1978] On the lattice of sublattices of a modular lattice, Nanta Math. 11, 17–21.

Welsh, D. [1976] *Matroid Theory*, Academic Press, London.

White, N. L. (ed.) [1986] *Theory of Matroids*, Encyclopedia of Mathematics and its Applications, Vol. 26, Cambridge Univ. Press, Cambridge.

White, N. L. (ed.) [1987] *Combinatorial Geometries*, Encyclopedia of Mathematics and its Applications, Vol. 29, Cambridge Univ. Press, Cambridge.

White, N. L. (ed.) [1992] *Matroid Applications*, Encyclopedia of Mathematics and its Applications, Vol. 40, Cambridge University Press, Cambridge.

Whitman, P. M. [1946] Lattices, equivalence relations, and subgroups, Bull. Amer. Math. Soc. 52, 507–522.

Whitney, H. [1935] On the abstract properties of linear dependence, Amer. J. Math. 57, 509–533.

1.8 Existence of Decompositions

Summary. We briefly review some fundamental notions and results concerning the existence of (meet) decompositions for elements of lattices that are not of finite length. These results include Birkhoff's subdirect-product theorem (the existence of meet decompositions in algebraic lattices) and Crawley's theorem on the existence of irredundant meet decompositions in strongly atomic algebraic lattices.

Irredundant finite meet decompositions and join decompositions were defined in Section 1.3. In a lattice of finite length, the existence of irredundant meet and join decompositions is guaranteed, that is, every lattice element has such decompositions. Under the assumption of the ascending chain condition (ACC, for short), each element of a lattice has an irredundant meet decomposition and every such decomposition is finite. Similarly, in a lattice satisfying the descending chain condition (DCC, for short), every element has an irredundant join decomposition and every such decomposition is finite. If the ACC or the DCC is dropped, the situation becomes more complicated. In general nothing can be said about the existence of decompositions without additional assumptions. These additional assumptions will consist of suitable weakenings of the ACC or the DCC. More precisely, we will replace the ACC by algebraicity and the DCC by strong atomicity.

First of all, however, the notions of meet-irreducible and join-irreducible element have to be generalized, which will be done within the class of complete lattices. In what follows we consider meet representations. In a similar way one obtains the corresponding notions for join representations.

An element q of a complete lattice L is said to be *completely meet-irreducible* if, for every subset S of L, $q = \bigwedge S$ implies $q \in S$. For every completely meet-irreducible element $q \neq 1$, there is a unique element $q^* \in L$ such that $q \prec q^*$ and $q < w$ if and only if $q^* \leq w$. Every completely meet-irreducible element is meet-irreducible, but not conversely. On the other hand, if a complete lattice is strongly atomic, then every meet-irreducible element is also completely meet-irreducible, that is, the two concepts are equivalent.

Next we define meet decompositions and irredundant meet decompositions for elements of a complete lattice. Let a be an element of a complete lattice L, and Q a set of completely meet-irreducible elements of L. A representation $a = \bigwedge Q$ is called a *meet decomposition* (briefly: $\wedge$-decomposition) of a. The $\wedge$-decomposition $a = \bigwedge Q$ is *irredundant* if $\bigwedge (Q - q) \neq a$ for all $q \in Q$.

Let now L be an algebraic lattice. The subdirect-product theorem of Birkhoff [1944] is equivalent to the assertion that if L is the congruence lattice of an algebra, then every element of L is a meet of completely meet-irreducible elements. Birkhoff & Frink [1948] proved that this holds in every complete, upper continuous lattice in which every element is the join of so-called inaccessible elements. These lattices are precisely what are now called algebraic lattices. Hence we have

Theorem 1.8.1   *Every element of an algebraic lattice has a meet decomposition.*

For a proof see Crawley & Dilworth [1973], pp. 43–44, or McKenzie et al. [1987].

Let us mention some results underlining the importance of algebraic lattices. For instance, the subalgebra lattice of any universal algebra is algebraic. Birkhoff & Frink [1948] showed that every algebraic lattice can be represented as the subalgebra lattice of some algebra. Moreover, the lattice of congruence relations of any universal algebra is algebraic. Hence algebraic lattices play a fundamental role in the decomposition theory of algebras (see Dilworth & Crawley [1960]). Grätzer & Schmidt [1963] proved that every algebraic lattice can be represented as the congruence lattice of some algebra.

The existence of meet decompositions, however, does not imply the existence of irredundant meet decompositions, as the following example illustrates (cf. Crawley & Dilworth [1973], p. 44). Let L be the lattice consisting of the rational numbers 1, $1/2, 1/3, \ldots, 0$ with respect to the natural ordering. L is an infinite chain satisfying the ACC. Each positive number in L is completely meet-irreducible, and 0 is the meet of any infinite set of positive numbers. However, 0 is not the meet of any finite set of positive numbers. Thus the element 0 has many meet decompositions but no irredundant meet decomposition.

Further assumptions are needed to guarantee the existence of irredundant meet decompositions. A sufficient condition is provided by strong atomicity.

Theorem 1.8.2 *If an algebraic lattice is strongly atomic, then every element has an irredundant meet decomposition.*

This result is due to Crawley [1961]. For a proof see also Crawley & Dilworth [1973], Theorem 6.4, p. 45.

The following example due to Erné [1983] shows that the existence of irredundant meet decompositions does not even imply that the lattice is weakly atomic. (A lattice is said to be *weakly atomic* if, for every pair a, b with $a > b$, there exist elements x, y such that $a \geq x \succ y \geq b$.)

Consider the subset $L = \{(x, y) \in [0, 1]^2 : x + y \leq 1\} \cup \{(1, 1)\}$ of the unit square $[0, 1]^2$, partially ordered componentwise by the usual $\leq$-relation. This is a complete semimodular lattice in which every element has a unique irredundant finite meet decomposition (into at most two coatoms). This lattice is not weakly atomic, since the only covering pairs are $(x, 1 - x) \prec (1, 1)$. The dual of this lattice is isomorphic to the closure system of all closed subintervals of the unit interval $[0, 1]$.

In the modular case, strong atomicity is also a necessary condition for the existence of irredundant meet decompositions (cf. Crawley & Dilworth [1973], Theorem 6.3, p. 44):

Theorem 1.8.3 *If every element of a modular algebraic lattice has an irredundant meet decomposition, then the lattice is strongly atomic.*

The lattice of Figure 1.29 (cf. Crawley & Dilworth [1973], Fig. 6-1, p. 48) shows that modularity cannot be replaced by semimodularity in the preceding theorem: this lattice is semimodular and satisfies the ACC, but it is not atomic; every element is the unique meet of at most two completely meet-irreducible elements. The lattice of Figure 1.29 arises, for example, as the lattice of so-called normal congruences

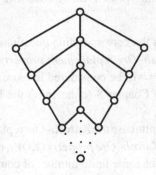

Figure 1.29

on the set of idempotents of certain inverse semigroups (see Eberhart & Seldon [1972]).

The preceding theorem also has a "local version" (due to Dilworth & Crawley [1960]) asserting the existence of irredundant meet decompositions for special elements:

Theorem 1.8.4   *If a is an element of a semimodular algebraic lattice and the interval* $[a, 1]$ *is atomic, then a has an irredundant meet decomposition.*

For a proof see also Crawley & Dilworth [1973], Theorem 6.9, pp. 47–48. Crawley & Dilworth [1973], p. 48, conjectured that the assumption of semimodularity is needed in Theorem 1.8.4. This conjecture was answered in the affirmative by Diercks [1982], who constructed a complete, atomic, lower but not upper semimodular lattice in which every element is compact, but the least element has no irredundant meet decomposition (see also Erné [1983]).

As in the finite-length case (see Section 1.6), we define the uniqueness property and the Kurosh–Ore replacement property for irredundant decompositions in complete lattices. A complete lattice is said to have *unique irredundant meet decompositions* if every lattice element has exactly one irredundant meet decomposition. A complete lattice is said to have the *Kurosh–Ore replacement property for meet decompositions* ($\wedge$-KORP) if each lattice element has at least one irredundant meet decomposition and the following property holds: If $a = \bigwedge Q = \bigwedge R$ are two irredundant meet decompositions of an element a, then for each $q \in Q$ there exists $r \in R$ such that $a = r \wedge \bigwedge (Q - q)$, and this resulting meet decomposition is irredundant. In Crawley & Dilworth [1973] lattices with $\wedge$-KORP are said to have *replaceable irredundant decompositions*.

Once the existence of irredundant meet decompositions is guaranteed (as in strongly atomic algebraic lattices), we may ask the following questions:

(i) When does the given lattice possess unique irredundant meet decompositions?

(ii) When does the given lattice possess the Kurosh–Ore replacement property for meet decompositions?

Similarly one formulates the corresponding questions for irredundant join decompositions and the *Kurosh–Ore replacement property for join decompositions* ($\vee$-KORP). These questions will be considered in more detail in Chapter 7 (concerning uniqueness) and in Chapter 8 (concerning the Kurosh–Ore replacement property).

We remarked that in the finite case the Kurosh–Ore replacement property (KORP) implies the corresponding *Kurosh–Ore property* (KOP), meaning that two irredundant decompositions have the same finite number of components (cf. Section 1.6, where this was mentioned for modular lattices). In general, however, it is not true that two irredundant meet decompositions of an element in a complete modular lattice have the same cardinality (see Crawley & Dilworth [1973], pp. 49–50, for an example).

Notes

Every algebraic lattice is weakly atomic and upper continuous, whereas the converse does not hold (see Crawley & Dilworth [1973], pp. 14–16). The fact that every element of an algebraic lattice has a meet decomposition (Theorem 1.8.1) may be extended to weakly atomic upper continuous lattices. The following results are all due to Erné [1983] and will be partly left as exercises.

If v covers u in L, then each maximal element of the set $M_u^v = \{w \in L : u = v \wedge w\}$ is completely meet-irreducible. Conversely, let q be a completely meet-irreducible element of a modular lattice L. Then $q \wedge y \prec q^* \wedge y$ holds for all $y \in L$ with $y \not\leq q$. If $u \prec v$ and $q \in M_u^v$, then q is maximal in M_u^v.

An immediate consequence is the following result: In order for each element of a complete lattice L to have a meet decomposition it is sufficient that L be weakly atomic and that for all $u \prec v$ the set M_u^v have a maximal element. In a modular lattice, this condition is also necessary for the existence of meet decompositions.

In semimodular lattices the second (but not the first) part of the existence criterion in the preceding result is also necessary. We also note that upper continuity implies the existence of maximal elements in M_u^v, but not conversely (for an example see Erné [1983]).

As a corollary we obtain: Suppose L is an upper continuous modular lattice or a (not necessarily complete) relatively pseudocomplemented lattice. Then L is weakly atomic if and only if every element of L has a meet decomposition.

We recall that a relatively pseudocomplemented lattice is a distributive lattice in which each of the sets M_u^v has a greatest element, denoted by $v * u$. Hence, if v covers u in a relatively pseudocomplemented lattice, then the relative pseudocomplement $v * u$ is completely meet-irreducible. Thus in weakly atomic relatively pseudocomplemented lattices, meet decompositions can be constructed without applying any maximal principle.

Avann [1964a] remarked that in a complete distributive lattice the existence of meet decompositions already implies upper continuity. Hence we have the following consequences:

(i) A complete distributive lattice is weakly atomic and upper continuous (or relatively pseudocomplemented) if and only if each of its elements has a meet decomposition.

(ii) For an upper and lower continuous lattice L, the following conditions and their duals are all equivalent:

(a) L is weakly atomic;
(b) L is algebraic;
(c) every element of L has a meet decomposition.

The last result extends a result of Bruns [1959] on distributive lattices and a result of Geissinger & Graves [1972]. Walendziak [1990a] further generalized some results of Erné [1983].

Exercises

1. Show that in a complete distributive lattice an element has at most one irredundant (meet or join) decomposition. (cf. e.g. Crawley & Dilworth [1973], Theorem 7.1, p. 49).

2. If an element a of a complete modular lattice has two meet decompositions $a = \bigwedge Q = \bigwedge R$, then for each $q \in Q$ there exists $r \in R$ such that $a = r \wedge \bigwedge (Q - q)$. Moreover, this resulting meet decomposition is irredundant if the meet decomposition $a = \bigwedge Q$ is irredundant (cf. Crawley & Dilworth [1973], Theorem 7.2, p. 49).

3. Prove that if an element u of a semimodular lattice has a meet decomposition, then for all $v \succ u$ the set M_u^v has a maximal element. (In fact, we may choose a completely meet-irreducible element q with $u \leq q$ and $v \not\leq q$; then $q \vee v$ covers q and thus $q \vee v = q^*$. Hence $v \leq w$ for all $w > q$, and q is maximal in M_u^v.)

4. For $x \in L$, set $M_x = \{y \in L : x = y \wedge z \text{ for some } z > x\}$. Note that $M_1 = \emptyset$ and $x \in M_x$ for all $x \neq 1$. Let x be an element of an upper continuous lattice such that each $y \in M_x$ is covered by another element. Show that then x has an irredundant meet decomposition (Erné [1983], Theorem 2). (The hypothesis on x is satisfied provided $[x, 1]$ is strongly atomic or $[x, 1]$ is atomic and semimodular.)

References

Avann, S. P. [1964a] Dependence of finiteness conditions in distributive lattices, Math. Z. 85, 245–256.

Birkhoff, G. [1944] Subdirect unions in universal algebra, Bull. Amer. Math. Soc. 50, 764–768.

Birkhoff, G. and O. Frink [1948] Representation of lattices by sets, Trans. Amer. Math. Soc. 64, 299–316.

Bogart, K. P., Freese, R., and Kung, J. P. S. (eds.) [1990] *The Dilworth Theorems. Selected Papers of Robert P. Dilworth*, Birkhäuser, Boston.

Bruns, G. [1959] Verbandstheoretische Kennzeichnung vollständiger Mengenringe, Arch. d. Math. 10, 109–112.

Crawley, P. [1961] Decomposition theory for non-semimodular lattices, Trans. Amer. Math. Soc. 99, 246–254.

Crawley, P. and R. P. Dilworth [1973] *Algebraic Theory of Lattices*, Prentice–Hall, Englewood Cliffs, NJ.

Diercks, V. [1982] Zerlegungstheorie in vollständigen Verbänden, Diplomarbeit, Univ. Hannover.

Dilworth, R. P. [1990a] Background to Chapter 3, in: Bogart et al. [1990], pp. 89–92.

Dilworth, R. P. and P. Crawley [1960] Decomposition theory for lattices without chain conditions, Trans. Amer. Math. Soc. 96, 1–22.

Eberhart, C. and J. Seldon [1972] One-parameter inverse semigroups, Trans. Amer. Math. Soc. 168, 53–66.

Erné, M. [1983] On the existence of decompositions in lattices, Algebra Universalis 16, 338–343.

Geissinger, L. and W. Graves [1972] The category of complete algebraic lattices, J. Combin. Theory (A) 13, 332–338.

Grätzer, G. and E. T. Schmidt [1963] Characterizations of congruence lattices of abstract algebras, Acta Sci. Math. (Szeged) 24, 34–59.

Jónsson, B. [1990] Dilworth's work on decompositions in semimodular lattices, in: Bogart et al. [1990], pp. 187–191.

McKenzie, R., G. F. McNulty, and W. Taylor [1987] *Algebras, Lattices, Varieties*, Vol. 1, Wadsworth & Brooks/Cole, Montcrey, Calif.

Walendziak, A. [1990a] Meet-decompositions in complete lattices, *Periodica Math. Hungar.* 21, 219–222.

1.9 The Jordan–Dedekind Chain Condition

Summary. The original Jordan–Hölder theorem for groups has been extended and generalized by many authors. We focus here on the result of Jordan that the principal series of a group have the same length. Already Dedekind [1900] recognized this as a special case of a result on modular lattices. This is why one speaks nowadays of the Jordan–Dedekind chain condition. However, the expression "Jordan–Hölder chain condition" is also used. Ore [1943a] obtained characterizations of the Jordan–Dedekind chain condition in finite partially ordered sets; see also Mac Lane [1943]. Their results imply results on semimodular lattices that were later rediscovered several times by several authors. We shall have a closer look here at some of these results. For the group-theoretic background cf. e.g. Hall [1959].

The *length* of a finite chain C is $|C| - 1$. The length of the chain C will be denoted by $l(C)$. We write $l(C) = \infty$ if C is an infinite chain. A poset P is said to be *of length n* [briefly: $l(P) = n$], where n is a natural number, if there is a chain in P of length n and all chains in P are of length $\leq n$. A poset P is of *finite length* if it is of length n for some natural number n. Otherwise a poset is said to be of *infinite length*, in which case we write $l(P) = \infty$.

A chain C in a poset P is called *maximal* if, for any chain D in P, $C \subseteq D$, implies that $C = D$. Using Zorn's lemma, one can show that every chain is contained in a maximal chain. In particular, in any poset there is at least one maximal chain between any pair of elements a, b with $a \leq b$. In general there is more than one maximal chain between such a pair of elements, and these maximal chains have different length.

The following restriction on the lengths of maximal chains between two elements of a poset is the *Jordan–Dedekind chain condition*, (JD) for short:

(JD) For all elements a, b with $a < b$, all maximal chains of the interval $[a, b]$ have the same length, that is, either all chains between a and b are infinite or the lengths of all chains equal the same finite number.

We define the *natural rank* (briefly: *rank* or *height*) $r(x)$ of an element x of a poset bounded below as the length of the interval $[0, x]$. The domain of values of the rank function r (also called the height function) consists of nonnegative numbers (including perhaps the symbol $+\infty$) with $r(x) = 0$ if and only if $x = 0$. No confusion will arise from the fact that in general we use the same symbol for the least element of a lattice as for the number 0.

A poset is said to be of *locally finite length* (or *discrete*) if each of its intervals is of finite length. If a poset with 0 (bounded below) is of locally finite length, then each of its elements has finite rank. If a poset is bounded, then it has locally finite length if and only if it has finite length. If a poset with 0 has locally finite length and satisfies the Jordan–Dedekind chain condition, then for each element x the rank $r(x)$ equals the common length of the maximal chains between 0 and x. In this case $x \prec y$ holds if and only if $x \le y$ and $r(x) + 1 = r(y)$.

More generally, let d be a function defined on a poset (not necessarily bounded from below) such that the domain of values of d consists of arbitrary integers and perhaps one or both of the symbols $+\infty$, $-\infty$. Then d is called a *dimension function* provided that $x \prec y$ holds if and only if $x \le y$ and $d(x) + 1 = d(y)$.

If a poset has a dimension function d, then we also say that the poset is *graded by* d. If a poset is graded by its rank function, we also say that it is *graded*. Let now P be a poset with 0 in which all chains are finite. Then P satisfies the Jordan–Dedekind chain condition if and only if it is graded by its rank function $r(x)$.

For the following result (which is also known as the Croisot–Szász theorem) we refer to Szász [1951–2] (see also Szász [1963], Theorem 41) and Dubreil-Jacotin et al. [1953], Theorem 1, p. 88. The proof is left as an exercise.

Theorem 1.9.1 *Any upper or lower semimodular lattice satisfies the Jordan–Dedekind chain condition.*

Corollary 1.9.2 *If a lattice of locally finite length is upper or lower semimodular, then it also satisfies the Jordan–Dedekind chain condition and has a dimension function.*

Independent subsets of atoms in a semimodular lattice were considered in Section 1.7. Let us note here the following characterization of finite independent subsets of atoms in a semimodular lattice in terms of the rank function.

Theorem 1.9.3   *Let* $X = \{p_1, \ldots, p_n\}$ *be a set of n atoms of a semimodular lattice. Then the following conditions are equivalent:*

(i) *X is independent;*
(ii) $(p_1 \vee \cdots \vee p_i) \wedge p_{i+1} = 0$ *for* $i = 1, 2, \ldots, n - 1;$
(iii) $r(p_1 \vee \cdots \vee p_n) = n.$

For a proof see e.g. Grätzer [1978], Theorem 4, pp. 174–175, or Crawley & Dilworth [1973], Theorem 6.8, p. 47.

We have already given examples of graded lattices being neither upper nor lower semimodular, such as the lattice of Figure 1.4(a) (Section 1.2). Another such example is shown in Figure 1.30.

The lattice of Figure 1.30 occurs as the *face lattice of a square*. More generally, let Δ be a finite simplicial or polyhedral complex with all maximal faces of the same dimension. Then the faces of Δ (including the improper faces $\emptyset$ and Δ) form a graded lattice $L(\Delta)$ with respect to set inclusion. The lattice $L(\Delta)$ is called the *face lattice* of Δ. For more details on face lattices see Björner et al. [1982].

Another source of graded lattices comes from group theory. Let G be a group, and denote by $L(G)$ its subgroup lattice (for a formal definition see Section 8.3). We have the following characterization of the Jordan–Dedekind chain condition for the subgroup lattice of a finite group.

Theorem 1.9.4   *Let G be a finite group. Then the subgroup lattice $L(G)$ satisfies the Jordan–Dedekind chain condition if and only if G is supersolvable.*

This theorem is due to Iwasawa [1941] (see e.g. Birkhoff [1967], Theorem 21, p. 177). For the notion of supersolvability in groups we refer to Hall [1959] and to

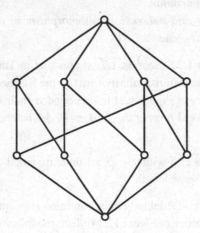

Figure 1.30

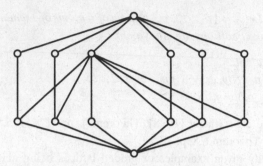

Figure 1.31

Section 8.3. A group G whose subgroup lattice $L(G)$ satisfies the Jordan–Dedekind chain condition is also called a *J-group* (Suzuki [1956]).

A specific example of such a subgroup lattice is given by the group H_{20}, the metacyclic, supersolvable holomorph of the cyclic group of order 5, with the defining relations $a^5 = b^4 = 1$, $b^{-1}ab = a^2$ (cf. Birkhoff [1967], p. 177). The lattice $L(H_{20})$ of subgroups of H_{20} is shown in Figure 1.31.

A group with a lower semimodular subgroup lattice of finite order is always a J-group, and so it is supersolvable by Theorem 1.9.4. Thus finite nilpotent groups (in particular, finite p-groups) are supersolvable (see Section 8.3 for more details). We have the following characterization due to Ito [1951] and Jones [1946] (see also Suzuki [1956] and Schmidt [1994]).

Theorem 1.9.5 *Let G be a finite group and $L(G)$ its lattice of subgroups. Then the following two conditions are equivalent:*

(i) *$L(G)$ is lower semimodular;*
(ii) *G is supersolvable and induces an automorphism of prime order in every factor of a principal series.*

The lattice of Figure 1.21 (Section 1.7) shows that in Theorem 1.9.1 it is not possible to replace upper semimodularity (Sm) by the Birkhoff condition (Bi). On the other hand, the lattice of Figure 1.21 has a bounded chain of infinite length and satisfies the following *weak form of the Jordan–Dedekind chain condition*, (WJD) for short:

(WJD) For all elements a, b with $a < b$, all finite maximal chains of the interval $[a, b]$ have the same length.

It is clear that the Jordan–Dedekind chain condition (JD) implies the weak form (WJD). Moreover, the lattice of Figure 1.21 reflects the following general situation:

Theorem 1.9.6 *Birkhoff's condition (Bi) implies the weak form (WJD) of the Jordan–Dedekind chain condition.*

For a proof – which is left as an exercise – see Rudeanu [1964], Lemma 5 (cf. also Dubreil-Jacotin et al. [1953], pp. 64–65). In upper semimodular lattices of locally finite length all definable dimension functions satisfy the inequality of submodularity (for a proof see e.g. Szász [1963], Theorem 42):

Theorem 1.9.7 *If L is a semimodular lattice of locally finite length, then every dimension function d definable on L is submodular, that is, $d(a) + d(b) \geq d(a \wedge b) + d(a \vee b)$ holds for all $a, b \in L$.*

One also says in this case that d is a *semimodular function*. If L is a lower semimodular lattice of locally finite length, then every dimension function d definable on L is *supermodular*, that is, for all $a, b \in L$ one has $d(a) + d(b) \leq d(a \wedge b) + d(a \vee b)$. One also says in this case that d is a *lower semimodular function*.

Corollary 1.9.8 *If a lattice L of locally finite length is both upper and lower semimodular, then for every dimension function d definable on L the dimension equation $d(a) + d(b) = d(a \wedge b) + d(a \vee b)$ holds for all $a, b \in L$.*

We leave it as an exercise to show that lattices satisfying the assumptions of this corollary are modular (Birkhoff [1933], Theorem 10.2). The next result (also due to Birkhoff [1933]) gives two further characterizations of upper semimodularity in lattices of finite length.

Theorem 1.9.9 *For a lattice L of finite length the following conditions are equivalent:*

  (i) *L is upper semimodular [condition (Sm)];*
  (ii) *L satisfies Birkhoff's condition (Bi);*
  (iii) *at least one dimension function can be defined on L, and every dimension function d of L is a submodular function.*

Sketch of proof.   (i) $\Rightarrow$ (iii): If L is a finite semimodular lattice, then it satisfies the Jordan–Dedekind chain condition (Theorem 1.9.1) and at least one dimension function can be constructed on L (for details of this construction see e.g. Szász [1963], Section 9). Now the assertion follows from Theorem 1.9.7.

(iii) $\Rightarrow$ (ii): Let $x, y, a \in L$ such that $x \succ a$, $y \succ a$, $x \neq y$. Then we have by (iii) for any dimension function d of L the relation $d(x \vee y) - d(x) \leq d(y) - d(x \wedge y) = d(y) - d(a) = 1$. Here $d(x \vee y) - d(x) = 0$ is not possible, and thus $d(x \vee y) - d(x) = 1$, that is, $x \vee y \succ x$. Similarly we get $x \vee y \succ y$, that is, (Bi) holds.

(ii) $\Rightarrow$ (i): This implication will be shown to hold for a class of lattices properly containing the class of lattices of finite length (see Theorem 3.1.7).   ■

Corollary 1.9.10 *A lattice of finite length graded by its rank function r is upper semimodular if and only if r is submodular.*

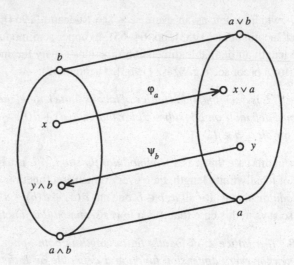

Figure 1.32

Corollary 1.9.11   *If* $\{x_0, x_1, \ldots, x_n\}$ *is a connected chain in a semimodular lattice of finite length, then the range of the mapping* $\varphi_a : x_i \to x_i \vee a$ *is a (possibly shorter) connected chain.*

Let us now establish a connection between modular pairs and the rank function. We start with the following result known as *Schwan's lemma* (see Birkhoff [1967], p. 73) involving the mappings visualized in Figure 1.32.

Lemma 1.9.12   *In any lattice, the maps* $\varphi_a : x \to x \vee a$ *from* $[a \wedge b, b]$ *(the lower transpose) to* $[a, a \vee b]$ *(the upper transpose) and* $\psi_b : x \to x \wedge b$ *from* $[a, a \vee b]$ *to* $[a \wedge b, b]$ *are isotone and satisfy*

$(+)\quad \varphi_a \psi_b \varphi_a = \varphi_a$ *on* $[a \wedge b, b]$

and

$(++)\quad \psi_b \varphi_a \psi_b = \psi_b$ *on* $[a, a \vee b]$.

Proof.   By the modular inequality we have $\varphi_a \psi_b(x) \leq x$ for all $x \in [a, a \vee b]$ and $y \leq \psi_b \varphi_a(y)$ for all $y \in [a \wedge b, b]$. We get $\psi_b \varphi_a \psi_b(x) \leq \psi_b(x)$ and $\psi_b \varphi_a \psi_b(x) = b \wedge (a \vee (b \wedge x)) \geq b \wedge (b \wedge x) = \psi_b(x)$, which shows that $(++)$ holds. Similarly $(+)$ holds.   ∎

Modular pairs can be characterized by means of the order-preserving maps φ_a and ψ_b (cf. e.g. Reeg & Weiß [1991], Lemma 6.2.6, p. 67).

Lemma 1.9.13   *With* φ_a *and* ψ_b *as defined above, the relation* $a \, M \, b$ *holds if and only if*

$(+++)$ $x = (x \vee a) \wedge b = \psi_b \varphi_a(x)$

holds for all $x \in [a \wedge b, b]$, *that is,* $\psi_b \varphi_a = \mathrm{id}_E$.

Proof. For $x \in [a \wedge b, b]$ we have $x \vee (a \wedge b) = x$, and hence $a\ M\ b$ implies $(+++)$. Conversely, if $(+++)$ holds and $t \in L$, $t \leq b$, then $x = t \vee (a \wedge b) \in [a \wedge b, b]$ and thus $x = t \vee (a \wedge b) = $ [because of $(+++)$] $[t \vee (a \wedge b) \vee a] = (t \vee a) \wedge b$, that is, $a\ M\ b$. ∎

We can now derive the following characterization of modular pairs.

Lemma 1.9.14 *In any lattice the following statements are equivalent:*

(i) $a\ M\ b$;

(ii) *the map* φ_a *of the lower transpose to the upper transpose is one-to-one;*

(iii) *the map* ψ_b *of the upper transpose to the lower transpose is onto.*

Proof. (i) $\Rightarrow$ (ii) and (i) $\Rightarrow$ (iii): By Lemma 1.9.13, condition (i) implies that $\psi_b \varphi_a$ is the identity map and thus (ii) and (iii) follow from (i).

(ii) $\Rightarrow$ (i): If $a\ M\ b$ does not hold, then there exists an element $x \in [a \wedge b, b]$ such that $x < (x \vee a) \wedge b = z$. Since $\varphi_a(z) = \varphi_a(x)$, it follows that φ_a is not one-to-one.

(iii) $\Rightarrow$ (i): If (iii) holds, then for all $x \in [a \wedge b, b]$ there exists $y \in [a, a \vee b]$ such that (applying Lemma 1.9.12) $x = \psi_b(y) = \psi_b \varphi_a \psi_b\ (y) = \psi_b \varphi_a(x)$. Hence $a\ M\ b$ follows from Lemma 1.9.13. ∎

Since $a\ M\ b$ implies that φ_a is one-to-one, we get in particular: if C is a chain of length n in $[a \wedge b, b]$, then $\varphi_a(C) = \{\varphi_a(c) : c \in C\}$ is a chain of the same length in $[a, a \vee b]$. Hence if $[a \wedge b, a \vee b]$ is an interval of finite length in a lattice that satisfies the Jordan–Dedekind chain condition, then $a\ M\ b$ implies $r(b) - r(a \wedge b) \leq r(a \vee b) - r(a)$. Thus we have proved

Lemma 1.9.15 *In a lattice of finite length that satisfies the Jordan–Dedekind chain condition, $a\ M\ b$ implies $r(a) + r(b) \leq r(a \wedge b) + r(a \vee b)$.*

In the semimodular case we have the following characterization of modular pairs.

Lemma 1.9.16 *In a semimodular lattice of finite length $a\ M\ b$ holds if and only if $r(a) + r(b) = r(a \wedge b) + r(a \vee b)$.*

Proof. Let L be a semimodular lattice of finite length and $a, b \in L$. Then $r(a) + r(b) \geq r(a \wedge b) + r(a \vee b)$ (by Theorem 1.9.9). If $a\ M\ b$ holds, then we also have (by Lemma 1.9.15) $r(a) + r(b) \leq r(a \wedge b) + r(a \vee b)$. Thus equality holds.

Conversely, assume that $a\ M\ b$ fails. Then by Lemma 1.9.14, φ_a maps two distinct elements x and y ($y \nleq x$) of $[a \wedge b, b]$ onto the same element $z = \varphi_a(x)$ of $[a, a \vee b]$. It follows that there is a connected chain C from $a \wedge b$ to

b containing the two distinct elements x and $x \vee y$ that have the same image under φ_a. By Corollary 1.9.11, the map φ_a carries connected chains of $[a \wedge b, b]$ onto connected chains of $[a, a \vee b]$. Hence by assumption, $\varphi_a(C)$ is a shorter connected chain from a to $a \vee b$ through z. Therefore $r(b) - r(a \wedge b) > r(a \vee b) - r(a)$. ∎

We have already noted that $a \, M \, b$ does not imply $b \, M \, a$ in general (take for example the pentagon). A lattice L is called *M-symmetric* if $a \, M \, b$ implies $b \, M \, a$ for arbitrary $a, b \in L$.

Lemma 1.9.17   *In any lattice, M-symmetry implies upper semimodularity.*

Proof.   Assume that a lattice L is not upper semimodular. Then there exist elements $x, y, z \in L$ such that $x \wedge y \prec x$ and $y < z < x \vee y$. It follows that $x \succ x \wedge z = x \wedge y$ and hence $z \, M \, x$ (cf. Maeda & Maeda [1970], Lemma 7.5, p. 31). On the other hand, we have $x \, \bar{M} \, z$, and thus L is not M-symmetric. ∎

The converse implication is not true in general, as the lattice of Figure 1.33 (taken from Croisot [1951], p. 242) shows. In Figure 1.33 the arrows indicate a chain isomorphic to the chain of the positive integers. It is clear that this lattice is upper semimodular. On the other hand we have $b \, M \, d$ but $d \, \bar{M} \, b$, and thus the lattice is not M-symmetric.

The concepts of modular pairs and M-symmetric lattices were introduced by Wilcox [1938], [1939]. The preceding lemma is due to Wilcox [1944], who also showed that, for lattices of finite length, upper semimodularity means exactly the same as M-symmetry:

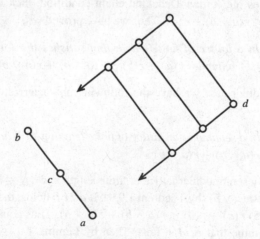

Figure 1.33

Theorem 1.9.18 *A lattice of finite length is upper semimodular if and only if it is M-symmetric.*

The equivalence of upper semimodularity and M-symmetry will be shown for a class of lattices properly containing the lattices of finite length (see Theorem 3.1.7). Let us note, however, that the lattice of Figure 1.33 satisfies the ACC and hence it is algebraic. It follows that M-symmetry and upper semimodularity are not equivalent for algebraic lattices. Thus the lattice of Figure 1.33 anticipated a negative answer to Grätzer [1978], Problem IV.16, p. 225. M-symmetric lattices will be considered in more detail in Chapter 2.

Referring to the survey given by Monjardet [1981], we close this section with a metric characterization of semimodularity. Let X be a set and f a nonnegative real-valued function defined on the set X^2 of ordered pairs of X. We shall have a look at the following properties that can be satisfied by f:

(1) For every $x \in X$, $f(x, x) = 0$.
(2) For all $x, y \in X$, $f(x, y) = 0$ implies $x = y$.
(3) For all $x, y \in X$, $f(x, y) = f(y, x)$.
(4) For all $x, y, z \in X$, $f(x, y) \le f(x, z) + f(z, y)$.
(5) For all $x, y, z \in X$, $f(x, y) \le \max[f(x, z), f(z, y)]$.

The function f is called a semimetric, quasimetric, metric, ultrametric if it satisfies the properties [(1), (2), (3)], [(1), (3), (4)], [(1), (2), (3), (4)], [(1), (2), (3), (5)], respectively. An ultrametric is a metric, but not conversely.

In what follows, the ground set X will always be assumed to be a finite poset and will be denoted by $(P, \le)$ or briefly by P. The *covering graph* $G(P)$ of P is the graph whose vertices are the elements of P and whose edges are the pairs (x, y) with $x, y \in P$ satisfying $x \prec y$ or $y \prec x$. Thus $G(P)$ is the *unoriented graph* underlying the Hasse diagram of P. For more on diagrams and covering graphs of finite posets see Section 5.1, where also examples of covering graphs are given. A *path* from x to y is a path from x to y in $G(P)$, that is, a sequence $p_{xy} : x = x_0, x_1, \ldots, x_n = y$ such that $(x_0, x_1), \ldots, (x_{n-1}, x_n)$ are edges in $G(P)$; the *length* of such a path is n. The poset P is *connected* if for all $x, y \in P$ there exists a path from x to y.

Let P now be a finite poset with greatest element 1, and v a real-valued map defined on P. For an edge $e = (x, y)$ in the covering graph $G(P)$ of P we define the *weight* of e by $w(e) = |v(x) - v(y)|$. For $x, y \in P$ and p_{xy}, a path from x to y, we define the weight of p_{xy} by $w(p_{xy}) = \sum_{e \in p_{xy}} w(e)$, and we write $\delta_v(x, y)$ for the minimum weight of a path from x to y.

Clearly, δ_v is a quasimetric on P. We call such a quasimetric a *minimum-weighted-path quasimetric*. If v is the dimension function of a poset P graded by v, one obtains the classical *shortest path metric* (relative to v) on $G(P)$. In this case δ_v is the *shortest path distance* in $G(P)$, which will be denoted by δ. The shortest path distance δ can be used to characterize semimodularity:

Theorem 1.9.19   *A finite lattice is semimodular if and only if it is graded by the shortest path distance δ and $\delta(x, y) = \delta(x, x \vee y) + \delta(x \vee y, y)$ holds for all x, y.*

In fact, this result can be proved in a more general setting for semilattices (see Monjardet [1977]). For more on metrics on posets see Barbut & Monjardet [1970], Haskins & Gudder [1972], and the survey Monjardet [1981] (with a comprehensive bibliography containing more than 50 items). We shall use theorem 1.9.19 in Section 5.2.

Notes

As sources for a treatment of the group-theoretical background of the Jordan–Dedekind chain we refer to Barbilian [1946] and Felscher [1961]. Felscher's paper is a detailed investigation of the validity of Jordan–Dedekind-type statements for posets. The abstraction from lattices to posets was first done by Ore [1943a], who introduced the so-called quadrilateral condition. Felscher [1961] extends and unifies the results of Ore [1943a], Croisot [1951], Szász [1951–2], and others.

Let us make some more remarks on the Jordan–Dedekind chain condition. Consider first the following finiteness conditions:

(FL)   The lattice is of finite length.

(LFL)   The lattice is of locally finite length (discrete).

(BCF)   All bounded chains of the lattice are finite.

Then (FL) $\Rightarrow$ (LFL) $\Rightarrow$ (BCF), while the reverse implications are not true (this is left as an exercise). We used (JD) as shorthand for the Jordan–Dedekind chain condition and (WJD) for a weak form of the Jordan–Dedekind chain condition. Let us note, however, that Birkhoff [1948] calls (WJD) the Jordan–Dedekind chain condition. In the terminology of Dubreil-Jacotin et al. [1953] the Jordan–Dedekind chain condition is the logical conjunction of (JD) and (BCF) [or of (WJD) and (BCF)]. For a survey on the above-mentioned properties and related ones we refer to Rudeanu [1964].

Szász [1955] showed that if we define the length of an infinite chain as the power of the set of its elements, then even in distributive lattices (JD) does not hold. However, Grätzer & Schmidt [1957] established an analogue of condition (JD) for infinite chains in distributive lattices.

Nieminen [1985b] characterized the Jordan–Dedekind chain condition for finite lattices by means of special ideals and gave a similar characterization for semimodularity. Nieminen [1990] investigated the connection between the Jordan–Dedekind chain condition and annihilators in finite lattices. He defined a concept of "weak semimodularity" differing from Birkhoff's condition (Bi) and proved that a finite weakly semimodular lattice satisfies the Jordan–Dedekind chain condition.

The set Γ of bases of a matroid on a finite set E may be characterized as a nonempty antichain in the power set 2^E of E that enjoys the following interpolation property: if A, B are in Γ and $A \subseteq Y$, $Y \supseteq X$, $X \subseteq B$ (where $X, Y \subseteq E$), then $X \subseteq C \subseteq Y$ for some C in Γ. This characterization (along with two others) was used by Pezzoli [1981] to extend matroid theory to situations where 2^E is replaced by a more general lattice or even by a poset. Pezzoli's approach differs from other extensions of matroids such as that of Dunstan et al. [1972]. Following Pezzoli's line of research, Higgs [1985] introduced the notion of a nicely graded lattice and showed that two classes of lattices are nicely graded: semimodular lattices in which bounded chains are finite, and lattices of independent sets (together with a top element) in finite matroids. The face lattice of a square (cf. Figure 1.30) is nicely graded, but the face lattices of n-gons for $n > 4$ are not. The property of being nicely graded is self-dual. Several results that are known to be true for semimodular lattices in which bounded chains are finite hold more generally for nicely graded lattices. An example of such a result is the following one (see Duffus & Rival [1977], Theorem 5.3): If a, b, c are elements of a semimodular lattice such that $\delta(a, b) = \delta(a, c) + \delta(c, b)$, then c is in $[a \wedge b, a \vee b]$ with $\delta(x, y)$ denoting the distance between x and y in the covering graph of the lattice.

Kolibiar [1965] introduced the notion of a *line* in a lattice, which is a natural generalization of the notion of a chain. He proved that an analogue of the Jordan–Dedekind chain condition is true for lines in a modular lattice. He has also provided an example showing that this analogue is not true for lines in a semimodular lattice. Gedeonová [1971], [1972] proved that there are nonmodular semimodular lattices satisfying this analogue; in fact she showed that this analogue is valid for lattices that are semimodular and p-modular.

The condition (Sm) of upper semimodularity and related conditions such as Birkhoff's condition (Bi), M-symmetry, and others have been extended in several natural ways to semilattices and to posets by many authors. We refer to Rival [1982b] for a bibliography containing papers that appeared up to that time.

We have already remarked that the importance of matroids has long been recognized and has led to much work on these structures. The weaker notion of closure spaces satisfying the Jordan–Dedekind chain condition (also called Jordan–Dedekind spaces) has also increasingly been found important. Investigations of closure spaces satisfying the Jordan–Dedekind chain condition (but not possessing the Steinitz–Mac Lane exchange property) can be found in the theory of polar spaces (see e.g. Tits [1974]) and in the incidence structures of Buekenhout [1979]. In view of the many characterizations of matroids, it is natural to ask for characterizations of the weaker Jordan–Dedekind spaces. Two such characterizations were given by Batten [1983], [1984].

Exercises

1. Prove that any upper or lower semimodular lattice satisfies the Jordan–Dedekind chain condition (Theorem 1.9.1).
2. Prove that Birkhoff's condition (Bi) implies the weak form (WJD) of the Jordan–Dedekind chain condition (Theorem 1.9.6).
3. Show that a poset P satisfies the ACC if and only if every nonempty subset has a maximal element. (Hint: If there is an infinite ascending chain in P, then this chain obviously has no maximal element. For the converse use the axiom of choice.)
4. A poset P has no infinite chains if and only if it satisfies both the ACC and the DCC. (Hint: If P has no infinite chains, then it obviously satisfies both the ACC and the DCC; for the converse use the preceding exercise.)
5. Show that lattices satisfying the assumptions of Corollary 1.9.8 are modular. (Birkhoff [1933], Theorem 10.2).
6. Let L be a lattice in which all bounded chains are finite, and suppose that the Jordan–Dedekind chain condition holds. Then L is of locally finite length (Rudeanu [1964], Lemma 3).
7. A finite lattice L satisfies the Jordan–Dedekind chain condition if and only if, for any $a \in L$, the dual ideal $[a)$ is a $2n$-lattice (Nieminen [1985b]; for the notion of a $2n$-lattice see Exercise 6 of Section 1.7).

References

Barbilian, D. [1946] Metrisch-konkave Verbände, Disquisitiones Math. Phys. (Bucureşti), V (1–4), 1–63.

Barbut, M. and B. Monjardet [1970] *Ordre et classification*, Vols. 1, 2, Hachette, Paris.

Batten, L. M. [1983] A rank-associated notion of independence, in: *Finite Geometries, Proc. Conf. in Pullman, Washington, April 1981*, Marcel Dekker, New York, pp. 33–46.

Batten, L. M. [1984] Jordan–Dedekind spaces, Quart. J. Math. Oxford 35, 373–381.

Birkhoff, G. [1933] On the combination of subalgebras, Proc. Camb. Phil. Soc. 29, 441–464.

Birkhoff, G. [1948] *Lattice Theory* (2nd edition), Amer. Math. Soc. Colloquium Publications, Vol. 25, New York.

Birkhoff, G. [1967] *Lattice Theory*, Amer. Math. Soc. Colloquium Publications, Vol. 25, Providence, R. I.; reprinted 1984.

Björner, A., A. Garsia, and R. Stanley [1982] An introduction to Cohen–Macaulay partially ordered sets, in: Rival [1982a], pp. 583–615.

Buekenhout, F. [1979] Diagrams for geometries and groups, J. Combin. Theory (A) 27, 121–151.

Crawley, P. and R. P. Dilworth [1973] *Algebraic Theory of Lattices*, Prentice-Hall, Englewood Cliffs, N.J.

Croisot, R. [1951] Contribution à l'étude des treillis semi-modulaires de longueur infinie, Ann. Sci. Ecole Norm. Sup. (3) 68, 203–265.

Dedekind, R. [1900] Über die von drei Moduln erzeugte Dualgruppe, Math. Ann. 53, 371–403; *Gesammelte Werke*, Vol. 2, pp. 236–271.

Dubreil-Jacotin, M. L., L. Lesieur, and R. Croisot [1953] *Leçons sur la théorie des treillis, des structures algébriques ordonnées et des treillis geometrique*, Gauthier-Villards, Paris.

Duffus, D. and I. Rival [1977] Path length in the covering graph of a lattice, Discrete Math. 19, 139–158.

Dunstan, F. D. J., A.W. Ingleton, and D. J. A. Welsh [1972] Supermatroids, in: *Proc. Conf. Combinatorial Math., Math. Inst. Oxford*, pp. 72–122.

Faigle, U. [1986a] Lattices, in: White [1986], Chapter 3.

Felscher, W. [1961] Jordan–Hölder Sätze und modular geordnete Mengen, Math. Z. 75, 83–114.

Gedeonová, E. [1971] Jordan–Hölder theorem for lines, Acta F.R.N. Univ. Comen. – Math., Memorial Volume, pp. 23–24.

Gedeonová, E. [1972] Jordan–Hölder theorem for lines, Mat. Časopis 22, 177–198.

Grätzer, G. [1978] *General Lattice Theory*, Birkhäuser, Basel.

Grätzer, G. and E. T. Schmidt [1957] On the Jordan–Dedekind chain condition, Acta Sci. Math. (Szeged) 7, 52–56.

Hall, M. [1959] *The Theory of Groups*, Macmillan, New York.

Haskins, L. and S. Gudder [1972] Heights on posets and graphs, Discrete Math. 2, 357–382.

Higgs, D. [1985] Interpolation antichains in lattices, in: *Universal Algebra and Lattice Theory, Proc. Conf. Charleston, July 11–14, 1984*, Lecture Notes in Mathematics 1149, Springer-Verlag, Berlin, pp. 142–149.

Ito, N. [1951] Note on (LM-)groups of finite order, Kodai Math. Sem. Rep., 1–6.

Iwasawa, K. [1941] Über die endlichen Gruppen und die Verbände ihrer Untergruppen, J. Univ. Tokyo 4(3), 171–199.

Jones, A. W. [1946] Semi-modular finite groups and the Burnside basis theorem (Abstract), Bull. Amer. Math. Soc. 52, 418.

Kolibiar, M. [1965] Linien in Verbänden, Analele Ştiinţifice Univ. Iaşi 11, 89–98.

Mac Lane, S. [1943] A conjecture of Ore on chains in partially ordered sets, Bull. Amer. Math. Soc 49, 567–568.

Maeda, F. and S. Maeda [1970] *Theory of Symmetric Lattices*, Springer-Verlag, Berlin.

Monjardet, B. [1977] Caractérisation métrique des ensembles ordonnées semi-modulaires, Math. Sci. Hum. 56, 77–87.

Monjardet, B. [1981] Metrics on partially ordered sets – a survey, Discrete Math. 35, 173–184.

Nieminen, J. [1985b] A characterization of the Jordan–Hölder chain condition, Bull. Inst. Math. Acad. Sinica 13, 1–4.

Nieminen, J. [1990] The Jordan–Hölder chain condition and annihilators in finite lattices, Tsukuba J. Math 14, 405–411.

Ore, O. [1943a] Chains in partially ordered sets, Bull. Amer. Math. Soc. 49, 558–566.

Pezzoli, L. [1981] Sistemi di independenza modulari, Boll. Un. Mat. Ital. 18-B, 575–590.

Reeg, S. and W. Weiß [1991] Properties of finite lattices, Diplomarbeit, Darmstadt.

Rival, I. (ed.) [1982a] *Ordered Sets, Proc. NATO Advanced Study Inst. Conf. held at Banff, Canada, Aug 28–Sep 12, 1981*.

Rival, I. [1982b] A bibliography (on ordered sets), in: Rival [1982a], pp. 864–966.

Rudeanu, S. [1964] Logical dependence of certain chain conditions in lattice theory, Acta Sci. Math. (Szeged) 25, 209–218.

Schmidt, R. [1994] *Subgroup Lattices of Groups*, de Gruyter, Berlin.

Suzuki, M. [1956] *Structure of a Group and the Structure of Its Lattice of Subgroups*, Ergebnisse, Vol. 10, Springer-Verlag, Berlin.

Szász, G. [1951–2] On the structure of semi-modular lattices of infinite length, Acta Sci. Math. (Szeged) 14, 239–245.

Szász, G. [1955] Generalization of a theorem of Birkhoff concerning maximal chains of a certain type of lattices, Acta Sci. Math. (Szeged) 16, 89–91.

Szász, G. [1963] *Introduction to Lattice Theory*, Akadémiai Kiadó, Budapest.

Tits, J. [1974] Buildings of spherical type and finite BN-pairs, *Lecture Notes in Mathematics*, Vol. 386, Springer-Verlag, Berlin.

Wilcox, L. R. [1938] Modularity in the theory of lattices, Bull. Amer. Math. Soc. 44, 50.

Wilcox, L. R. [1939] Modularity in the theory of lattices, Ann. of Math. 40, 490–505.

Wilcox, L. R. [1944] Modularity in Birkhoff lattices, Bull. Amer. Math. Soc. 50, 135–138.

Zassenhaus, H. [1958] *The Theory of Groups* (2nd edition), New York.

2

M-Symmetric Lattices

2.1 Modular Pairs and Modular Elements

Summary. Modular pairs were defined in Section 1.2, and later several properties were given, including a characterization via certain mappings (see Section 1.9). Here we give another characterization in terms of forbidden pentagons and some consequences. We present the parallelogram law, which is an extension of the isomorphism theorem (Dedekind's transposition principle) for modular lattices.

Blyth & Janowitz [1972], Theorem 8.1, p. 72, provided the following characterization of modular pairs in terms of relative complements, that is, excluding certain pentagon sublattices.

Theorem 2.1.1 *Let a, b be elements of a lattice L. Then $a \, M \, b$ holds if and only if L does not possess a pentagon sublattice $\{a \wedge b, a, x, y, a \vee x = a \vee y\}$ with $x < y \leq b$ (see Figure 2.1).*

Proof. If $a \, M \, b$ and there exists a sublattice of the indicated form, then $x < y \leq b$ and thus $y = (x \vee a) \wedge y = (x \vee a) \wedge b \wedge y = x \vee (a \wedge b) \wedge y = x \wedge y = x$, a contradiction. If $a \, M \, b$ fails, then we can find an element $t < b$ such that $t \vee (a \wedge b) < (t \vee a) \wedge b$. Setting $x = t \vee (a \wedge b)$ and $y = (t \vee a) \wedge b$, we get $a \wedge b \leq a \wedge y = a \wedge [(t \vee a) \wedge b] = a \wedge b$ and hence $a \wedge b = a \wedge y = a \wedge x$. Dually we get $a \vee x = a \vee y = a \vee t$. Thus we obtain a pentagon sublattice of the required form. ∎

Corollary 2.1.2 *The relation $a \, M \, b$ holds in a lattice if and only if it holds in $[a \wedge b, a \vee b]$. The corresponding statement for dual modular pairs also holds.*

Theorem 2.1.1 also implies the characterization of modularity by forbidden pentagons (see Theorem 1.2.2):

Corollary 2.1.3 *A lattice is modular if and only if no element admits distinct but comparable complements in an interval sublattice.*

73

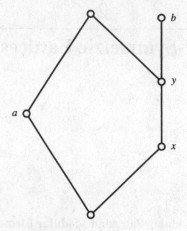

Figure 2.1

The next result – sometimes also referred to as the *parallelogram law* – is a transposition principle for modular pairs. In the modular case it yields the isomorphism theorem (Theorem 1.6.1).

Theorem 2.1.4   *Let L be a lattice and $a, b \in L$. If both a M b and b M^* a hold, then the intervals $[a \wedge b, b]$ and $[a, a \vee b]$ are isomorphic via the isomorphisms $\varphi_a : x \to x \vee a \ (x \in [a \wedge b, b])$ and $\psi_b : y \to y \wedge b \ (y \in [a, a \vee b])$.*

Proof.   Under the assumptions of the theorem, φ_a and ψ_b are inverses of each other (see Lemma 1.9.13). The assertion now follows from Lemma 1.9.14.   ∎

The term *parallelogram law* is employed in a different sense for orthomodular lattices (see e.g. Kalmbach [1983], p. 74).

In a lattice L, an element b is called a *right modular element* (briefly: *modular element*) if $x M b$ holds for every $x \in L$. By Mod(L) we denote the set of all right modular elements of L. The elements 0, 1 and every atom, if they exist, are right modular elements. Recall that an element b of a lattice L is said to be a *left modular* element if $b M x$ holds for every $x \in L$ (cf. Section 1.7).

Corollary 2.1.5   *Let L be an M-symmetric lattice (in particular, a semimodular lattice of finite length), and let $b \in$ Mod(L). Then the intervals $[x \wedge b, b]$ and $[x, x \vee b]$ are isomorphic for all $x \in L$.*

Proof.   The relation $b \in$ Mod(L) means that $x M b$ holds for all $x \in L$. This implies by M-symmetry that $b M x$ holds for all $x \in L$. But then also $b M^* x$ for all $x \in L$ (see Exercise 1). Now $x M b$ and $b M^* x$ together yield by Theorem 2.1.4 that the intervals $[x \wedge b, b]$ and $[x, x \vee b]$ are isomorphic.   ∎

Notes

General references are Maeda & Maeda [1970] and Blyth & Janowitz [1972]. In Annexe 1 to Duquenne & Monjardet [1982] one finds a concise exposition concerning modular pairs. Note that in the French terminology a distinction is made between *couple modulaire* (used for elements a, b with $a \, M \, b$) and *paire modulaire* (used for elements a, b satisfying simultaneously $a \, M \, b$ and $b \, M \, a$).

Exercises

1. Let L be a lattice and $a \in L$. Then $a \, M \, x$ holds for all $x \in L$ if and only if $a \, M^* \, x$ holds for all $x \in L$ (cf. Maeda & Maeda [1970], Lemma 1.2).

2. Two elements a, b of a lattice L with 0 are called *perspective*, denoted by $a \sim b$, if there exists an element $x \in L$ such that $a \vee x = b \vee x$ and $a \wedge x = b \wedge x = 0$. Show that in a relatively complemented lattice with 0 and 1 the relation $a \sim b$ holds if and only if a and b have a common complement.

3. Prove that, in a modular lattice with 0, $a \sim b$ implies that the intervals $[0, a]$ and $[0, b]$ are isomorphic (cf. Blyth & Janowitz [1972], p. 73).

4. Prove that in a relatively complemented lattice, the following three properties are equivalent:

   (i) $a \, M^* \, b$;
   (ii) b is a maximal relative complement of a in the interval $[a \wedge b, a \vee b]$;
   (iii) the relations $c \in [b, a \vee b]$ and $b \prec c$ imply $b \wedge a < c \wedge a$.

   Remark. For the equivalence of conditions (i) and (ii) see Crapo & Rota [1970 b]; condition (iii) was given in Aigner [1979] for partition lattices (cf. Duquenne & Monjardet [1982], Proposition 6).

5. Let L be an M-symmetric lattice, and let $a \, M \, x$ ($x \in L$ arbitrary) hold for a given $a \in L$, that is, a is a left modular element. Show that if $b \leq a$ and $b \, M \, e$ holds in $[0, a]$ for every $e \in [0, a]$ then $b \, M \, x$ holds (Sachs [1961], Theorem 1).

References

Aigner, M. [1979] *Combinatorial Theory*, Springer-Verlag, Berlin.

Blyth, T. S. and M. F. Janowitz [1972] *Residuation Theory*, Pergamon, Oxford.

Crapo, H. H. and G.-C. Rota [1970b] Geometric lattices, in: *Trends in Lattice Theory* (ed. J. C. Abbott), van Nostrand-Reinhold, New York, pp. 127–172.

Duquenne, V. and B. Monjardet [1982] Relations binaires entre partitions, Math. Sci. Hum. 80, 5–37.

Kalmbach, G. [1983] *Orthomodular Lattices*, Academic Press, London.

Maeda, F. and S. Maeda [1970] *Theory of Symmetric Lattices*, Springer-Verlag, Berlin.

Sachs, D. [1961] Partition and modulated lattices, Pacific J. Math. 11, 325–345.

2.2 Distributive, Standard, and Neutral Elements

Summary. Given the importance of distributive lattices, it is natural to consider distributivity conditions for elements and ideals in nondistributive lattices. We recall several distributivity conditions and clarify the relationship between distributive elements, modular elements, and M-symmetry.

If not stated otherwise, in this section we mean by L an arbitrary lattice. The following notions all involve some kind of distributivity for triples of elements.

A triple (a, b, c) of elements of L is called a *distributive triple* and we write $(a, b, c)D$ if $(a \vee b) \wedge c = (a \wedge c) \vee (b \wedge c)$. A triple (a, b, c) of elements of L is called a *dually distributive triple* and we write $(a, b, c)D^*$ if $(a \wedge b) \vee c = (a \vee c) \wedge (b \vee c)$. An element $d \in L$ is said to be *distributive* if $d \vee (x \wedge y) = (d \vee x) \wedge (d \vee y)$ holds for all $x, y \in L$, that is, if $(x, y, d)D^*$. An element $d \in L$ is called *standard* if $(x \vee d) \wedge y = (x \wedge y) \vee (d \wedge y)$ holds for all $x, y \in L$, that is, if $(x, d, y)D$. An element $t \in L$ is called *neutral* if $(x, y, d)D$ and $(x, y, d)D^*$ hold for all $x, y \in L$ and for all permutations of the triple (x, y, d).

Dualizing these notions, one defines *dually distributive elements* and *dually standard elements*. The concept of neutrality is self-dual. The three types of elements (distributive, standard, and neutral elements) were introduced by Ore [1935], Grätzer [1959], and Birkhoff [1940b], respectively. Ore [1935] studied neutral elements in modular lattices. The concept was later extended to general lattices by Birkhoff [1940b]. The above notions are interrelated in the following way (cf. Grätzer & Schmidt [1961]).

Theorem 2.2.1

(i) *Every neutral element is standard.*

(ii) *Every standard element is distributive.*

(iii) *Every standard element that is dually distributive is a neutral element.*

For a proof see also Grätzer [1978], Theorem 5, p. 142. For the following characterizations of standard and neutral elements we recall that an element $d \in L$ is said to be *separating* if $d \vee x = d \vee y$ and $d \wedge x = d \wedge y$ together imply $x = y$. The following characterization of standard elements is due to Grätzer & Schmidt [1961], Theorem 1, p. 28:

Theorem 2.2.2   *An element of a lattice is standard if and only if it is both distributive and separating.*

The following characterization of neutral elements is due to Birkhoff [1940b]:

Theorem 2.2.3   *An element is neutral if and only if it is distributive, dually distributive, and separating.*

An early result was that in a modular lattice any distributive (or dually distributive) element is in fact neutral and thus also standard, that is, all three concepts coincide. Grätzer & Schmidt [1961] proved that these concepts also coincide in lattices that are weakly modular in the sense of Grätzer & Schmidt [1958a] (see also Grätzer [1978], Theorem 6, p. 142). The concept of weak modularity in that sense (which will not be used here) must not be confused with weak modularity as defined in Section 1.7 and used here throughout.

Following Jones [1983a], we call an element $d \in L$ *weakly separating* if $d \vee x = d \vee y$, $d \wedge x = d \wedge y$, and $x \leq y$ together imply $x = y$. One of the elementary relationships we shall prove below is that a weakly separating element is left modular. A source of weakly separating elements is provided by the following result due to Jones [1983a], which can be obtained by modifying Jónsson's proof that a lattice with a "type two" representation is modular (see Grätzer [1978], Theorem IV.4.8, p. 197). Consider $\Pi(X)$, the partition lattice of the set X. Let the symbol $\circ$ denote composition, that is, if α, $\beta \in \Pi(X)$, then $(a, b) \in \alpha \circ \beta$ if $a \equiv c(\alpha)$ and $c \equiv b(\beta)$ for some $c \in X$.

Lemma 2.2.4   *Let $d \in L$, and suppose that L has a representation $\varphi : L \to \Pi(X)$ such that $(a \vee d)\varphi = a\varphi \circ d\varphi \circ a\varphi$ for all $a \in L$. Then d is weakly separating.*

Clearly, if $\mathbf{A} = (A, \ldots)$ is an algebra and $\rho \in \mathrm{Con}(A)$ has the property that for all $\tau \in \mathrm{Con}(A)$ one has $\rho \vee \tau = \rho \circ \tau \circ \rho$, then ρ weakly separates $\mathrm{Con}(A)$. We now recall some relationships between concepts mentioned above. For the following results we refer again to Jones [1983a].

Lemma 2.2.5

(i) *Any weakly separating element is left modular.*
(ii) *Any dually distributive element is right modular.*
(iii) *Any distributive, left modular element is separating.*

The proof of (ii) is immediate from the definition. The proofs of (i) and (iii) are left as exercises. Observing that left modularity is self-dual (see Section 2.1, Exercise 1), we get

Corollary 2.2.6   *If d is distributive or dually distributive, the following conditions are equivalent:*

(i) *d is weakly separating;*
(ii) *d is left modular;*
(iii) *d is separating.*

In what follows we consider some results concerning M-symmetric lattices.

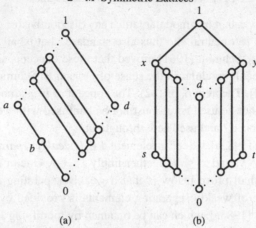

Figure 2.2

Theorem 2.2.7   *In an M-symmetric lattice any dually distributive element is sepa-rating. Hence any element that is both distributive and dually distributive is neutral.*

Proof.   If d is dually distributive, then by Lemma 2.2.5(ii), d is right modular and therefore by M-symmetry left modular. The result now follows from Corollary 2.2.6 and from Birkhoff's characterization of neutral elements (Theorem 2.2.3).

∎

The lattice of Figure 2.2(a) shows that M-symmetry cannot be replaced by the weaker condition of upper semimodularity. In this lattice the intervals $[a, 1]$ and $[0, d]$ are isomorphic dense bounded chains and b is the only other element. The lattice is semimodular and the element d is dually distributive but not separating.

In the lattice of Figure 2.2(b) each interval $(0, s]$, $(0, t]$, $(d, x]$, and $(d, y]$ is supposed to be isomorphic with the chain of natural numbers ordered by $1 > 2 > 3 > \cdots$. This lattice is M-symmetric, and the element d is standard, but not dually distributive, since $d \wedge (s \vee t) = d$ whereas $(d \wedge s) \vee (d \wedge t) = 0$. Hence this example shows that in M-symmetric lattices "standard implies neutral" is not true in general.

The lattice of Figure 2.2(b) satisfies the ACC but not the DCC. A similar example can be given for an M-symmetric lattice satisfying the DCC but not the ACC. However, if such a lattice is moreover assumed to be algebraic, then the concepts of standard and neutral elements coincide (Jones [1983a]):

Theorem 2.2.8   *In an algebraic, M-symmetric lattice with DCC, any standard element is neutral.*

Proof.   Let L be an algebraic, M-symmetric lattice with DCC. Assume that d is a standard element of L that is not neutral. By Theorem 2.2.1(iii), d is not dually distributive and thus there exist $x, y \in L$ such that

$$(+) \quad d \wedge (x \vee y) > (d \wedge x) \vee (d \wedge y) = b.$$

Since d is separating (Theorem 2.2.2), it is left modular [by Lemma 2.2.5(i)], and hence it is also right modular by M-symmetry. In particular $x \, M \, d$ and $y \, M \, d$, and since $b \in [d \wedge x, d]$ and $b \in [d \wedge y, d]$, we have

$$(++) \quad b = d \wedge (x \vee b) = d \wedge (y \vee b).$$

Thus $d \wedge \{(x \vee b) \vee (y \vee b)\} \geq d \wedge (x \vee y) > b = \{d \wedge (x \vee b)\} \vee \{d \wedge (y \vee b)\}$, so that in $(+)$ we may assume without loss of generality that $x, y \geq b$. In fact, since $d > b$, the inequality $(+)$ is valid for some $x, y, d \in [b)$, and thus we may assume from now on $b = 0$. Hence we have for some x and y the relation $d \wedge (x \vee y) > 0$ whereas $d \wedge x = d \wedge y = 0$.

Using the DCC, we may choose $z \in L$ minimally with respect to the following property: there is an element $p \in L$ for which $d \wedge (z \vee p) > 0$ while $d \wedge z = d \wedge p = 0$. On the other hand, by the maximum principle there is an element $m \in [p, z \vee p]$ maximal such that $d \wedge m = 0$. Note that $m < z \vee p$, for otherwise $d \wedge (z \vee p) = 0$. Thus $z \not\leq m$.

We show now that z covers $z \wedge m$. Let $w \in [z \wedge m, z)$. From the minimality of z (and noting that $d \wedge w = d \wedge m = 0$) it follows that $d \wedge (w \vee m) = 0$. But $w \vee m \in [m, p \vee z]$, and so from the maximality of m we deduce that $w \vee m = m$, so that $w = m \wedge z$. Hence $z \succ z \wedge m$. From M-symmetry (which implies semimodularity) it is now immediate that $z \vee p \succ m$. But if we put $c = d \wedge (z \vee p)$, then $c \vee m \in [m, z \vee p] = [m, z \vee m]$; thus $c \vee m = m$ or $c \vee m = z \vee m$. As in the equation $(++)$, using $m \, M \, d$, $d \wedge m = 0$, and $c \leq d$, we obtain $c = d \wedge (c \vee m)$. Thus if $c \vee m = m$, then $c = d \wedge m = 0$, a contradiction. Otherwise $d \vee m \geq c \vee m \geq z \vee m \geq z$, in which case we obtain (using for d the defining property of standard elements) $z = z \wedge (d \vee m) = (z \wedge d) \vee (z \wedge m) = z \wedge m$, contradicting $z \not\leq m$. It follows that d is dually distributive and therefore neutral. ∎

For lattices of finite length, M-symmetry and semimodularity are equivalent (cf. Theorem 1.9.18). Thus the preceding theorem implies the following result due to Eberhart & Williams [1978].

Corollary 2.2.9 *In a semimodular lattice of finite length every standard element is neutral.*

Notes

Grätzer & Schmidt [1961] give a detailed motivation as well as the historical background for introducing distributive, neutral, and standard elements and ideals into lattice theory. Additional information can be found in Janowitz [1965a].

The *center* of a bounded lattice is the set of all neutral elements that possess a complement. The elements of the center are called *central elements*. Neutral and central elements play a significant role in the decomposition theory of lattices. For example, a neutral element a in a bounded lattice L provides a representation $x \to (x \wedge a, x \vee a)$ of L as a subdirect product of $[0, a] \times [a, 1]$. The relationship of central element a in a bounded lattice L to the direct decomposition $[0, a] \times [a, 1]$ of L will be used in Chapter 5 (cf. Section 5.3).

Exercises

1. Prove that an element s of a lattice is a standard element if and only if the following relation Θ_s is a congruence relation: $x \equiv y(\Theta_s) \Leftrightarrow (x \wedge y) \vee s_1 = x \vee y$ holds for some $s_1 \leq s$ (Grätzer & Schmidt [1961], Theorem 1, p. 28).
2. Let s be a standard element and let a be an arbitrary element of a lattice.
   (a) Prove that then $a \wedge s$ is a standard element of the principal ideal $(a]$.
   (b) Give an example showing that the preceding conclusion is not valid for distributive elements. (Hint: the lattice L_{15} – see Section 1.3 – may be used.)
   (Grätzer & Schmidt [1961], Lemma 9, p. 38.)
3. Prove conditions (i) and (iii) of Lemma 2.2.5 (Jones [1983a], Lemma 1.3).

References

Birkhoff, G. [1940b] Neutral elements in general lattices, Bull. Amer. Math. Soc .46, 702–705.

Eberhart, C. and W. Williams [1978] Semimodularity in the lattice of congruences, J. Algebra 52, 75–87.

Grätzer, G. [1959] Standard ideálok (Hungarian), Magyar Tud. Akad. III. Oszt. Közl. 9, 81–97.

Grätzer, G. [1978] *General Lattice Theory*, Birkhäuser, Basel.

Grätzer, G. and E. T. Schmidt [1958a] Ideals and congruence relations in lattices, Acta Math. Acad. Sci. Hungar. 9, 137–175.

Grätzer, G. and E. T. Schmidt [1961] Standard ideals in lattices, Acta Math. Acad. Sci. Hungar. 12, 17–86.

Janowitz, M. F. [1965a] A characterisation of standard ideals, Acta Math. Acad. Sci. Hungar. 16, 289–301.

Jones, P. R. [1983a] Distributive, modular and separating elements in lattices, Rocky Mountain J. Math. 13, 429–436.

Maeda, F. and S. Maeda [1970] *Theory of Symmetric Lattices*, Springer-Verlag, Berlin.

Ore, O. [1935] On the foundations of abstract algebra I, Ann. of Math. 36, 406–437.

2.3 *M*-Symmetry and Related Concepts

Summary. In Chapter 1 we defined *M*-symmetric lattices [condition (Ms)] and stated that it implies upper semimodularity [condition (Sm)] but not conversely. Here we briefly consider some other notions of symmetry deriving from modular pairs or dual modular pairs such as M^*-symmetry [dual to (Ms)], $\perp$-symmetry [weaker than (Ms)], and cross-symmetry [stronger than (Ms)]. We have a look at the interrelationships between these concepts. One of our sources is the monograph Maeda & Maeda [1970] to which we refer for proofs not given here and further details.

For the notion of modular pair and dual modular pair see Section 1.2. We recall that a lattice L is M^*-*symmetric* if $a\,M^*\,b$ implies $b\,M^*a$ for all $a, b \in L$. The condition of M^*-symmetry will be briefly denoted by (Ms*). The pentagon shows that the relation of being a modular pair (or dual modular pair) is not symmetric in general. This example also shows that modular pairs need not be dual modular pairs. We have seen that any interval of an M^*-symmetric lattice is likewise M^*-symmetric (Corollary 2.1.2).

Let us first have a look at AC lattices, which were defined in Section 1.7. An M-symmetric atomistic lattice is always an AC lattice, since M-symmetry [condition (Ms)] implies upper semimodularity [condition (Sm)], which in turn yields the atomic covering property [condition (C)].

Maeda & Maeda [1970], Problem 2, p. 54, ask if there is an AC lattice that is not M-symmetric. This problem was solved affirmatively and in a constructive manner by Janowitz (see Supplement to Maeda & Maeda [1970]). The following simple example is due to Maeda & Kato [1974]:

Let E be an infinite set, and let $a, b \in E$ $(a \neq b)$. Put $A = E - \{a, b\}$, and consider $L = \{E, A\} \cup \mathbf{F}$, where $\mathbf{F}$ is the set of all finite subsets of E. It is clear that, ordered by set inclusion, L forms a complete lattice where the meet of elements of L coincides with their intersection, and $A \vee \{a\} = A \vee \{b\} = E$. It is not difficult to verify that L is an AC lattice. The pair $(\{a, b\}, A)$ is obviously modular. On the other hand, $(A, \{a, b\})$ is not modular, since $(\{a\} \vee A) \wedge \{a, b\} = E \wedge \{a, b\} = \{a, b\} \neq \{a\} = \{a\} \vee (A \wedge \{a, b\})$.

It is therefore natural to ask the following question: What does it mean for AC lattices to be M-symmetric? Similarly one may ask about the meaning of M-symmetry for other classes of lattices. In Section 2.5 and Section 2.6 we return to the question of M-symmetry in atomistic algebraic lattices and orthomodular lattices, respectively.

Like M-symmetry [condition (Ms)], the weaker concept of $\perp$-symmetry (for lattices with 0) was introduced by Wilcox [1939]. A lattice L with 0 is called $\perp$-*symmetric* if it satisfies the condition

(Ms$_0$) $a\,M\,b$ and $a \wedge b = 0$ together imply $b\,M\,a$.

In other words, $\perp$-symmetry means that the condition of M-symmetry holds for disjoint pairs. To be more consistent with the terminology "M-symmetry" we might also use the expression "M_0-symmetry" in place of $\perp$-symmetry. In lattices with 0 we clearly have

$$M\text{-symmetry } [(Ms)] \Rightarrow \perp\text{-symmetry } [(Ms_0)]$$
$$\Rightarrow \text{ atomic covering property } [(C)].$$

The reverse implications do not hold in general. Consider for example the lattice in Figure 2.3 (cf. Maeda & Maeda [1970], Exercise 1.2, p. 5), where the intervals $[0, a]$ and $[0, b]$ are supposed to be isomorphic to the chain $\{0, 1/n\}$ $(n = 1, 2, \ldots)$ taken with the natural order. This lattice is (complemented and) $\perp$-symmetric but not M-symmetric.

The above-mentioned example of Janowitz (cf. Supplement to Maeda & Maeda [1970]) also shows that in an atomistic lattice the atomic covering property [condition (C)] does not imply $\perp$-symmetry [condition (Ms_0)]. This example moreover shows that upper semimodularity does not imply $\perp$-symmetry, while Figure 2.3 shows that $\perp$-symmetry does not imply upper semimodularity.

For later use (see Section 3.5) we also mention the following sufficient condition for a $\perp$-symmetric lattice with 1 to be M-symmetric (cf. Maeda & Maeda [1970], Theorem 1.14). This result also follows from Theorem 3.5.26.

Theorem 2.3.1   *A $\perp$-symmetric lattice L with 1 is M-symmetric if it satisfies the following condition:*

(+) Every element $a \in L$ has a complement $\bar{a}$ such that $a\, M\, \bar{a}$ and $\bar{a}\, M\, a$.

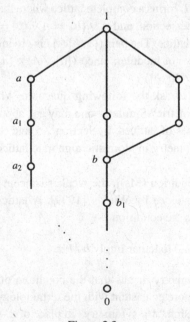

Figure 2.3

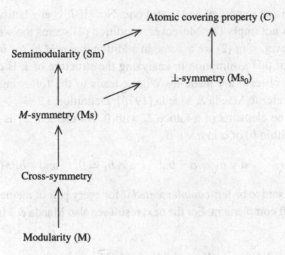

Figure 2.4

The concept of cross-symmetry is located between modularity and M-symmetry. A lattice L is called *cross-symmetric* (*dual cross-symmetric*) if $a \, M \, b$ implies $b \, M^* \, a$ ($a \, M^* \, b$ implies $b \, M \, a$) in L. We have (cf. Maeda & Maeda [1970], Theorem 1.9):

Theorem 2.3.2   *Any cross-symmetric lattice is M-symmetric, and any dual cross-symmetric lattice is M^*-symmetric.*

For a proof we refer to Exercise 3. The centered hexagon S_7 is M-symmetric but not cross-symmetric. For an example of a cross-symmetric lattice that is not modular see Maeda & Maeda [1970], Theorem 34.10, p. 157.

The interrelationships mentioned above are visualized in Figure 2.4.

We still note that, for a semilattice S, $\mathrm{Con}(S)$ is not only upper semimodular but even M-symmetric (this follows from Theorem 9.1.1). More examples will be given in the subsequent sections of this chapter and in the next chapter.

In the theory of modular lattices (in particular, in the theory of continuous geometries of von Neumann) the assumption of complementation has proved extremely powerful. Wilcox [1942] has found that some analogue of this assumption is usually necessary in the study of nonmodular lattices. Let us first make the following observation (the proof of which is left as an exercise): If a modular lattice with 0 and 1 has the property

(1) each $a \in L$ has a complement,

then it can be proved that

(2) $a, b \in L$ implies the existence of $b_1 \, P \leq b$ such that $a \vee b_1 = a \vee b$ and $a \wedge b_1 = 0$.

Property (2) is an important and useful one. Now if L is any lattice with 0 and 1, then (1) does not imply (2). Moreover, condition (2) seems too weak for most purposes. However, if in (2) we assume in addition that $a \, M \, b_1$ (or $b_1 \, M \, a$), then the usefulness of this assumption in analyzing the structure of L is considerably increased. This observation made by Wilcox leads to the following notion (for which we also refer to Maeda & Maeda [1970], Definition 3.7).

Let a and b be elements of a lattice L with 0. An element b_1 is called a *left complement* (within b) of a in $a \vee b$ if

$$b_1 \leq b, \qquad a \vee b_1 = a \vee b, \qquad a \wedge b_1 = 0, \quad \text{and} \quad b_1 \, M \, a.$$

The lattice L is said to be *left-complemented* if for every pair of elements $a, b \in L$ there exists a left complement. For the next result see also Maeda & Maeda [1970], Theorem 3.9.

Theorem 2.3.3   *A left-complemented lattice is M-symmetric.*

Sketch of proof.   Suppose $a \, M \, b$, and let b_1 be a left complement (within b) of a in $a \vee b$. Then by Exercise 6 we have $b = b_1 \vee (a \wedge b)$. Since $(a \wedge b) \, M \, a, b_1 \, M \, a$, and $b_1 \wedge a \leq a \wedge b$, it follows from Exercise 2 that $b \, M \, a$ holds. Hence the lattice is M-symmetric. ∎

Wilcox [1942] noted that lattices of flats of affine geometries are left-complemented. This is generalized by the following result due to Sachs [1961] (see also Maeda & Maeda [1970], Theorem 7.15, $(\alpha) \Rightarrow (\delta)$).

Theorem 2.3.4   *A matroid lattice is left-complemented.*

Sketch of proof.   Let a and b be elements of a matroid lattice L and consider the set S of all $x \in L$ such that $x \leq b$, $x \wedge a = 0$, and $x \, M \, a$. The set S is nonempty (since $0 \in S$), and it is partially ordered by the order inherited from L. Applying Zorn's lemma, it follows that there exists a maximal element b_1 such that $b_1 \leq b$, $a \wedge b_1 = 0$, and $b_1 \, M \, a$. It remains to show that b_1 is a left complement (within b) of a in $a \vee b$. To see this, we need only to show that $a \vee b_1 = a \vee b$. Assume on the contrary that $a \vee b_1 < a \vee b$. This implies (since L is atomistic) the existence of an atom p such that $p \leq b$ and $p \not\leq a \vee b_1$. Thus $p \wedge (b_1 \vee a) = 0$, and from Theorem 1.7.2 it follows that $p \, M \, (b_1 \vee a)$. Therefore we obtain by Exercise 2 that $(p \vee b_1) \, M \, a$ and $(p \vee b_1) \wedge a = b_1 \wedge a = 0$. This, however, contradicts the maximality of b_1, since $b_1 < p \vee b_1 \leq b$. Hence L is left-complemented. ∎

Corollary 2.3.5   *A matroid lattice is relatively complemented.*

For geometric lattices (i.e. matroid lattices of finite length) this was already stated in Section 1.7, Exercise 4. The proof of the more general assertion of the

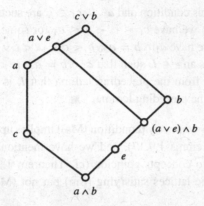

Figure 2.5

preceding corollary is likewise left as an exercise. Ramalho [1994] proved that *M*-symmetry can be characterized in terms of existing pentagons:

Lemma 2.3.6 *A lattice L is M-symmetric if and only if, for any $a, b, c \in L$ with $a \wedge b < c < a < c \vee b$, there exists an $e \in L$ such that $a \wedge b < e < (a \vee e) \wedge b$.*

Proof. (For an illustration see Figure 2.5.) Assume that L is M-symmetric and that the sublattice generated by a, b, c is a pentagon. Then $b \, \bar{M} \, a$, since $c \vee (b \wedge a) = c \neq a = (c \vee b) \wedge a$. M-symmetry implies that then also $a \, \bar{M} \, b$. Hence $z \vee (a \wedge b) < (z \vee a) \wedge b$ for some $z < b$. Now set $e = z \vee (a \wedge b)$. We obtain

$$e = e \vee (a \wedge b) = z \vee (a \wedge b) < (z \vee a) \wedge b = (e \vee a) \wedge b = (a \vee e) \wedge b.$$

This implies $a \wedge b \neq e$ and thus $a \wedge b < e$.

To prove the converse, assume that L satisfies the above condition and that $b \, \bar{M} \, a$. Then, for some $x < a$, we have $x \vee (b \wedge a) < (x \vee b) \wedge a$. From this we deduce that $a \wedge b < x \vee (a \wedge b) < (x \vee b) \wedge a$. Thus $\{x \vee (a \wedge b), (x \vee b) \wedge a, b\}$ generates a pentagon. By assumption, there exists an e such that $a \wedge b < e < \{[(x \vee b) \wedge a] \vee e\} \wedge b$. From this we conclude that $e < b$, and that $e \vee (a \wedge b) = e < (a \vee e) \wedge b = (e \vee a) \wedge b$, since $\{[(x \vee b) \wedge a] \vee e\} \wedge b \leq (a \vee e) \wedge b$. Hence $a \, \bar{M} \, b$. ∎

For weakly atomic lattices we shall make use of the following version of the preceding characterization of *M*-symmetry.

Lemma 2.3.7 *A weakly atomic lattice L is M-symmetric if and only if $a \wedge b < c \prec a < c \vee b$, implies the existence of an $e \in L$ such that $a \wedge b < e < (a \vee e) \wedge b$ for any $a, b, c \in L$.*

Proof. If L satisfies this condition and $a_1, b, c_1 \in L$ are such that $a_1 \wedge b < c_1 < a_1 < c_1 < c_1 \vee b$, then we have $c_1 \leq c \prec a \leq a_1$ for some $c, a \in L$, by weak atomicity. Moreover, we have $a_1 \wedge b = a \wedge b < c \prec a < c \vee b = c_1 \vee b$, and, by assumption, there exists an $e \in L$ such that $a_1 \wedge b = a \wedge b < e < (a \vee e) \wedge b \leq (a_1 \vee e) \wedge b$. It follows from the preceding lemma that L is M-symmetric. The converse is clear from the preceding lemma. ■

We have seen that M-symmetry [condition (Ms)] implies upper semimodularity [condition(Sm)] (see Lemma 1.9.17), and we have mentioned that, for lattices of finite length, the two concepts coincide (cf. Theorem 1.9.18). On the other hand there are algebraic lattices satisfying (Sm) but not (Ms) (cf. Figure 1.33, Section 1.9).

We shall consider to what degree it is possible to relax the property of finite length and still be able to prove that (Sm) implies (Ms), that is, that the two conditions are equivalent. We will replace "finite length" first by "lower continuous + strongly coatomic" (Theorem 2.3.8 below) and then by "upper and lower continuous + weakly atomic" (Theorem 2.3.9 below). Later (in Chapter 3), we replace "finite length" by "upper continuous + strongly atomic" (cf. Theorem 3.1.7). Note that in Theorem 3.1.7 we do not simply dualize Theorem 2.3.8, because we do not change (Sm) and (Ms) to (Sm*) and (Ms*), respectively. Ramalho [1994] obtained the following result.

Theorem 2.3.8 *For lower continuous strongly coatomic lattices, upper semimodularity implies M-symmetry and thus the two concepts are equivalent.*

Proof. We shall use the characterization of M-symmetry given in Lemma 2.3.6. Let L be a lower continuous strongly coatomic lattice, and $a, b, c \in L$ such that $a \wedge b < c < a < c \vee b$. Putting $Y = \{y \in L : a \wedge b \leq y = (a \vee y) \wedge b$ and $a \vee y = c \vee y\}$, we have $Y \neq \emptyset$ (since $b \in Y$) and $a \wedge b \notin Y$. Lower continuity implies that Y is dually inductive (down directed). Thus by the dual to Zorn's lemma, Y contains a minimal element y_m. Since L is strongly coatomic, there exists an element $e \in L$ such that $a \wedge b \leq e \prec y_m$.

Obviously $e \leq (a \vee e) \wedge b$. Assuming that $e = (a \vee e) \wedge b$, we get $e = (a \vee e) \wedge y_m$ and hence also $e = (c \vee e) \wedge y_m$. Using (Sm), it follows that $a \vee e \prec c \vee y_m = a \vee y_m$ and $c \vee e \prec c \vee y_m = a \vee y_m$. Since $c \vee e \leq a \vee e$, this implies $c \vee e = a \vee e$. Thus $e \in Y$, contradicting the minimality of y_m. Therefore $e < (a \vee e) \wedge b$, which also implies $a \wedge b < e$. From Lemma 2.3.6 it follows that L is M-symmetric.
　　　　　　　　　　　　　　　　　　　　　　　　　　　　　　　　■

Using arguments as in the proof of the preceding theorem, Ramalho [1994] provided an alternative proof of the following result due to Malliah & Bhatta [1986].

Theorem 2.3.9 *In a weakly atomic upper and lower continuous lattice, upper semimodularity implies M-symmetry and hence the two concepts are equivalent.*

Proof. We shall use the characterization of M-symmetry for weakly atomic lattices as given in Lemma 2.3.7. Let L be a weakly atomic, upper and lower continuous lattice, and $a, b, c \in L$ such that $a \wedge b < c \prec a < c \vee b$. As in the proof of Theorem 2.3.8, lower continuity implies the existence of a minimal element y_m in the set $Y = \{y \in L : a \wedge b \le y = (a \vee y) \wedge b$ and $a \vee y = c \vee y\}$. Note that $\bigvee(z_i : i \in I) < y_m$ is not possible and hence the set $Z = \{z \in L : a \wedge b < z < y_m\}$ is not empty. We show that for some $z \in L$ we have $z < (a \vee z) \wedge b$. Contrary to this, assume that $z = (a \vee z) \wedge b$ holds for all $z \in L$. Then $c \vee z < a \vee z$ for all $z \in L$, since $z \notin Y$. This implies $a \wedge (c \vee z) \ne a$, which together with $c \le a \wedge (c \vee z) \le a$ and $c \prec a$ yields $c = a \wedge (c \vee z)$. Let now $(z_i : i \in I)$ be any maximal chain in Z. If $\bigvee(z_i : i \in I) = y_m$, then (by upper continuity) we get $a = a \wedge (a \vee y_m) = a \wedge (c \vee y_m) = a \wedge (c \vee \bigvee z_i) = \bigvee[a \wedge (c \vee z_i)] = c$, a contradiction.

If $\bigvee(z_i : i \in I) = y_m$, then $e = \bigvee(z_i : i \in I) \in Z$ and $a \wedge b < e \prec y_m$. As in the proof of the preceding theorem, we conclude $e < (a \vee e) \wedge b$, a contradiction. ∎

In lattices of finite length M-symmetry and M^*-symmetry together imply modularity. On the other hand, the lattice of closed subspaces of an infinite-dimensional Hilbert space (see Section 2.5) shows that M-symmetry and M^*-symmetry together do not imply modularity in general. Wilcox [1944] asked what can be said generally of nonmodular lattices that together with their duals are M-symmetric. A partial answer to this question was given by Croisot [1952], Lemme 1:

Lemma 2.3.10 *If a nonmodular lattice is both M-symmetric and M^*-symmetric, then there exist both an infinite ascending chain and an infinite descending chain.*

Proof. (See Figure 2.6 for an illustration.) Since the lattice is assumed to be nonmodular, there exist elements x_1, z_1, y_1, s_1, t_1 such that $y_1 > s_1$, $t_1 = x_1 \wedge y_1 = x_1 \wedge s_1$, $z_1 = x_1 \vee y_1 = x_1 \vee s_1$. Since $(s_1 \vee x_1) \wedge y_1 = y_1 \ne s_1$, we have $x_1 \, \bar{M} \, y_1$. Since the lattice is M-symmetric, it follows that also $y_1 \, \bar{M} \, x_1$. Hence there exists an element x_2 $(t_1 < x_2 < x_1)$ such that for $x_2 \vee y_1 = z_2$ (z_2 may equal z_1) one has $z_2 \wedge x_1 = r_2 > x_2$. Since $(r_2 \wedge y_1) \vee x_2 = x_2 \ne r_2$, we have $y_1 \, \bar{M}^* \, x_2$. Since the lattice is M^*-symmetric, it follows that $x_2 \, \bar{M}^* \, y_1$. Hence there exists an element y_2 $(y_1 < y_2 < z_2)$ such that for $y_2 \wedge x_2 = t_2$ (t_2 may equal t_1) one has $t_2 \vee y_1 = s_2 < y_2$.

The reasoning concerning the pair (y_1, x_1) can be repeated with the pair (y_2, x_2), yielding the existence of an element x_3 such that $x_2 > x_3 > t_2$. The reasoning concerning the pair (x_2, y_1) can similarly be repeated with the pair (x_3, y_2), showing

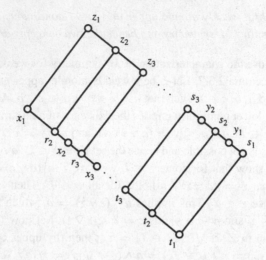

Figure 2.6

the existence of an element y_3 such that $y_2 < y_3 < z_3$. Continuing in this way, we obtain an infinite descending chain $x_1 > x_2 > x_3 > \cdots$ and an infinite ascending chain $y_1 < y_2 < y_3 < \cdots$. ∎

As a consequence we get the following result (Croisot [1952], Théorème 5), which implies that a lattice of finite length is modular if and only if it is both upper and lower semimodular.

Corollary 2.3.11   *A lattice satisfying the* ACC *or the* DCC *is modular if and only if it is both M-symmetric and M*-symmetric.*

Let us finally have a look at the preservation of *M*-symmetry. *M*-symmetry is obviously preserved by ideals, but not by sublattices (any nonmodular *M*-symmetric lattice has a pentagon sublattice, and this is not *M*-symmetric). It is also easy to see that if *L* is the direct product of lattices L_α ($\alpha \in I$), then *L* is *M*-symmetric if and only if L_α is *M*-symmetric for every $\alpha \in I$ (cf. Maeda & Maeda [1970], Lemma 1.17, p. 5). We also have (Jones [1978], Proposition 5.1)

Lemma 2.3.12   *A subdirect product of M-symmetric lattices is M-symmetric.*

Proof.   Let *L* be a subdirect product of *M*-symmetric lattices L_i, $i \in I$. There exist epimorphisms $\varphi_i : L \to L_i$ such that $a\varphi_i = b\varphi_i$ for all $i \in I$ implies $a = b$. If $a \, M \, b$ holds for $a, b, \in L$, then $a\varphi_i \, M \, b\varphi_i$ holds for all $i \in I$, and thus $b\varphi_i \, M \, a\varphi_i$ for all $i \in I$ by *M*-symmetry. Hence if $c \in [a \wedge b, a]$ and therefore $c\varphi_i \in [a\varphi_i \wedge b\varphi_i, a\varphi_i]$ for all i, then $(a \wedge (b \vee c))\varphi_i = a\varphi_i \wedge (b\varphi_i \vee c\varphi_i) = c\varphi_i$ for all i, whence $a \wedge (b \vee c) = c$. Hence $b \, M \, a$, and therefore *L* is *M*-symmetric. ∎

Using this lemma and a result of Kogalovskiĭ, Jones [1978] obtained the following result.

Corollary 2.3.13 *A homomorphic image of an M-symmetric lattice is not M-symmetric in general.*

Sketch of proof. Kogalovskiĭ [1965] has shown that any class of algebras closed under subdirect products and homomorphic images is a variety (see Grätzer [1979], Theorem 3, p. 153) and thus closed under the formation of subalgebras. However, the class of M-symmetric lattices is not closed under the formation of sublattices. Hence by Lemma 2.3.12 and by Kogalovskiĭ's result, the class of M-symmetric lattices is not closed under the formation of homomorphic images. ∎

This result contributes to the solution of Problem 13 in Birkhoff [1967], p. 109: Find necessary and sufficient conditions on a lattice for all its epimorphs to be semimodular. Is every epimorph of a semimodular lattice semimodular? The preceding corollary answers the second part of this problem negatively in a nonconstructive way. Jones [1978] also constructed an example of an M-symmetric lattice whose homomorphic image is not M-symmetric (see also Jones [1979]).

Notes

The above-mentioned constructive approach of Jones [1978], [1979] is via semimodular inverse semigroups (an inverse semigroup is called semimodular if its lattice of full inverse subsemigroups is semimodular). It is not possible here to go into more details concerning properties (including semimodularity and M-symmetry) of the subsemigroup lattices of certain semigroups. Rather we refer here to the papers of Johnston & Jones [1984] and Jones [1978], [1979]. For the vast field of subsemigroup lattices of semigroups we refer to the surveys Shevrin & Ovsyannikov [1983] and Jones [1990]. A comprehensive treatment can be found in the monograph Shevrin & Ovsyannikov [1990], [1991].[1]

Dedekind's construction of the real numbers by cuts of rational numbers (Dedekind [1872]) was generalized by MacNeille [1937] to arbitrary quasiordered sets. (A *quasiordered set* (*quoset*, for short) is a set endowed with a reflexive and transitive relation $\leq$.)

Let A be a subset of a poset P. By $A^{\uparrow}$ and $A^{\downarrow}$ we denote the sets of all upper and lower bounds of A, respectively. A cut is a pair (A, B) such that $A = B^{\downarrow}$ and $B = A^{\uparrow}$. The collection of all cuts, ordered by

$$(A, B) \leq (C, D) \quad \Leftrightarrow \quad A \subseteq C \quad \Leftrightarrow \quad B \subseteq D,$$

---

[1]Cf. also *Semigroups and their Subsemigroup Lattices* by L. N. Shevrin and A. J. Ovsyannikov. Translated and revised from the 1990/1991 Russian originals by the authors. Kluwer, Dordrecht (1996).

is a complete lattice, called the *completion by cuts* (or *Dedekind–MacNeille completion* or *normal completion*) of P.

It is well known that the completion by cuts of a modular lattice is not necessarily modular (see Birkhoff [1967], Example 9, p. 127). Maeda & Maeda [1970], Problem 4, p. 55, raised the question whether the completion by cuts of an M-symmetric lattice is M-symmetric. A negative answer to this question was given by Maeda & Kato [1974], who constructed an atomistic M-symmetric lattice whose completion by cuts is not M-symmetric.

Exercises

1. Prove the following three statements for elements a, b, c of an arbitrary lattice:

    (i) If $a \, M \, b$ and $(a \wedge b) \, M \, c$ then $a_1 \, M \, (b \wedge c)$ for any $a_1 \in [a \wedge c, a]$.
    (ii) If $a \, M \, b$ then $a_1 \, M \, b_1$ for any $a_1 \in [a \wedge b, a]$ and $b_1 \in [a \wedge b, b]$.
    (iii) If $a \, M \, b$ and $a \wedge b = 0$ then $a_1 \, M \, b_1$ for any $a_1 \leq a$ and $b_1 \leq b$.

    (Maeda & Maeda [1970], Lemma 1.5, p. 2).

2. If $a \, M \, b, c \, M \, (a \vee b)$, and $c \wedge (a \vee b) \leq a$, then $(c \vee a) \, M \, b$ and $(c \vee a) \wedge b = a \wedge b$ (Maeda & Maeda [1970], Lemma 1.6, p. 2).

3. Use Exercise 1(ii) and Corollary 2.1.2 to prove that any cross-symmetric lattice is M-symmetric (Theorem 2.3.2). (Cf. Maeda & Maeda [1970], Proof of Theorem 1.9.)

4. Prove the following strengthening of Theorem 2.3.2 for locally finite lattices (i.e. lattices in which all bounded chains are finite): A locally finite cross-symmetric lattice is modular (Fort [1974]).

5. Show that if a and b are elements of a complemented modular lattice, then there exists an element $b_1 \leq b$ such that $a \vee b_1 = a \vee b$ and $a \wedge b_1 = 0$. (Cf. Maeda & Maeda [1970], Remark 3.10, p. 12.) Give an example showing that the preceding assertion does not hold for nonmodular complemented lattices.

6. In a lattice L with 0, if $a \, M \, b$ and if b_1 is a left complement within b of a in $a \vee b$, then b_1 is a left complement of $a \wedge b$ in b. (Cf. Maeda & Maeda [1970], Lemma 3.8.)

7. Using the fact that a matroid lattice is left-complemented (Theorem 2.3.4), show that it is also relatively complemented (Corollary 2.3.5). (Cf. Maeda & Maeda [1970], p. 34.)

8. Show that if, in an M-symmetric lattice, b covers a and $a \, M \, y$, then $b \wedge y = a \wedge y$ or $b \wedge y$ covers $a \wedge y$ (Sachs [1961], Lemma 19).

References

Birkhoff, G. [1967] *Lattice Theory*, Amer. Math. Soc. Colloquium Publications, Vol. 25, Providence, R. I. (reprinted 1984).

Croisot, R. [1952] Quelques applications et propriétés des treillis semi-modulaires de longueur infinie, Ann. Fac. Sci. Univ. Toulouse 16, 11–74.

Dedekind, R. [1872] *Stetigkeit und irrationale Zahlen* (7th edition, 1969), Vieweg, Braunschweig.

Fort, A. [1974] Una caratterizzazione dei reticoli modulari a catene limitate-finite, Rend. Sem. Mat. Univ. Padova 51, 269–273.

Grätzer, G. [1978] *General Lattice Theory*, Birkhäuser, Basel.

Grätzer, G. [1979] *Universal Algebra* (2nd edition), Springer-Verlag, New York.

Johnston, K. G. and P. R. Jones [1984] The lattice of full regular subsemigroups of a regular semigroup, Proc. Roy. Soc. Edinburgh 89A, 203–204.

Jones, P. R. [1978] Semimodular inverse semigroups, J. London Math. Soc. 17, 446–456.

Jones, P. R. [1979] A homomorphic image of a semimodular lattice need not be semimodular: an answer to a problem of Birkhoff, Algebra Univsalis 9, 127–130.

Jones, P. R. [1990] Inverse semigroups and their lattices of inverse subsemigroups, in: *Lattices, Semigroups, and Universal Algebra* (ed. J. Almeida et al.), Plenum, New York, pp. 115–127.

Kogalovskiî, S. R. [1965] On the theorem of Birkhoff (Russian), Uspehi Mat. Nauk 20, 206–207.

MacNeille, H. M. [1937] Partially ordered sets, Trans. Amer. Math. Soc. 42, 416–460.

Maeda, S. and Y. Kato [1974] The completion by cuts of an M-symmetric lattice, Proc. Japan. Acad. 50, 356–358.

Maeda, F. and S. Maeda [1970] *Theory of Symmetric Lattices*, Springer-Verlag, Berlin.

Malliah, C. and S. P. Bhatta [1986] Equivalence of M-symmetry and semimodularity in lattices, Bull. London Math. Soc. 18, 338–342.

Ramalho, M. [1994] On upper continuous and semimodular lattices, Algebra Universalis 32, 330–340.

Sachs, D. [1961] Partition and modulated lattices, Pacific J. Math. 11, 325–345.

Shevrin, L. N. and A. J. Ovsyannikov [1983] Semigroups and their subsemigroup lattices, Semigroup Forum 27, 1–154.

Shevrin, L. N. and A. J. Ovsyannikov [1990] *Semigroups and Their Subsemigroup Lattices, Part I: Semigroups with Certain Types of Subsemigroup Lattices and Lattice Characterizations of Semigroup Classes.* (Russian), Ural State Univ. Publishers, Sverdlovsk.

Shevrin, L. N. and A. J. Ovsyannikov [1991] *Semigroups and their subsemigroup lattices, Part II: Lattice isomorphisms.* (Russian), Ural State Univ. Publishers, Sverdlovsk.

Wilcox, L. R. [1939] Modularity in the theory of lattices, Ann. of Math. 40, 490–505.

Wilcox, L. R. [1942] A note on complementation in lattices, Bull. Amer. Math. Soc. 48, 453–458.

Wilcox, L. R. [1944] Modularity in Birkhoff lattices, Bull. Amer. Math. Soc. 50, 135–138.

2.4 Wilcox Lattices

Summary. Weakly modular lattices were defined in Section 1.7, where we also noted that these lattices are not M-symmetric in general. Here we briefly mention an important subclass of weakly modular lattices, viz. Wilcox lattices. These lattices, which turn out to be M-symmetric, have their origin in the investigations of Wilcox [1938], [1939] on the lattice-theoretic treatment of the relationships between projective and affine geometries.

The relationships between projective and affine geometries may be stated lattice-theoretically as follows (Wilcox, unpublished manuscript; cf. also Wilcox [1955]):

(1) If a hyperplane and its nonzero subelements are deleted from a projective geometry Λ (regarded as a complemented modular lattice), what remains is an affine geometry L.

(2) Any affine geometry may be embedded into a projective geometry by appending suitable "ideal" elements.

   A few pertinent properties of L in (1) are the following:

(3) L is a lattice with the ordering, zero, and unit of Λ;

(4) L is weakly modular and M-symmetric;

(5) L is complemented;

(6) joins in L agree with joins in Λ;

(7) meets in L agree with meets in Λ for nonparallel elements;

(8) pairs of elements of L having a nonzero meet (in L) are modular.

In (2), points at infinity may be defined as sets of mutually parallel lines; other ideal elements are defined as certain sets of ideal points. Wilcox first studied a class of subsystems L of a complemented modular lattice Λ that have properties (3)–(6). If parallelism is properly defined, (7) will hold, while (8) will not be required, although it plays a later role. The lattices L are then characterized by their intrinsic properties, that is, an embedding of abstract lattices with these properties into complemented modular lattices is obtained. While the procedure remotely resembles that of (2) and contains the geometric case as an instance, it possesses much more generality. For example, indecomposability or finite dimensionality of Λ is not required, nor is the existence of points, lines, hyperplanes, and so on. Moreover, even in the geometric case Wilcox's theory treats a much wider class of subsystems than the affine geometries.

   More formally, we have the following result, which gives rise to the definition of Wilcox lattices.

Theorem 2.4.1 *Let Λ be a given complemented modular lattice with the lattice operations $a \sqcap b$ and $a \sqcup b$ and let $\underline{S}$ be a fixed subset of $\Lambda - \{0, 1\}$ with the following two properties:*

 (i) *$a \in \underline{S}$ and $0 < b \leq a$ imply $b \in \underline{S}$;*
 (ii) *$a, b \in \underline{S}$ implies $a \sqcup b \in \underline{S}$.*

If in the set $L \equiv \Lambda - \underline{S}$ we give the same order as in Λ, then L is a weakly modular M-symmetric lattice where the lattice operations $a \vee b$ and $a \wedge b$ satisfy

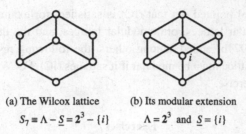

(a) The Wilcox lattice   (b) Its modular extension

$S_7 \equiv \Lambda - \underline{S} = 2^3 - \{i\}$ $\Lambda = 2^3$ and $\underline{S} = \{i\}$

Figure 2.7

the following conditions:

$$a \vee b = a \sqcup b \quad \text{for all } a, b \in L \quad \text{and} \quad a \wedge b = \begin{cases} a \sqcap b & \text{if } a \sqcap b \in L, \\ 0 & \text{if } a \sqcap b \in \underline{S}. \end{cases}$$

For a proof see Maeda & Maeda [1970], Theorem 3.11. When a weakly modular M-symmetric lattice L arises from a complemented modular lattice Λ in the manner described in the above theorem, we call L a *Wilcox lattice* (see Maeda & Maeda [1970]) and Λ the *modular extension* of L. An element in $\underline{S}$ is called an *imaginary element* for L, and if $\underline{S}$ has a greatest element i, then it is called the *imaginary unit* for L.

For example, $L = S_7$ is a Wilcox lattice that is not atomistic and not left-complemented; its modular extension is $\Lambda = 2^3$ with $\underline{S} = \{i\}$. This Wilcox lattice together with its modular extension is shown in Figure 2.7(a), (b).

Another example of a Wilcox lattice is the lattice of affine subsets of a vector space (cf. Maeda & Maeda [1970]). This lattice is atomistic and left-complemented.

Notes

Wilcox first considered the case where the resulting lattice is always left-complemented. For more details we refer to the original papers Wilcox [1939], [1942]. Wilcox wrote a comprehensive manuscript on his theory. This manuscript was never published; only the summary Wilcox [1955] appeared in print. A treatment of Wilcox's theory can be found in Maeda & Maeda [1970]. Some properties of atomistic Wilcox lattices will be dealt with in Section 2.5. Generalizations of Wilcox's construction have been given by Fujiwara [1965] and Bennett [1989–90]. For a thorough study of modularity relations in lattices see also Mihalek [1960].

Janowitz [1965b] introduced the concept of independent complement in the following way: Let L be a lattice with 0 and 1. A complement $\bar{a}$ of an element $a \in L$ is called an *independent complement* or an *I-complement* when $\bar{a} \, M \, a$ and $a \, M^* \, \bar{a}$. Consider the following condition on a lattice with 0 and 1:

(IC) every element has an independent complement.

Janowitz [1965b] pointed out that (IC) is satisfied, for example, by complemented orthomodular lattices, orthomodular lattices, and the duals of partition lattices. Maeda [1974b] added, among others, the following results: a partition lattice or Wilcox lattice must be modular if it satisfies (IC). For Wilcox lattices we leave this as an exercise.

Exercises

1. A Wilcox lattice $L \equiv \Lambda - \underline{S}$ is left-complemented if and only if $\underline{S}$ satisfies the following condition: If $b \leq a$ in Λ and if $a \notin \underline{S}$ then there exists $c \notin \underline{S}$ such that $a = b \sqcup c$ and $b \sqcap c = 0$. (Maeda & Maeda [1970], Lemma 3.13, p. 14.)
2. Let $L \equiv \Lambda - \underline{S}$ be a Wilcox lattice. An element $a \in L$ has an independent complement (see Notes above) if and only if a has a complement and $x \, M \, a$ holds for all $x \in L$. It follows that a Wilcox lattice satisfying (IC) is modular. (Maeda [1974b], Theorem 3.)

References

Bennett, M. K. [1989–90] Rectangular products of lattices, Discrete Math. 79, 235–249.
Fujiwara, S. [1965] The generalization of Wilcox lattices, Res. Bull. Fac. Liberal Arts Oita Univ. 2, 1–6.
Janowitz, M. F. [1965b] IC-lattices, Portugal. Math. 24, 115–122.
Maeda, F. and S. Maeda [1970] *Theory of Symmetric Lattices*, Springer-Verlag, Berlin.
Maeda, S. [1974b] Independent complements in lattices, in: *Coll. Math. Soc. J. Bolyai, 14. Lattice Theory*, Szeged, Hungary, 1974, pp. 215–226.
Mihalek, R. J. [1960] Modularity relations in lattices, Proc. Amer. Math. Soc. 11, 9–16.
Wilcox, L. R. [1938] Modularity in the theory of lattices, Bull. Amer. Math. Soc. 44, 50.
Wilcox, L. R. [1939] Modularity in the theory of lattices, Ann. of Math. 40, 490–505.
Wilcox, L. R. [1942] A note on complementation in lattices, Bull. Amer. Math. Soc. 48, 453–458.
Wilcox, L. R. [1944] Modularity in Birkhoff lattices, Bull. Amer. Math. Soc. 50, 135–138.
Wilcox, L. R. [1955] Modular extensions of semi-modular lattices (abstract), Bull. Amer. Math. Soc. 61, 542.

2.5 Finite-Modular and Weakly Modular AC Lattices

Summary. AC lattices, that is, atomistic lattices satisfying the atomic covering property (C), were introduced in Section 1.7 as important instances of semimodular lattices. Here we mention some more facts, in particular concerning the subclasses of finite-modular AC lattices, strongly planar AC lattices, and weakly modular AC lattices.

We recall that in arbitrary lattices upper semimodularity [condition (Sm)] implies the atomic covering property [condition (C)] and this in turn implies the Steinitz–Mac Lane exchange property [condition (EP)], while the reverse implications do not hold in general (see Section 1.7, in particular Figure 1.22). On the other hand, in atomistic lattices the Steinitz–Mac Lane exchange property implies upper semimodularity, that is, all three conditions are equivalent (cf. Theorem 1.7.2).

Moreover, in atomistic lattices M-symmetry [condition (Ms)] implies $\perp$-symmetry [condition (Ms$_0$)] and this in turn implies the atomic covering property [condition (C)]. Indeed, these implications hold in arbitrary lattices bounded below – see Section 2.3. As already noted, the atomic covering property and upper semimodularity [condition (Sm)] are equivalent in atomistic lattices.

If, in addition, an atomistic lattice is algebraic, then all four conditions (Ms), (Ms$_0$), (C), and (Sm) are equivalent. A direct proof of this can be found in Maeda & Maeda [1970], Theorem 7.15, p. 35. This equivalence also follows from Corollary 3.1.8 below.

We have also seen that an AC lattice has "many" left-modular elements. Indeed, if p is an atom in a lattice L with 0 satisfying the covering property (C), then for every $x \in L$ we have $p \, M \, x$ (that is, every atom is a left-modular element) and $p \, M^* \, x$ (see Theorem 1.7.2). There are AC lattices with "many more" (left and right) modular elements. Above all, we mention the class of finite-modular AC lattices, at which we shall now have a closer look.

An element of a lattice L with 0 is called a *finite element* if it is either 0 or the join of finitely many atoms. The set of all finite elements of L is denoted by $F(L)$. An atomistic lattice is *finite-modular* if every finite element is a right modular element (the concept "finite-modular" was introduced in Maeda [1967]). For the following characterization of finite-modularity in AC lattices see Maeda & Maeda [1970], Lemma 9.2 and Theorem 9.5.

Theorem 2.5.1 *Let L be an AC lattice. Then the following conditions are equivalent:*

(i) *L is finite-modular;*

(ii) *L is M^*-symmetric;*

(iii) *L is lower semimodular [condition (Sm*)];*

(iv) *If p and q are atoms and $p \leq q \vee a$ in L ($a \neq 0$), then there exists an atom $r \in L$ such that $p \leq q \vee r$ and $r \leq a$;*

(v) *$a \in F(L)$ implies $a \, M \, x$ and $x \, M \, a$ for every $x \in L$, that is, every finite element is both a right- and a left-modular element.*

For a further equivalent condition see Exercise 1. Although a finite-modular AC lattice is always M^*-symmetric by the preceding theorem, it is not M-symmetric [condition (Ms)] in general. Indeed, an example provided by Janowitz (cf. Supplement to Maeda & Maeda [1970]) exhibits a finite-modular AC lattice lacking $\perp$-symmetry [condition (Ms$_0$)]. On the other hand, if a finite-modular AC lattice is of finite length, then it is modular and thus M-symmetric. However, M-symmetric finite-modular AC lattices are not modular in general (cf. the example below derived from an infinite-dimensional Hilbert space).

The dual of an AC lattice is not an AC lattice, in general. However, there are nonmodular AC lattices having this property, and they are important enough to deserve a name of their own: a lattice L with 0 and 1 is called a *DAC lattice* if both L and its dual L^* are AC lattices.

Any upper continuous modular lattice is obviously a DAC lattice. We leave it as an exercise to show that any upper continuous DAC lattice is modular. For the following result see Maeda & Maeda [1970], Theorem 27.6.

Theorem 2.5.2 *Any* DAC *lattice is* M*-symmetric and* M**-symmetric. Moreover, it is finite-modular.*

This result follows from the fact that any DAC lattice L is lower semimodular and hence is M^*-symmetric and finite-modular by Theorem 2.5.1. Dual arguments show that L also M-symmetric.

By Maeda & Maeda [1970], Theorem 34.8, the set $L_c(H)$ of all closed subspaces of a Hilbert space H forms, with respect to set inclusion, a complete orthomodular AC lattice and hence a DAC lattice. (Note that an orthocomplemented AC lattice is always a DAC lattice.) Moreover, a Hilbert space H is finite-dimensional if and only if the lattice $L_c(H)$ is modular (this follows from Maeda & Maeda [1970], Theorem 32.17).

A weakening of condition (iv) in Theorem 2.5.1 leads to the concept of strongly planar lattices. A lattice L with 0 is called *strongly planar* if it satisfies the following condition:

(+) If p, q, and r are atoms and if $p \leq q \vee a$ and $r \leq a$, then there exists an atom $s \in L$ such that $p \leq q \vee r \vee s$ and $s \leq a$.

For a motivation and the geometric background of this concept cf. Jónsson [1959]. Strong planarity in AC lattices is related to finite-modularity in the following way:

Lemma 2.5.3 *An* AC *lattice* L *is strongly planar if and only if the principal dual ideal* $[p)$ *(which is likewise an* AC *lattice) is finite-modular for every atom* $p \in L$.

If in the preceding lemma the lattice L is algebraic (i.e. a matroid lattice), then strong planarity just means that for every atom $p \in L$ the interval $[p, 1]$ is modular (see Maeda & Maeda [1970], Theorem 14.1); moreover, the interval $[0, p]$ is distributive. This observation was generalized by Wille [1967], who defined *incidence geometries of grade n* and characterized them lattice-theoretically as matroid lattices having the property that for every element b of rank n the interval $[0, b]$ is distributive and the interval $[b, 1]$ is modular. Thus the geometries of grade 0 are just the projective geometries. For $n = 1$ we obtain the strongly planar geometries (cf. Sasaki [1952–3] and Jónsson [1959]); if in addition Euclid's parallel axiom is satisfied, we get the affine geometries. Möbius geometries (see Hoffman [1951]) are geometries of grade 2.

Let us have a look now at some further properties of the set of all finite elements $F(L)$ of an AC lattice L. First we note that clearly $a \in F(L)$ if and only if $r(a) < \infty$. Moreover, $F(L)$ is an ideal of L and for $a, b \in F(L)$, $a < b$ implies $r(a) < r(b)$ (cf. Maeda & Maeda [1970], Lemma 8.8). If $a < b$ in $F(L)$, then all maximal chains between a and b have the same length $r(b) - r(a)$, since $F(L)$ satisfies the Jordan–Dedekind chain condition.

We shall be interested in several properties of the ideal $F(L)$ of an AC lattice L, in particular in conditions for $F(L)$ to be a standard ideal in the sense of Grätzer & Schmidt [1961]. Let us recall that an ideal of a lattice L is called *distributive, standard,* or *neutral* if it is distributive, standard, or neutral, respectively, as an element of the ideal lattice $\mathbf{I}(L)$ (distributive, standard, and neutral elements were defined in Section 2.2). We shall use the following characterization of standard ideals due to Grätzer & Schmidt [1961].

Theorem 2.5.4 *An ideal S of a lattice L is standard if and only if $S \vee (x] = \{s \vee x_1 : s \in S, \ x_1 \leq x\}$ holds for every principal ideal $(x]$ of L.*

For a proof and for more information on standard ideals we refer to Grätzer & Schmidt [1961] and Janowitz [1965a] (see also Grätzer [1978], Chapter III, pp. 146–150). For further use we also note two other types of ideals and their relationship with standard ideals (for details see Janowitz [1965a]).

An ideal J of a lattice L is said to be *homomorphism kernel* if there exists a congruence relation Θ of L such that L/Θ has a zero element and $J = \{a \in L : a/\Theta = 0/\Theta\}$. An ideal J of a lattice L is called a *p-ideal* if L is bounded below and J is closed under perspectivity of elements (as defined in Exercise 2 of Section 2.1). In an arbitrary lattice one has the following string of implications:

neutral ideal $\Rightarrow$ standard ideal $\Rightarrow$ distributive ideal $\Rightarrow$ homomorphism kernel,

and if the lattice is bounded below one also has

homomorphism kernel $\Rightarrow$ p-ideal.

Counterexamples exist that show that all implications are irreversible (see Janowitz [1965a] and Schmidt [1965]).

In what follows let L be an AC lattice. If L is of finite length, then $F(L)$ is trivially a standard ideal. In an arbitrary AC lattice, however, the ideal $F(L)$ is not standard: consider, for instance, an infinite partition lattice, that is, the lattice of equivalence relations on an infinite ground set. Ore [1942] has proved that partition lattices are *simple,* that is, they have no nontrivial congruence relations; hence in an infinite partition lattice L, $F(L)$ is not the kernel of a homomorphism and thus not a standard ideal. Applying Theorem 2.5.4 we now give a necessary and sufficient condition for $F(L)$ to be a standard ideal in an arbitrary AC lattice L (cf. Stern [1991a], Corollary 41.4).

Theorem 2.5.5   *For an* AC *lattice L, the following two conditions are equivalent:*

(i) $F(L)$ *is a standard ideal.*

(ii) *If, for* $b, x \in L$, *the interval* $[x, b \vee x]$ *is of finite length, then the lower transposed interval* $[b \wedge x, b]$ *is of finite length, too.*

Proof.   Let $F(L)$ be a standard ideal of an AC lattice L, and suppose that the interval $[x, b \vee x]$ is of finite length. Since L is atomistic, there exists an $a \in F(L)$ such that $x \leq b \vee x = a \vee x$ and thus $b \in F(L) \vee (x]$. From Theorem 2.5.4 it follows that there are elements $x_1 \leq x$ and $a_1 \in F(L)$ such that $b = x_1 \vee a_1$. Hence the length of the interval $[x_1, b] = [x_1, x_1 \vee a_1]$ is finite. Since $x_1 \leq x \wedge b \leq b$, it follows that the length of the interval $[b \wedge x, b]$ is also finite. Conversely let now $b \leq x \vee a$ with $a \in F(L)$. Then the interval $[x, x \vee a]$ is of finite length. Since $x \leq b \vee x = a \vee x$, the interval $[x, b \vee x]$ is also of finite length. Hence by condition (ii), the interval $[b \wedge x, b]$ is of finite length, too. Since L is atomistic, there exists an $a_1 \in F(L)$ such that $b = (x \wedge b) \vee a_1$. From Theorem 2.5.4 it follows that $F(L)$ is standard. ∎

Corollary 2.5.6   *For an* AC *lattice L, the following two conditions are equivalent:*

(i) $F(L)$ *is a standard ideal;*

(ii) *if, for* $z, y \in L$, *one has* $z \prec z \vee y$, *then the lower transpose* $[z \wedge y, y]$ *is of finite length.*

The proof is left as an exercise. Let us now turn to some applications. Condition (ii) of Corollary 2.5.6. is, of course, satisfied in any lower semimodular AC lattice. Since for AC lattices lower semimodularity is equivalent to finite-modularity (see Theorem 2.5.1), we obtain the following result of Janowitz [1970], Theorem 4.6.

Corollary 2.5.7   *If L is a finite-modular* AC *lattice, then* $F(L)$ *is a standard ideal.*

Before giving further applications to other special classes of AC lattices, let us recall a notion that abstracts the geometric concept of parallelism. Let L be a lattice with 0 and $a, b \in L$ ($a \neq 0, b \neq 0$). We write $a <| b$ if $a \wedge b = 0$ and $b \prec a \vee b$. If both $a <| b$ and $b <| a$ hold, we say that a and b are *parallel* and write $a \parallel b$.

Lemma 2.5.8   *In an* AC *lattice L consider the following conditions:*

(i) $F(L)$ *is a standard ideal;*

(ii) $F(L)$ *is a homomorphism kernel;*

(iii) $F(L)$ *is a p-ideal;*

(iv) $y <| z$ *implies* $y \in F(L)$;

(v) $y \parallel z$ *implies* $y \in F(L)$.

Then (i) $\Rightarrow$ (ii) $\Rightarrow$ (iii) $\Rightarrow$ (iv) $\Rightarrow$ (v).

The implications (i) $\Rightarrow$ (ii) $\Rightarrow$ (iii) follow from the above-mentioned relationships between the given types of ideals. The remaining implications are also easy consequences of the definitions and are left as exercises.

We now ask for classes of AC lattices in which some or all conditions of the preceding lemma are equivalent. This question can be answered affirmatively for section-complemented AC lattices, for strongly planar AC lattices, and for atomistic Wilcox lattices. The proofs of the following two lemmas are left as exercises.

Lemma 2.5.9 Let L be a section complemented AC lattice. $F(L)$ is standard if and only if, for $y, z \in L$, $y <| z$ implies $y \in F(L)$.

Extending the preceding result, Maeda [1977] provided similar necessary and sufficient conditions for an arbitrary ideal of a section-complemented lattice to be standard. Since a matroid lattice (i.e. an algebraic AC lattice) is relatively complemented, it is section-complemented, and thus the preceding result holds for matroid lattices. Our next application concerns strongly planar lattices.

Lemma 2.5.10 Let L be a strongly planar AC lattice. $F(L)$ is standard if and only if, for $y, z \in L$, $y <| z$ implies $y \in F(L)$.

Since strongly planar AC lattices are finite-modular, the preceding lemma implies once more Corollary 2.5.7. Let us next consider weakly modular AC lattices and, in particular, atomistic Wilcox lattices. Since any weakly modular AC lattice is strongly planar (Maeda & Maeda [1970], Section 14) and since an atomistic Wilcox lattice is a weakly modular AC lattice (cf. Maeda & Maeda [1970], Remark 20.1) we obtain furthermore the following result:

Lemma 2.5.11 Let L be a weakly modular AC lattice (in particular, an atomistic Wilcox lattice). Then $F(L)$ is standard if and only if, for $y, z \in L$, $y <| z$ implies $y \in F(L)$.

In the special case of atomistic Wilcox lattices further equivalent conditions can be added for the ideal $F(L)$ to be a standard:

Theorem 2.5.12 In an atomistic Wilcox lattice $L \equiv \Lambda - \underline{S}$ consider the following six conditions:

(i) $F(L)$ is a standard ideal;
(ii) $F(L)$ is a p-ideal;
(iii) $y, z \in L$, $y <| z$ implies $y \in F(L)$;
(iv) $y \parallel z$ implies $y \in F(L)$;

(v) $\underline{S} \subseteq F(\Lambda)$;

(vi) *the imaginary unit i is finite in* Λ.

Then conditions (i)–(v) *are equivalent. If L has an imaginary unit i, then all six conditions are equivalent.*

For a proof of the preceding result see Stern [1991a], Theorem 41.11. The equivalence of the conditions (ii)–(v) is an unpublished result of M. F. Janowitz.

Let now $L \equiv \Lambda - \underline{S}$ be an atomistic Wilcox lattice of infinite length having the property that the imaginary unit i is finite in the modular extension Λ. From Theorem 2.5.12 it follows that L is not a simple lattice: $F(L)$ is a standard ideal and hence induces a nontrivial congruence relation. We now ask for conditions on a special class of atomistic Wilcox lattices to be simple. Let us first recall the formal definition of this class of lattices.

A matroid lattice L of length ≥ 4 is said to be an *affine matroid lattice* if it is weakly modular and satisfies *Euclid's weak parallel axiom*: If g is a line (i.e. an element of rank 2) and if p is a point (atom) with $p \not\leq g$, then there exists *at most* one line k such that $g \parallel k$ and $p < k$. If, in the preceding sentence, *at most* is replaced by *exactly one*, we say that L satisfies *Euclid's strong parallel axiom*. For both Euclid's weak and Euclid's strong parallel axiom see Maeda & Maeda [1970], Section 18.

It follows from Maeda & Maeda [1970], Corollary 19.14, that a nonmodular affine matroid lattice is an atomistic Wilcox lattice with imaginary unit i. Maeda [1977] characterized the simplicity of these lattices.

Theorem 2.5.13 *Let L be a nonmodular affine matroid lattice. Then*

(i) *L is simple if and only if there exist elements* $a_1, \ldots, a_n \in L$ *such that* $a_1 \vee \cdots \vee a_n = 1$ *and each* a_k *is subprojective to an element* $x_k \sqcup i$ *with* $0 \neq x_k \in F(L)$;

(ii) *if i is dual-finite in the modular extension* Λ, *then L is simple.*

For the notions of subprojectivity and dual-finiteness, and for the proof, see Maeda [1977]. As a special case we get (cf. Stern [1991a], Corollary 41.14):

Corollary 2.5.14 *A nonmodular affine matroid lattice satisfying Euclid's strong parallel axiom is a simple lattice.*

Let us now have a look at complements in finite-modular AC lattices. Following a suggestion of S. Maeda, an AC lattice will be called *finite-complemented* if each finite element has a complement. Figure 2.8 shows the interrelationships between types of complementation for certain classes of finite-modular AC lattices. The implications (a), (b), (c), (d), and (f) are immediate from the definitions.

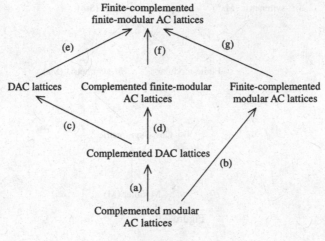

Figure 2.8

Implication (e) follows from Maeda & Maeda [1970], Theorem 27.6 and Theorem 27.10. Let us give some examples concerning the nonreversibility of these implications.

We have already mentioned that the lattice of all closed subspaces of an infinite-dimensional Hilbert space is a complemented DAC lattice. It is finite-modular but not modular. Thus the implications (a) and (g) are not reversible.

The lattice L of all closed subspaces of the Banach space $\Gamma(M)$ (where M is an uncountable set) is a DAC lattice, but it is not complemented (cf. Lindenstrauss [1968]). On the other hand, L is finite-complemented (cf. Maeda & Maeda [1970], Theorem 27.10). Hence the implications (c) and (f) are not reversible.

The next example is due to S. Maeda (personal communication). Let H be an infinite-dimensional Hilbert space, let K denote the lattice of all subspaces of H, and denote by L the set formed by removing from K every nonclosed subspace M for which $\dim(M^{\perp}) < \infty$, that is, the closure $cl(M)$ has finite codimension. (Here $X^{\perp}$ denotes the orthocomplement of a subspace X.) We have the set inclusions $L_c(H) \subset L \subset K$. It can be shown that the lattice L satisfies the conditions (15.15.2) and (15.15.3) of Maeda & Maeda [1970] (see Stern [1992a] for details). Hence by Maeda & Maeda [1970], Theorem 15.15, L is a complete finite-modular AC lattice. We leave it as an exercise to show that L is complemented but not coatomistic. The preceding example shows that the implications (d) and (e) in Figure 2.8 are not reversible.

Let us finally turn to the interrelationship between concepts connected with upper semimodularity in atomistic lattices. From Maeda & Maeda [1970] and Maeda [1981] we obtain Figure 2.9. In Maeda [1981] one finds examples showing that these implications are not reversible. Let us note the following ones:

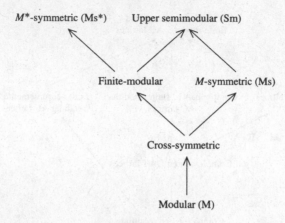

Figure 2.9

The atomistic lattice of all closed subspaces of an infinite-dimensional Banach space is cross-symmetric but not modular (Maeda & Maeda [1970], (32.14) and (32.17)).

The atomistic lattice given in the supplement of Maeda & Maeda [1970] is finite-modular but not M-symmetric (and hence not cross-symmetric).

The dual of a nonmodular matroid lattice is atomistic (Maeda & Maeda [1970], (13.1)), but it does not possess the atomic covering property (C) (Maeda & Maeda [1970], (14.1)) and hence it is not finite-modular. (Maeda [1981] proved that any finite-modular atomistic lattice has the covering property.)

Notes

For complete atomistic lattices the concept of *static* (*statisch*) lattices was introduced by Wille [1966] and extended by Janowitz [1970] to the more general notion of *finite-static* lattices. Maeda & Maeda [1970], p. 65, applies these concepts to arbitrary atomistic lattices without assuming completeness: An atomistic lattice is called *static* if $p \leq a \vee b$, p atom, implies the existence of finite elements a_1, b_1 such that $p \leq a_1 \vee b_1$ with $a_1 \leq a$ and $b_1 \leq b$.

In algebraic atomistic lattices the atoms are compact and hence these lattices are static. Janowitz [1975] proved the following result: Let L be an AC lattice such that every infinite element dominates a finite element b having the property that the principal dual ideal $[b)$ is static. Then L is static. As a corollary one gets that any weakly modular AC lattice is static. Janowitz [1970] also proved that every static AC lattice is M-symmetric. It follows, in particular, that a weakly modular AC lattice is M-symmetric (note that this also follows Maeda & Maeda [1970], Corollary 15.13, p. 66).

Janowitz [1975] proposed to study AC lattices in which the principal dual ideal $[b)$ is a modular sublattice for elements of rank n. Such a lattice will be referred to as an AC lattice that is *modular of grade n*. The incidence geometries of grade n introduced by Wille [1967] share this property. From the results of Janowitz it follows that if an AC lattice is modular of grade n, then it is static.

The notions of static and finite-static atomistic lattices have been further extended by Janowitz [1976] and by Maeda et al. [1985] (the latter paper investigates these concepts within the framework of join semilattices).

Janowitz & Coté [1976] call an atomistic lattice L *finite-distributive* if every element of $F(L)$ is a standard element; they show, among other things, that an atomistic lattice is finite-distributive if and only if it is a finite-modular AC lattice having the property that each of its lines (i.e. elements of rank 2) contains exactly two points (atoms). Moreover they prove that a finite-distributive lattice is ∇-symmetric (i.e., the relation $a \nabla b$ is symmetric – cf. Maeda & Maeda [1970], p. 16) if and only if it the lattice is distributive. It follows that any nondistributive finite-distributive atomistic lattice provides an affirmative answer to Maeda & Maeda [1970], Problem 3, p. 55.

An atomistic lattice is said to be *biatomic* if $p \leq a \vee b$, p atom, $a \neq 0, b \neq 0$, implies the existence of atoms $q \leq a$ and $r \leq b$ such that $p \leq q \vee r$ (Bennett [1987]). Atomistic lattices with this property were also considered by Wille [1966] and Maeda [1981]. It is clear from the definitions that any biatomic lattice is static, and it is easy to see that the converse is not true in general. Wille [1966], Satz 3.11, proved that any modular atomistic lattice is biatomic. This implies that the lattice of all subspaces of a vector space is biatomic.

The latter lattice is an instance of a biatomic lattice satisfying the Steinitz–Mac Lane exchange property [condition (EP); see Section 1.7]. Another class of biatomic lattices arises from convex substructures (cf. Birkhoff & Bennett [1985]). Let us note here that lattices whose members are convex sets often enjoy the so-called *antiexchange property*

(AEP) $p \leq q \vee b$ and $p \not\leq b$ imply that $q \not\leq p \vee b$ (p, q atoms, b arbitrary).

For more details on convexity and the antiexchange property see Section 7.3. Let us also note that biatomic lattices satisfy neither (EP) nor (AEP) in general (cf. Bennett [1989–90]).

Exercises

1. Let L be an AC lattice. Show that condition (iv) of Theorem 2.5.1 implies the following condition:

   (∗) If p is an atom, b a finite element, and $p \leq a \vee b$ in L ($a \neq 0$, $b \neq 0$), then there exist atoms $q, r \in L$ such that $p \leq q \vee r$, $q \leq a$, and $r \leq b$.

   Show that this condition in turn implies condition (v) of Theorem 2.5.1 (cf. Maeda & Maeda [1970], Lemma 9.2).

2. Show that if L is an AC lattice, then the ideal $F(L)$ of the finite elements of L is left-complemented and M-symmetric. (Cf. Maeda & Maeda [1970], Theorem 8.11, p. 37.)

3. Let L be an algebraic lattice satisfying the dual covering property (C^*) and the descending chain condition for compact elements. Show that if each atom of L has a complement, then L is atomistic (Stern [1991b]).

4. Show that an atomistic lattice is finite-distributive if and only if each of its atoms is a standard element (Janowitz & Coté [1976]).

5. Prove the following statements:

   (a) An AC lattice L is static if and only if its ideal of the finite elements $F(L)$ is dually distributive.

   (b) In an AC lattice L, the ideal $F(L)$ is neutral if and only if L is static and $F(L)$ is standard.

   (c) If L is an AC lattice that is modular of grade n, then the ideal $F(L)$ of the finite elements of L is standard if and only if it is neutral.

6. Prove that for a nonmodular affine matroid lattice L one can add in Theorem 2.5.12 the statement "$F(L)$ is a neutral ideal" as a further equivalent condition.

7. Show that any cross-symmetric atomistic lattice is finite-modular (Maeda [1981]).

8. Let L be a complete atomic lattice that is both upper and lower semimodular. Show that each atom of L has a complement if and only if each finite element of L has a complement (Stern [1992a]).

References

Bennett, M. K. [1987] Biatomic lattices, Algebra Universalis 24, 60–73.

Bennett, M. K. [1989–90] Rectangular products of lattices, Discrete Math. 79, 235–249.

Birkhoff, G. and M. K. Bennett [1985] The convexity lattice of a poset, Order 2, 223–242.

Grätzer, G. [1978] *General Lattice Theory*, Birkhäuser, Basel.

Grätzer, G. and E. T. Schmidt [1961] Standard ideals in lattices, Acta Math. Acad. Sci. Hungar. 12, 17–86.

Hoffman, A. J. [1951] On the foundations of inversion geometry, Trans. Amer. Math. Soc. 71, 218–242.

Janowitz, M. F. [1965a] A characterisation of standard ideals, Acta Math. Acad. Sci. Hungar. 16, 289–301.

Janowitz, M. F. [1970] On the modular relation in atomistic lattices, Fund. Math. 66, 337–346.

Janowitz, M. F. [1975] Examples of statisch and finite-statisch AC-lattices, Fund. Math. 89, 225–227.

Janowitz, M. F. [1976] On the "del" relation in certain atomistic lattices, Acta Math. Acad. Sci. Hungar. 28, 231–240.

Janowitz, M. F. and N. H. Coté [1976] Finite-distributive atomistic lattices, Portugal. Math. 35, 80–91.

Jónsson, B. [1959] Lattice-theoretic approach to projective and affine geometry, in: *The Axiomatic method* (ed. L. Henkin, P. Suppes, and A. Tarski), Studies in Logic, Amsterdam, pp. 188–203.

Lindenstrauss, J. [1968] On subspaces of Banach spaces without quasicomplement, Israel J. Math 6, 36–38.

Maeda, F. and S. Maeda [1970] *Theory of Symmetric Lattices*, Springer-Verlag, Berlin.

Maeda, S. [1967] On atomistic lattice with the covering property, J. Sci. Hiroshima Univ. Ser. A-I 31, 105–121.

Maeda, S. [1977] Standard ideals in Wilcox lattices, Acta Math. Acad. Sci. Hungar. 29, 113–118.

Maeda, S. [1981] On finite-modular atomistic lattices, Algebra Universalis 12, 76–80.

Maeda, S., N. K. Thakare, and M. P. Wasadikar [1985] On the "del" relation in join-semilattices, Algebra Universalis 20, 229–242.

Ore, O. [1942] Theory of equivalence relations, Duke Math. J. 9, 573–627.

Sasaki, U. [1952–3] Semi-modularity in relatively atomic upper continuous lattices, J. Sci. Hiroshima Univ. Ser. A 16, 409–416.

Schmidt, E. T. [1965] Remark on a paper of M. F. Janowitz, Acta Math. Acad. Sci. Hungar. 16, 435.

Stern, M. [1991a] *Semimodular Lattices*, B. G. Teubner, Stuttgart.

Stern, M. [1991b] Complements in certain algebraic lattices, Archiv d. Math. 56, 197–202.

Stern, M. [1992a] On complements in lattices with covering properties, Algebra Universalis 29, 33–40.

Wille, R. [1966] Halbkomplementäre Verbände, Math. Z. 94, 1–31.

Wille, R. [1967] Verbandstheoretische Charakterisierung n-stufiger Geometrien, Archiv d. Math. 18, 465–468.

2.6 Orthomodular M-Symmetric Lattices

Summary. We have a look at upper semimodularity and M-symmetry for orthocomplemented and for orthomodular lattices. Examples are given for both the atomistic and the nonatomistic case.

For an orthocomplemented lattice of finite length, upper semimodularity [condition (Sm)] is equivalent to modularity, since by orthocomplementation upper semimodularity implies lower semimodularity [condition (Sm*)].

Let now L be an arbitrary orthocomplemented lattice. Since $a \to a^{\perp}$ is a dual automorphism of L,

(1) $a\,M^{*}\,b$ is equivalent to $a^{\perp}\,M\,b^{\perp}$,

and

(2) L is M-symmetric if and only if L is M^{*}-symmetric.

For both properties see Maeda & Maeda [1970], Remark 29.6. Whereas for general lattices bounded below $\perp$-symmetry [condition (Ms$_0$)] is weaker than M-symmetry [condition (Ms)] (see Section 2.3), for orthomodular lattices the two concepts coincide and are equivalent to M^{*}-symmetry (see Maeda & Maeda [1970], Theorem 29.17):

Theorem 2.6.1   *If L is an orthomodular lattice, then the following three statements are equivalent:*

(i) *L is $\perp$-symmetric [condition (Ms_0)];*
(ii) *L is M-symmetric [condition (Ms)];*
(iii) *L is M^*-symmetric [condition (Ms^*)].*

Sketch of proof.   The equivalence of (ii) and (iii) follows from the above remark (2). (ii) $\Rightarrow$ (i) is trivial; (i) $\Rightarrow$ (ii) by Theorem 1.2.4 and by Theorem 2.3.1.   ∎

For orthocomplemented lattices there is a notion of symmetry, called O-symmetry, which is stronger than M-symmetry (in fact, it turns out to be equivalent to cross-symmetry): An orthocomplemented lattice is said to be O-*symmetric* if $a \, M \, b$ implies $b^\perp \, M \, a^\perp$. For the following result see Maeda & Maeda [1970], Theorem 29.8.

Theorem 2.6.2   *If an orthocomplemented lattice is O-symmetric, then it is M-symmetric and M^*-symmetric. Moreover, the four relations $a \, M \, b$, $b \, M \, a$, $a \, M^* \, b$, and $b \, M^* \, a$ are equivalent.*

Sketch of proof.   If an orthocomplemented lattice is O-symmetric, then it is cross-symmetric by the above remark (1) and thus it is M-symmetric by Theorem 2.3.2. By the above remark (2) it is also M^*-symmetric. Moreover O-symmetry implies that $a \, M \, b$ and $b \, M^* \, a$ are equivalent.   ∎

In fact, it can be shown that in an orthocomplemented lattice, cross-symmetry, dual cross-symmetry, and O-symmetry are all equivalent (cf. Piziak [1990]). Schreiner [1966] introduced O-symmetry in an orthomodular lattice and explored some basic properties; in particular, he proved that every O-symmetric lattice is M-symmetric (cf. Schreiner [1969]).

An example is provided by the projection lattice of a von Neumann algebra. Let H be a Hilbert space, and denote by $B(H)$ the $*$-algebra of all bounded linear operators on H. The set $P(B(H))$ of all projection operators forms a complete orthomodular lattice. This lattice is isomorphic to the lattice $L_c(H)$ of closed subspaces of H (for which see Section 1.2). For any subset S of $B(H)$ we put

$$\mathbf{S}' = \{T \in B(H) : TS = ST \text{ for every } S \in \mathbf{S}\}.$$

A *von Neumann algebra* $\mathbf{A}$ is a $*$-subalgebra of $B(H)$ such that $\mathbf{A}'' = \mathbf{A}$.

Let now a and b be elements of an orthomodular lattice. We say that a *commutes with* b and we write $a \, C \, b$ if $a = (a \wedge b) \vee (a \wedge b^\perp)$. It can be shown that $a \, C \, b$ if and only if $b \, C \, a$ (cf. Maeda & Maeda [1970], Theorem 36.1).

For a subset S of an orthomodular lattice L, the set $C(S) = \{a \in L : a \, C \, b \text{ for every } b \in S\}$ is an orthocomplemented sublattice of L, and we have $S \subseteq CC(S)$. If

$S = CC(S)$, then S is called a *C-closed sublattice* of L. A C-closed sublattice of an orthomodular lattice is itself orthomodular, since condition (iv) of Theorem 1.2.4 holds in S.

Let $P(A)$ denote the set of projection operators of a von Neumann algebra A. It can be shown that $P(A)$ is a C-closed sublattice of $P(B(H))$. Thus $P(A)$, the *projection lattice* of the von Neumann algebra A, is an orthomodular lattice. Topping [1967] proved that the projection lattice of any von Neumann algebra is O-symmetric. Generalizing Topping's approach, S. Maeda extended the preceding result to the projection lattice of any Baer *-ring satisfying the "square-root axiom" (see Maeda & Maeda [1970], Theorem 37.14). Thus by Theorem 2.6.2 the projection lattice of any von Neumann algebra is an M-symmetric orthomodular lattice.

Let us now turn to the atomistic case. Recall that in lattices with 0, M-symmetry [condition (Ms)] implies $\perp$-symmetry [condition (Ms$_0$)] and this, in turn, implies the atomic covering property [condition (C)]. Moreover these implications are strict. On the other hand, in atomistic lattices condition (C) is equivalent to upper semimodularity. If an atomistic lattice is orthocomplemented, all these concepts coincide and are equivalent to M^*-symmetry (cf. Maeda & Maeda [1970], Theorem 30.2):

Theorem 2.6.3 *Let L be an orthocomplemented atomistic lattice. The following four conditions are equivalent:*

(i) *L has the atomic covering property [condition (C)];*
(ii) *L is $\perp$-symmetric [condition (Ms$_0$)];*
(iii) *L is M-symmetric [condition (Ms)];*
(iv) *L is M^*-symmetric.*

Sketch of proof. (iii) $\Rightarrow$ (ii) $\Rightarrow$ (i): As already noted, these implications hold in arbitrary lattices with 0; the equivalence of (iii) and (iv) holds by the above remark (2).

(i) $\Rightarrow$ (iii): If an orthocomplemented atomistic lattice satisfies (C), then it is a DAC lattice and therefore it is M-symmetric (by Theorem 2.5.2). ∎

Thus an orthocomplemented AC lattice is always M-symmetric. In particular, the preceding result holds for atomistic orthomodular lattices, for example, for $L_c(H)$.

Maeda & Maeda [1970], Problem 7, p. 135, ask the following questions:

Is there an orthocomplemented AC lattice that is not O-symmetric?
Is there an orthomodular AC lattice that is not O-symmetric?

Saarimäki [1982] constructed examples of anisotropic quadratic spaces (k, E, Φ) such that $L_c(k, E, \Phi)$ is not O-symmetric. Since $L_c(k, E, \Phi)$ is an orthocomplemented AC lattice, this provides a negative answer to the first one

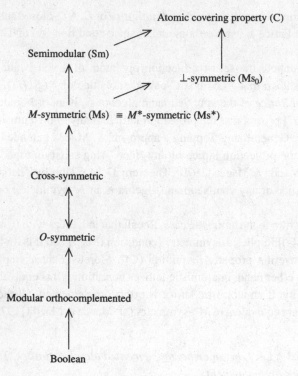

Figure 2.10

of the preceding two questions. The second question seems to be still open – see Piziak [1991], who also asks more generally:

Does an *M*-symmetric orthomodular lattice have to be *O*-symmetric?

Figure 2.10 shows the interrelationships in orthocomplemented lattices.

Notes

For general information on orthomodular lattices we refer to Holland [1970], Kalmbach [1983], [1986], Beran [1984], and Maeda & Maeda [1970]. For relations to physics (quantum logics) see for example Greechie & Gudder [1973], Beltrametti & Cassinelli [1981], Cohen [1989], and Pták & Pulmannová [1991].

Exercises

1. Give an example of an orthomodular lattice that is not *M*-symmetric.

2. If $c\,C\,a$ and $c\,C\,b$ hold for elements a, b, c of an orthomodular lattice, then
$(a \vee c)\,M\,(b \vee c)$, $(a \wedge c)\,M\,(b \wedge c)$, $a\,M\,(b \wedge c)$, $(a \vee c)\,M\,b$, $(a \vee c)$

M $(b \wedge c)$, and $(a \wedge c)$ M $(b \vee c)$, but the pairs $(a, b \vee c)$ and $(a \wedge c, b)$ are not modular pairs in general (Schreiner [1966]; cf. Kalmbach [1983], p. 101).

3. Show that if H is a finite-dimensional Hilbert space, then the lattice of closed subspaces, $L_c(H)$, is a modular lattice. (Cf. Kalmbach [1983], p. 71.)

4. Recall that two elements a and b (of an orthomodular lattice) are perspective, denoted by $a \sim b$, if and only if they have a common complement, that is, if there exists an element x such that $a \vee x = b \vee x = 1$ and $a \wedge x = b \wedge x = 0$ (cf. Section 2, Exercise 2). Elements a and b (of an orthomodular lattice) are called *strongly perspective*, denoted by $a \sim_s b$, if they have a common complement in the interval $[0, a \vee b]$.

   (a) Show that strong perspectivity implies perspectivity, and give an example showing that the converse is not true.

   (b) Show that an orthomodular lattice is modular if and only if perspectivity implies strong perspectivity.

   (Cf. Kalmbach [1983], p. 74, Theorem 3.)

References

Beltrametti, E. G. and G. Cassinelli [1981] *The Logic of Quantum Mechanics*, Encyclopedia of Mathematics and its Applications, Vol. 15, Addison-Wesley, Reading, Mass.

Beran, L. [1984] *Orthomodular Lattices, Algebraic Approach*, Reidel, Dordrecht.

Cohen, D. W. [1989] *An Introduction to Hilbert Space and Quantum Logic*, Springer-Verlag, New York.

Greechie, R. J. and S. Gudder [1973] Quantum logics, in: *Contemporary Research in the Foundations and Philosophy of Quantum Theory* (ed. C. A. Hooker), Reidel, Dordrecht, pp. 143–173.

Holland, S. S., Jr. [1970] The current interest in orthomodular lattices, in: *Trends in Lattice Theory* (ed. J. C. Abbott), van Nostrand Reinhold, New York, pp. 41–126.

Kalmbach, G. [1983] *Orthomodular Lattices*, Academic Press, London.

Kalmbach, G. [1986] *Measures and Hilbert Lattices*, World Scientific, Singapore.

Maeda, F. and S. Maeda [1970] *Theory of Symmetric Lattices*, Springer-Verlag, Berlin.

Piziak, R. [1990] Lattice theory, quadratic spaces, and quantum proposition systems, Found. Phys. 20, 651–665.

Piziak, R. [1991] Orthomodular lattices and quadratic spaces: a survey, Rocky Mountain J. Math. 21, 951–992.

Pták, P. and S. Pulmannová [1991] *Orthomodular Structures as Quantum Logics*, Kluwer Academic, Dordrecht.

Saarimäki, M. [1982] Counterexamples to the algebraic closed graph theorem, J. London Math. Soc. 26, 421–424.

Schreiner, E. A. [1966] Modular pairs in orthomodular lattices, Pacific J. Math. 19, 519–528.

Schreiner, E. A. [1969] A note on O-symmetric lattices, Caribbean J. Sci. and Math. 1, 40–50.

Topping, D. M. [1967] Asymptoticity and semimodularity in projection lattices, Pacific J. Math. 20, 317–325.

3

Conditions Related to Semimodularity, 0-Conditions, and Disjointness Properties

3.1 Mac Lane's Condition

Summary. In Chapter 2 we dealt with M-symmetry, a condition related to semimodularity in the sense that the two conditions are equivalent in lattices of finite length. Wilcox's concept of M-symmetry is one important approach to the question of replacing upper semimodularity by a condition that is nontrivially satisfied in lattices having continuous chains. Another approach is that of Mac Lane [1938], at which we cast a glance here. His condition will be given in two equivalent forms. We study the interrelationships between Mac Lane's condition, semimodularity, Birkhoff's condition, and M-symmetry and show, among other things, that these conditions are equivalent in upper continuous strongly atomic lattices.

In atomistic lattices, the Steinitz–Mac Lane exchange property [condition (EP)] is equivalent to the atomic covering property [condition (C)] and thus to upper semimodularity [condition (Sm)] (see Theorem 1.7.3). On the other hand, the conditions (Sm), (EP), and (C) are trivially satisfied in lattices with continuous chains. It was this criticism of (Sm) that led Wilcox [1938], [1939] to introduce M-symmetric lattices (cf. Chapter 2).

Mac Lane [1938] was led in his investigations to an"exchange axiom," denoted by him as condition (E_5), which is stronger than (Sm) and goes in a different direction than M-symmetry. Like M-symmetry, Mac Lane's condition does not involve any covering relation. Mac Lane's point of departure was the following condition due to Menger [1936] : If p is an atom and a, c are arbitrary elements, then $p \not\leq a \vee c$ implies $(a \vee p) \wedge c = a \wedge c$. Mac Lane called this condition *Menger's exchange axiom* and denoted it by (E_3).

To modify Menger's condition (E_3), Mac Lane replaced the atom p by an arbitrary element b and the conclusion $(a \vee p) \wedge c = a \wedge c$ by the statement that $(a \vee b_1) \wedge c = a \wedge c$ holds for some nontrivial part b_1 of b. No generality is lost if one requires $a < c$: the hypothesis $p \not\leq a \vee c$ or $p \wedge (a \vee c) = 0$ then becomes $b \wedge c < a$, and Mac Lane's modified statement is the atomless condition

(E$_5$) $b \wedge c < a < c < b \vee c$ implies that there exists b_1 such that $b \wedge c < b_1 \leq b$ and $(a \vee b_1) \wedge c = a$.

The statement that this law or other similar laws called (E$_6$) and (E$_7$) (for which see below) hold is to mean that they hold for all elements a, b, and c of the lattice.

The three atomless conditions (E$_5$), (E$_6$), and (E$_7$) do not, like the exchange property (EP) [called (E$_1$) by Mac Lane], hold trivially in lattices with continuous chains. Anyone of these three conditions is a weakening of the modular law. To see this, for example, for (E$_5$), we note that in any lattice the modular law is equivalent to the following assertion:

(*) $b \wedge c < a < c < b \vee c$ implies $(a \vee b) \wedge c = a$.

Thus condition (E$_5$) is a direct consequence of the modular law.

In a modular lattice it is impossible to find a sublattice isomorphic with the lattice of Figure 3.1(a).

Consider now a lattice L in which it is possible to find a sublattice isomorphic with the lattice in Figure 3.1(a). For an arbitrary element $s \in L$ one has $x \leq (x \vee s) \wedge z$. Specializing this property, we formulate

Definition 1. *A lattice L satisfies Mac Lane's condition* (Mac$_1$) *if, for any* $x, y, z \in L$ *such that*

(+) $y \wedge z < x < z < y \vee x$,

there exists an element $t \in L$ such that $y \wedge z < t \leq y$ and $x = (x \vee t) \wedge z$.

For an illustration see Figure 3.1(b). In Definition 1 we may replace the single occurring $\leq$ by $<$, since it is not possible to choose $t = y$ [because $(x \vee y) \wedge z = (y \vee z) \wedge z = z$]. From Figure 3.1(b) we see that the centered hexagon S_7 satisfies (Mac$_1$). Instead of Definition 1 we shall also use the following equivalent form, which is just Mac Lane's above-mentioned condition (E$_5$):

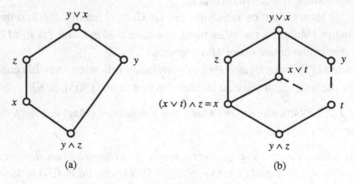

(a)　　　　　　　　　(b)

Figure 3.1

Definition 2.   *A lattice L satisfies Mac Lane's condition* (Mac$_2$) *if for any* x, y, $z \in L$ *such that*

$(++)$ $y \wedge z < x < z < y \vee z$

there exists an element $t \in L$ *such that* $y \wedge z < t \leq y$ *and* $x = (x \vee t) \wedge z$.

Property $(++)$ in Definition 2 is a weakening of property $(+)$ in Definition 1: if $y \wedge z < x < z < y \vee x$, then $y \vee x \leq y \vee z \leq y \vee x$ and thus $y \vee x = y \vee z$. In contrast to the situation in Definition 1, the single $\leq$ in Definition 2 cannot be replaced by $<$ [since $(y \vee x) \wedge z = x$ may occur].

We now show that the two preceding definitions are indeed equivalent:

Lemma 3.1.1   *Condition* (Mac$_1$) *holds in a lattice if and only if condition* (Mac$_2$) *holds.*

Proof.   If the assumptions of (Mac$_1$) are satisfied, then so are the assumptions of (Mac$_2$). Hence if (Mac$_2$) is true, then so is (Mac$_1$).

Suppose now that condition $(++)$ of Definition 2 holds, that is, the assumptions of (Mac$_2$) are satisfied. If $y \vee x = y \vee z$, then also condition $(+)$ of Definition 1 holds, that is, the assumptions of (Mac$_1$) are satisfied. Thus, in this case, if (Mac$_1$) is true, then (Mac$_2$) is also true.

If $y \vee x \neq y \vee z$, then $y \vee x \ngeq z$ and thus $y \vee x$ and z are incomparable. Set $(y \vee x) \wedge z = z_1$. Then $z_1 < z$. If $z_1 = x$, then, taking $t = y$, we see that the assertion holds true. If $z_1 \neq x$, then x, y, z satisfy the assumptions of (Mac$_1$). Hence there exists a t_1 such that $y \wedge z = y \wedge z_1 < t_1 \leq y$ and $(x \vee t_1) \wedge z_1 = x$. From $(x \vee t_1) \wedge z \leq (x \vee y) \wedge z = z_1$ we obtain therefore $(x \vee t_1) \wedge z = [(x \vee t_1) \wedge z] \wedge z_1 = (x \vee t_1) \wedge z_1 = x$. Taking $t = t_1$, we see that the assertion holds true also in this case.   ∎

If a lattice satisfies (Mac$_1$) or, equivalently, (Mac$_2$) we briefly say that the lattice satisfies *Mac Lane's condition* (Mac). Some authors (e.g. Croisot, Szász) call a lattice semimodular if it satisfies (Mac).

Figure 3.2 shows a lattice satisfying the DCC. It is readily checked that this lattice satisfies (Mac); on the other hand, we have $c \, M \, a_1$, whereas $a_1 \, M \, c$ does not hold, that is, the lattice is not M-symmetric.

Mac Lane [1938] investigated further conditions following from his condition (Mac). For the next result see also Dubreil-Jacotin et al. [1953], p. 87.

Lemma 3.1.2   *A lattice satisfying Mac Lane's condition* (Mac) *has the following two properties:*

(i)   *The relations* $y \wedge z < x < z < y \vee x$ *imply the existence of an element* t *such that* $y \wedge z < t \leq y$ *and* $(x \vee t) \wedge z < z$. *(This is condition* (E$_7$) *of Mac Lane [1938].)*

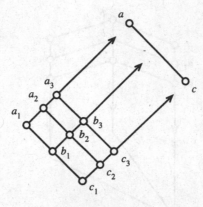

Figure 3.2

(ii) *The relations $y \wedge z < x < z < y \vee z$ imply the existence of an element t such that $y \wedge z < t \leq y$ and $(x \vee t) \wedge z < z$. (This is condition (E$_6$) of Mac Lane [1938].)*

Moreover, (i) *and* (ii) *are equivalent in any lattice.*

Proof. It is clear that (Mac$_1$) implies (i) and that (Mac$_2$) implies (ii), since (according to the corresponding definitions) $(x \vee t) \wedge z = x$ implies $(x \vee t) \wedge z < z$. It is also obvious that (ii) implies (i).

Let now (i) be satisfied and suppose that $y \wedge z < x < z < y \vee z$. If $y \vee x = y \vee z$, then the premises of (i) are satisfied and thus the common conclusion is true. If $y \vee x \neq y \vee z$, then $y \vee x \not\geq z$ and thus $(y \vee x) \wedge z < z$, showing that the conclusion of (ii) is true for $t = y$. ∎

The lattice of Figure 3.3 is taken from Dubreil-Jacotin et al. [1953], p. 100. In this lattice the chains $\{a_i\}_{i=1,2,3,\dots}$, $\{b_j\}_{j=1,2,3,\dots}$, $\{c_k\}_{k=1,2,3,\dots}$ are assumed to be isomorphic with the order dual of the chain of positive integers. This lattice satisfies (E$_6$) and equivalently (E$_7$), but it does not satisfy Mac Lane's condition (Mac).

Lemma 3.1.3 *In an arbitrary lattice, condition (E$_6$) implies upper semimodularity [condition (Sm)].*

Proof. If (Sm) does not hold in a lattice L, then there exist elements $x, y, z \in L$ such that $x \wedge y \prec y$ and $x < z < y \vee x$. Then $x \wedge y = y \wedge z$, and the premises of condition (i) in Lemma 3.1.2 are satisfied, whereas the conclusion does not hold [since $y \succ x \wedge y$ and $(x \vee y) \wedge z = z$]. Thus condition (i) of this proposition is not satisfied. ∎

The lattice of Figure 3.6(a) in Section 3.2 (where the chains $\{a_i\}_{i=1,2,3,\dots}$, $\{b_k\}_{j=1,2,3,\dots}$ are isomorphic with the order dual of the chain of positive integers)

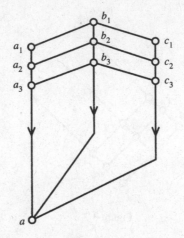

Figure 3.3

is trivially upper semimodular, but it does not satisfy Mac Lane's condition (Mac). Note that this lattice satisfies the ACC. Ramalho [1994] gives an example of a lower continuous lattice with ACC satisfying (Sm) but not (Mac).

Let us consider now some other properties. An arbitrary sublattice of a lattice satisfying Mac Lane's condition (Mac) does not satisfy (Mac) in general. We leave it as an exercise to show that condition (Mac) is inherited by convex sublattices and by direct products.

A homomorphic image of a lattice satisfying (Mac) need not satisfy (Mac). For an example of such a lattice L, which even satisfies the DCC, see Dubreil-Jacotin et al. [1953], p. 101, Figure 47. This example also shows that (Mac) together with the dual condition, that is, (Mac) for L^*, does not imply modularity. Similarly, in lattices satisfying the ACC, Mac Lane's condition (Mac) together with its dual does not imply modularity. On the other hand we have (see Croisot [1952], Théorème 11)

Theorem 3.1.4   *Let L be a lattice satisfying Mac Lane's condition* (Mac) *and the ACC. Then every homomorphic image K of L satisfies* (Mac).

Proof.   We use capital letters for elements of K. Let X, Y, Z be elements of K such that $X \wedge Y < X < Z < X \vee Y$. We have to show that there exists an element U such that $Z \wedge Y < U \leq Y$ and $(U \vee X) \wedge Z = X$. Put $A = Z \wedge Y = X \wedge Y$ and $B = Z \vee Y = X \vee Y$. Let a be an element of L that is maximal among the preimages of A (such an element exists because of the ACC, since the join of two preimages of A is again a preimage of A). Let x and y be preimages of X and Y, respectively, such that $x > a$ and $y > a$ (it is easy to see that such elements x and y exist). By the definition of a we have $x \wedge y = a$, since $x \wedge y \geq a$ is mapped on $X \wedge Y = A$. Let z be a preimage of Z such that $z > x$. We have $z \wedge y = a$, since

$z \wedge y \geq a$ is mapped on $Z \wedge Y = A$. Thus the elements $x, y, z \in L$ have been chosen such that $z \wedge y < x < z < z \vee y$. Since L satisfies (Mac), there exists an element $u \in L$ such that $a < u \leq y$ and $(u \vee x) \wedge z = x$. Let U denote the element of K on which u is mapped. Then $Z \wedge Y \leq U \leq Y$ and $(U \vee X) \wedge Z = X$. Finally we note that $Z \wedge Y \neq U$, since otherwise u would be mapped on A and $u > a$, in contradiction to the choice of a. ∎

Let us show now the equivalence of Mac Lane's condition with M-symmetry, semimodularity, and Birkhoff's condition in upper continuous strongly atomic lattices. We first note that, in strongly atomic lattices, upper semimodularity [condition (Sm)] and Mac Lane's condition (Mac) are equivalent. This follows from Lemma 3.1.2, Lemma 3.1.3, and

Lemma 3.1.5 *In strongly atomic lattices upper semimodularity implies Mac Lane's condition* (Mac).

Proof. Let L be a strongly atomic lattice such that (Sm) holds. Let $a, b, c \in L$, and assume that $a \wedge b < c < a < c \vee b$. Since L is strongly atomic, there exists an element $d \in L$ such that $a \wedge b \prec d \leq b$. We have $a \wedge b = c \wedge d \prec d$, and from (Sm) we conclude that $c \prec c \vee d$. Since $d \not\leq a$, we get that $c = a \wedge (c \vee d)$. It follows that (Mac) holds. ∎

The preceding lemma (cf. Dubreil-Jacotin et al. [1953], Lemme 4, p. 97, or Ramalho [1994], Lemma 1) strengthens the corresponding assertion for lattices satisfying the DCC proved in Croisot [1951], p. 235.

We have already noted that Mac Lane's condition (Mac) and M-symmetry [condition (Ms)] are independent of each other, that is, neither of these conditions implies the other one. On the other hand, in a lattice satisfying the ACC, (Mac) implies M-symmetry (see Croisot [1951], p. 225). The following result of Ramalho [1994] shows that the ACC may be replaced by the weaker condition of upper continuity:

Lemma 3.1.6 *In an upper continuous lattice Mac Lane's condition* (Mac) *implies M-symmetry.*

Proof. We use the characterization of M-symmetry given in Lemma 2.3.6. Suppose that L is an upper continuous lattice and $a, b, c \in L$ with $a \wedge b < c < a < c \vee b$. By condition (Mac$_1$) there exists an $y \in L$ such that $a \wedge b < y < b$ and $c = a \wedge (c \vee y)$. Set $Y = \{y \in L : a \wedge b < y \leq b$ and $c = a \wedge (c \vee y)\}$. Note that $c = a \wedge (c \vee y)$ together with $y \leq b$ yields $y < b$. Now Y is directed: if $\{y_\alpha : \alpha \in A\}$ is a chain, then upper continuity implies $a \wedge [c \vee (\bigvee y_\alpha)] = a \wedge [\bigvee (c \vee y_\alpha)] = c$ and thus $\bigvee Y_\alpha \in Y$. Zorn's lemma guarantees the existence of a maximal element $y_m \in Y$. We have $c \vee y_m < a \vee y_m$, since $c \vee y_m = a \vee y_m$ would imply $c = a \wedge (c \vee y_m) = a \vee (a \wedge y_m) = a$. We show that $y_m \neq (a \vee y_m) \wedge b$

and thus $y_m < (a \vee y_m) \wedge b$. Assuming $y_m = (a \vee y_m) \wedge b$, it is easy to see that $\{c \vee y_m, a \vee y_m, b\}$ generates a pentagon. By (Mac$_1$) there exists a $t \in L$ such that $y_m < t < b$ and $c \vee y_m = (a \vee y_m) \wedge (c \vee y_m \vee t) = (a \vee y_m) \wedge (c \vee t)$. From this we obtain $c = a \wedge (c \vee y_m) = a \wedge (a \vee y_m) \wedge (c \vee t) = a \wedge (c \vee t)$, contradicting the maximality of y_m in Y. Thus we have found that the condition of Lemma 2.4.6 characterizing M-symmetry is satisfied. ∎

Now we can prove the following result of Ramalho [1994], Theorem 1.

Theorem 3.1.7   *In upper continuous strongly atomic lattices, upper semimodularity [condition* (Sm)*], Birkhoff's condition* (Bi), *Mac Lane's condition* (Mac), *and M-symmetry [condition* (Ms)*] are equivalent.*

Sketch of proof.   (Mac) implies (Ms) by Lemma 3.1.6. (Ms) implies (Sm) by Lemma 1.9.17. (Sm) implies (Mac) by Lemma 3.1.5. (Sm) and (Bi) are equivalent by Ramalho [1994], Lemma 3 (the equivalence of these two conditions for algebraic strongly atomic lattices is proved in Crawley & Dilworth [1973], Theorem 3.7; the same arguments work also for upper continuous strongly atomic lattices). ∎

It follows in particular that for lattices of finite length the conditions (Sm), (Bi), (Mac), and (Ms) are equivalent. For lattices of finite length some of these equivalences were already stated earlier; for example, the equivalence of (Ms) and (Sm) was mentioned in Section 1.9.

Let us now make a digression to AC lattices. AC lattices (see Section 2.5) satisfy (Sm) and are strongly atomic. An atomistic lattice is upper continuous if and only if it is algebraic (see Maeda & Maeda [1970], Theorem 7.13). As a consequence of Theorem 3.1.7 we obtain the following characterization of matroid lattices (i.e. algebraic AC lattices).

Corollary 3.1.8   *For algebraic atomistic lattices upper semimodularity [condition* (Sm)*], Mac Lane's condition* (Mac), *and M-symmetry [condition* (Ms)*] are equivalent [and they are equivalent to the atomic covering property* (C)*].*

The following special case of Theorem 3.1.7 shows that Mac Lane's condition is a condition related to semimodularity.

Corollary 3.1.9   *A lattice of finite length is semimodular if and only if it satisfies Mac Lane's condition* (Mac).

We shall use this now to characterize modularity in semimodular lattices of finite length by means of forbidden cover-preserving centered hexagons. (Recall that a sublattice S of a lattice L is said to be cover-preserving in L if, for $a, b \in S$, $a \succ b$ in S implies $a \succ b$ in L.)

Theorem 3.1.10   *A semimodular lattice of finite length is modular if and only if it contains no cover-preserving sublattice isomorphic to the centered hexagon S_7.*

Proof.   Let L be a semimodular lattice of finite length that is not modular. Then L contains a nonmodular sublattice $\{a, b, c, b \wedge c, a \vee b\}$ isomorphic to N_5. Let S be such a sublattice of L having minimum length in L, and denote the length of S by $l(S)$. In view of Corollary 3.1.9 there exists an element $d \in L$ satisfying $b \wedge c < d \leq b$ and $(a \vee d) \wedge c = a$.

Let $c \vee d < a \vee b$. If $(c \vee d) \wedge b = d$, then $\{b, d, a \vee d, a \vee b, c \vee d\}$ is a sublattice of L isomorphic to N_5 and of length less than $l(S)$. Otherwise, $(c \vee d) \wedge b > d$, and $\{c, d, c \vee d, b \wedge c, (c \vee d) \wedge b\}$ is a sublattice of L isomorphic to N_5 and of length less than $l(S)$. Hence $c \vee d = a \vee b$. If $b \wedge (a \vee d) > d$, then $\{a, d, b \wedge c, a \vee d, b \wedge (a \vee d)\}$ is a sublattice of L isomorphic to N_5 and of length less than $l(S)$. We conclude that $b \wedge (a \vee d) = d$ and $\{a, b, c, d, b \wedge c, a \vee b, a \vee d\}$ is a sublattice of L isomorphic to the hexagon S_7. Assume now $c > e \succ a$. The minimality of $l(S)$ implies $a \vee b > d \vee e \succ a \vee d$ and $b \succ b \wedge (d \vee e) > d$. Then $\{a, e, b \wedge c, d \vee e, b \vee (d \vee e)\}$ is a sublattice of L isomorphic to N_5 and of length less than $l(S)$. Hence $c \succ a$ and, by symmetry, $b \succ d$. Semimodularity now implies $a \vee b \succ a \vee d$.

Let $a \succ f > b \wedge c$. The minimality of $l(S)$ now implies $a \vee d \succ d \vee f$ and $a \vee b \succ b \vee f$. If $c \wedge (b \vee f) > f$ [or $c \wedge (b \vee f) = f$], then $\{b \wedge c, f, c \wedge (b \vee f), b, b \vee f\}$ [or $\{a, c, f, b \vee f, a \vee b\}$, respectively] is a sublattice of L isomorphic to N_5 and of length less than $l(S)$. We conclude that $a \succ b \wedge c$ and, by symmetry, $d \succ b \wedge c$.

Finally, semimodularity implies $a \vee d \succ d$, $a \vee b \succ b$, $a \vee d \succ a$, and $a \vee b \succ c$, that is, $\{a, b, c, d, b \wedge c, a \vee b, a \vee d\}$ is a cover-preserving sublattice of L isomorphic to the centered hexagon S_7.   ∎

The preceding theorem follows from a result of Jakubík [1975a], who observed that the theorem is due to Šík (1972, unpublished) and can also be deduced from Vilhelm [1955]. The proof given above is due to Duffus & Rival [1977].

Notes

The following two conditions denoted by (B) and (F) (see Croisot [1951]) have been considered which are related to (E_6):

Condition (B).   We say that the ordered pair (x, y) has the property (b) if for every maximal chain $\{x_i\}$ (with $x \wedge y \leq x_i \leq x$) the chain $\{x_i \vee y\}$ (with $y \leq x_i \vee y \leq x \vee y$) is also maximal. We say that a lattice satisfies the condition (B) if each ordered pair of elements has the property (b).

Condition (F).   An ordered pair (x, y) of lattice elements has property (f) if for every maximal chain $\{(x_i, y_i)\}$ of the cardinal product $[x \wedge y, x] \times [x \wedge y, y]$ the

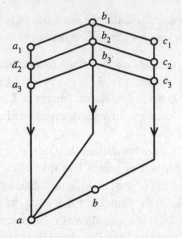

Figure 3.4

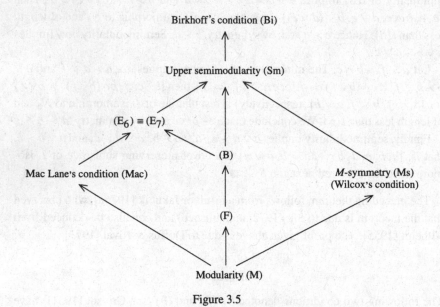

Figure 3.5

chain $\{x_i \vee y_i\}$ is maximal (in $[x \wedge y, x \vee y]$). A lattice is said to satisfy condition (F) if each ordered pair of elements has property (f).

It is easy to see that modularity implies condition (F), condition (F) implies condition (B), and condition (B) implies condition (E_6). The verifications of these implications are left as exercises. Moreover, in lattices of finite length, upper semimodularity and conditions (F) and (B) are equivalent (see Exercise 6).

The lattice in Figure 3.4 (cf. Croisot [1951], Fig. 43) consists of the three chains $\{a_i\}, \{b_j\}, \{c_k\}$ (each assumed to be isomorphic to the order dual of the chain of positive integers) and the elements a and b. This lattice does not satisfy Mac Lane's condition (Mac), since $a_1 \wedge b = a_1 \wedge c_1 = a$, $a_1 \vee b = a_1 \vee c_1 = b_1$, and for each element $x = a_i$ such that $a < x \leq a_1$ one has $(a_i \vee b) \wedge c_1 = b_i \wedge c_1 = c_i$. On the other hand, the lattice satisfies condition (B) and hence also condition (E_6). Incidentally we note that the lattice of Figure 3.4 is M-symmetric. Hence this example also shows that M-symmetry does not imply (Mac).

The interrelationships between some old and new conditions are visualized in Figure 3.5. This picture will be further refined in Figure 3.8 (Section 3.2).

Exercises

1. If a lattice satisfying Birkhoff's condition (Bi) is of locally finite length, then it also satisfies Mac Lane's condition (Mac) (Rudeanu [1964], Lemma 4).
2. Let L be a lattice satisfying (Mac) and the ACC. Then every homomorphic image of L satisfies (Mac) (Dubreil-Jacotin et al. [1953], Théorème 2, p. 95).
3. If an element r of a semicomplemented lattice L satisfying (Mac) has a maximal proper semicomplement m, then L has a greatest element and m is the complement of r (Szász [1957–58]).
4. Verify the following assertions: modularity implies condition (F) (cf. Birkhoff [1948], p. 101) but not conversely; condition (F) implies (B) (Croisot [1951], p. 217) but not conversely; condition (B) implies (E_6) (Croisot [1951], p. 219–220) but not conversely. [For a definition of (B) and (F) see the preceding Notes.]
5. Show that for lattices of finite length condition (F) is equivalent with upper semimodularity (cf. Birkhoff [1948], Theorem 2, p. 101).

References

Birkhoff, G. [1948] *Lattice Theory* (2nd ed.), Amer. Math. Soc. Colloquium Publications, Vol. 25, New York.

Crawley, P. and R. P. Dilworth [1973] Algebraic Theory of Lattices, Prentice-Hall, Englewood Cliffs, N.J.

Croisot, R. [1951] Contribution à l'étude des treillis semi-modulaires de longueur infinie, Ann. Sci. Ecole Norm. Sup. (3) 68, 203–265.

Croisot, R. [1952] Quelques applications et propriétés des treillis semi-modulaires de longueur infinie, Ann. Fac. Sci. Univ. Toulouse 16, 11–74.

Dubreil-Jacotin, M. L., L. Lesieur, and R. Croisot [1953] *Leçons sur la théorie des treillis, des structures algébriques ordonnées et des treillis géométriques*, Gauthier-Villards, Paris.

Duffus, D. and I. Rival [1977] Path length in the covering graph of a lattice, Discrete Math. 19, 139–158.

Jakubík, J. [1975a] Modular lattices of locally finite length, Acta Sci. Math. (Szeged) 37, 79–82.

Mac Lane, S. [1938] A lattice formulation for transcendence degrees and p-bases, Duke Math. J. 4, 455–468.

Maeda, F. and S. Maeda [1970] *Theory of Symmetric Lattices*, Springer-Verlag, Berlin.

Menger, K. [1936] New foundations of projective and affine geometry, Ann. of Math. 37, 456–482.

Ramalho, M. [1994] On upper continuous and semimodular lattices, Algebra Universalis 32, 330–340.

Rudeanu, S. [1964] Logical dependence of certain chain conditions in lattice theory, Acta Sci. Math. (Szeged) 25, 209–218.

Szász, G. [1957–58] Semicomplements and complements in semimodular lattices, Publ. Math. (Debrecen) 5, 217–221.

Vilhelm, V. [1955] The self-dual kernel of Birkhoff's condition in lattices with finite chains (Russian), Czech. Math. J. 5, 439–450.

Wilcox, L. R. [1938] Modularity in the theory of lattices, Bull. Amer. Math. Soc. 44, 50.

Wilcox, L. R. [1939] Modularity in the theory of lattices, Ann. of Math. 40, 490–505.

3.2 Conditions for the Ideal Lattice

Summary. We first have a look at lattices L that have the property that M-symmetry, semimodularity, or Birkhoff's condition holds in the lattice of ideals of L or in the lattice of dual ideals of L. Then we consider a condition introduced by Dilworth. This condition is, for a given lattice L, weaker than the requirement of semimodularity for the lattice of dual ideals of L but stronger than Mac Lane's condition for L. Finally we illustrate the interrelationships between the conditions for the lattices of ideals and dual ideals introduced in this section and previously introduced conditions.

Recall that for a lattice L we denote by $\mathbf{I}(L)$ the lattice of ideals of L and by $\mathbf{D}(L)$ the lattice of dual ideals of L. A lattice L of finite length is isomorphic to both the lattice of its ideals $\mathbf{I}(L)$ and the lattice of its dual ideals $\mathbf{D}(L)$ by means of the mappings $x \to (x]$ and $x \to [x)$, respectively, where $x \in L$.

Let (X) denote one of the conditions (Ms) [M-symmetry], (Sm) [upper semimodularity], or (Bi) [Birkhoff's condition]. By $(X)_i$ and $(X)_d$ we denote the condition (X) imposed on $\mathbf{I}(L)$ and $\mathbf{D}(L)$, respectively. Altogether we obtain in this way the following two sets of conditions:

$$(Ms)_i, (Sm)_i, (Bi)_i \quad \text{and} \quad (Ms)_d, (Sm)_d, (Bi)_d.$$

We say that a lattice L satisfies condition $(X)_i$ if $\mathbf{I}(L)$ satisfies (X). Similarly we say that a lattice L satisfies condition $(X)_d$ if $\mathbf{D}(L)$ satisfies (X). Thus, for a lattice L to satisfy $(Sm)_d$ means that (Sm) holds in $\mathbf{D}(L)$, that is, for dual ideals $\mathbf{a}, \mathbf{b} \in \mathbf{D}(L)$ the relation $\mathbf{a} \wedge \mathbf{b} \prec \mathbf{a}$ implies $\mathbf{b} \prec \mathbf{a} \vee \mathbf{b}$ [with meet, join, and covering relation as defined in $\mathbf{D}(L)$].

We first state the following implications, which are immediate from the definitions and from the known chain of implications $(Ms) \Rightarrow (Sm) \Rightarrow (Bi)$:

$$(M) \Rightarrow (Ms)_i \Rightarrow (Sm)_i \Rightarrow (Bi)_i \Rightarrow (Bi) \quad \text{and}$$
$$(M) \Rightarrow (Ms)_d \Rightarrow (Sm)_d \Rightarrow (Bi)_d \Rightarrow (Bi).$$

The implications $(Ms)_i \Rightarrow (Ms)$ and $(Ms)_d \Rightarrow (Ms)$ are left as exercises.

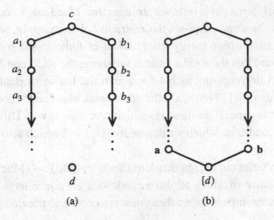

Figure 3.6

The lattice of Figure 3.6(a) (see Croisot [1951]) shows that (Ms) does not imply (Ms)$_d$. The elements of the lattice L of Figure 3.6(a) are the two chains $\{a_i\}$ and $\{b_j\}$, each isomorphic to the order dual of the chain of positive integers, and the elements c and d. Its lattice of dual ideals $\mathbf{D}(L)$ is shown in Figure 3.6(b); the elements $\mathbf{a}$ and $\mathbf{b}$ represent the dual ideals generated by the a_i and b_j, respectively. The lattice L does not satisfy (Bi)$_d$, since the dual ideals $\mathbf{a}$ and $\mathbf{b}$ cover $[d)$ while their join $[c)$ does not cover $\mathbf{a}$ and $\mathbf{b}$. It follows that L does not satisfy (Ms)$_d$. On the other hand, the lattice L is obviously M-symmetric.

For an example of a lattice satisfying (Ms) but not (Ms)$_i$ see Croisot [1951], Figure 45, p. 250. The implication (Sm)$_i$ $\Rightarrow$ (Sm) is again left as an exercise. Condition (Sm)$_d$ was characterized in the following way by Dilworth [1941b], Lemma 6. 2.

Lemma 3.2.1   *A lattice L satisfies* (Sm)$_d$ *[i.e.,* (Sm) *holds in* $\mathbf{D}(L)$*] if and only if* $\mathbf{a} \succ [x) \wedge \mathbf{a}$ *implies* $[x) \vee \mathbf{a} \succ [x)$ *for every dual ideal $\mathbf{a}$ and for every principal dual ideal* $[x)$.

Proof.   Let $\mathbf{a} \succ [x) \wedge \mathbf{a}$ imply $[x) \vee \mathbf{a} \succ [x)$ for every dual ideal $\mathbf{a}$ and for every x. Suppose that (Sm) does not hold in $\mathbf{D}(L)$. Then there exist dual ideals $\mathbf{a}$ and $\mathbf{b}$ such that $\mathbf{a} \succ \mathbf{a} \wedge \mathbf{b}$ but $\mathbf{a} \vee \mathbf{b} > \mathbf{c} > \mathbf{b}$. Since $\mathbf{b} \not\geq \mathbf{a}$, there exists an element x_1 such that $[x_1) \geq \mathbf{b}$ but $[x_1) \not\geq \mathbf{a}$. Similarly, since $\mathbf{b} \not\geq \mathbf{c}$, an element x_2 exists such that $[x_2) \geq \mathbf{b}$ but $[x_2) \not\geq \mathbf{c}$. Finally, since $\mathbf{b} \vee \mathbf{c} \not\geq \mathbf{a}$, there is an element x_3 such that $[x_3) \geq \mathbf{b}$ but $[x_3) \vee \mathbf{c} \not\geq \mathbf{a}$. Let $x = x_1 \wedge x_2 \wedge x_3$. Then $[x) \geq \mathbf{b}$, $[x) \not\geq \mathbf{a}$, and $[x) \vee \mathbf{c} \not\geq \mathbf{a}$. Now $\mathbf{a} \geq \mathbf{a} \wedge [x) \geq \mathbf{a} \wedge \mathbf{b}$ and $\mathbf{a} \neq \mathbf{a} \wedge [x)$. Hence $\mathbf{a} \wedge [x) = \mathbf{a} \wedge \mathbf{b}$ and thus $\mathbf{a} \succ [x) \wedge \mathbf{a}$. By hypothesis then $[x) \vee \mathbf{a} \succ [x)$. Now $[x) \vee \mathbf{a} \geq \mathbf{c} \vee [x) \geq [x)$ and $\mathbf{c} \vee [x) \neq [x)$. Hence $[x) \vee \mathbf{a} = \mathbf{c} \vee [x)$ which implies $[x) \vee \mathbf{c} \geq \mathbf{a}$ contrary to the definition of x. Thus (Sm) holds in $\mathbf{D}(L)$. ∎

Condition $(Sm)_d$ turns out to be much stronger than Mac Lane's condition (Mac). In fact, $(Sm)_d$ is for some purposes too restrictive. For example, when Dilworth extended his decomposition theory from lattices of finite length to larger classes of lattices, he found that there exist lattices satisfying the ACC and having unique irredundant meet decompositions but for which the lattice of dual ideals is not semimodular (Dilworth [1941b]). On the other hand, Mac Lane's condition (Mac) turned out to be too general for these investigations. This is why Dilworth isolated an intermediate condition, which we denote by (Dil) and which also involves dual ideals (see below).

We will briefly point out the implications $(Sm)_d \Rightarrow$ (Dil) $\Rightarrow$ (Mac) and mention an important instance in which all three conditions are equivalent. We shall also clarify other relationships between these new concepts and previously introduced ones.

In accordance with the terminology of Klein-Barmen [1937], Dilworth [1940], [1941a] called a lattice of finite length a *Birkhoff lattice* if it satisfies the condition of upper semimodularity (Sm), that is, if $a \succ a \wedge b$ implies $a \vee b \succ b$. Since both the ascending and the descending chain conditions were assumed to hold, (Sm) was never satisfied trivially. As already remarked in several places, in a sufficiently general lattice no covering relations may exist, and in these cases (Sm) holds in a trivial way. Therefore Dilworth [1941b] formulated a condition that reduces to (Sm) if the lattice satisfies the DCC. Dilworth first recalls the following equivalent form (Sm') of (Sm):

(Sm') $b \succ a$, $c \geq a$, and $c \not\geq b$ imply $b \vee c \succ c$.

If (Sm') is satisfied in a lattice L for a given a and any b and c, Dilworth says that the element a satisfies *Birkhoff's condition* in L. (Note that this differs from our usage of the concept of *Birkhoff's condition*.) Thus (Sm) holds in L if and only if every element of L satisfies Birkhoff's condition in the sense of Dilworth. Now Dilworth defines a lattice L to be a *Birkhoff lattice* if each element a of L satisfies Birkhoff's condition (in the sense of Dilworth) in the lattice of dual ideals $\mathbf{D}(L)$. We also note that Dilworth [1941b] uses the term "ideal" for what is now called "dual ideal." In other words, Dilworth introduced the following condition:

(Dil) For any principal dual ideal $[a]$ and for dual ideals $\mathbf{b}$ and $\mathbf{c}$, the relations $\mathbf{b} \succ [a]$, $\mathbf{c} \geq [a]$, $\mathbf{c} \not\geq \mathbf{b}$ imply $\mathbf{b} \vee \mathbf{c} \succ \mathbf{c}$.

A lattice L is never vacuously a Birkhoff lattice (in the sense defined by Dilworth), since covering ideals always exist (see Dilworth [1941b], Theorem 2.1). Moreover, if the DCC holds, then every dual ideal is principal and L is a Birkhoff lattice if and only if (Sm') holds in L. The verification of the

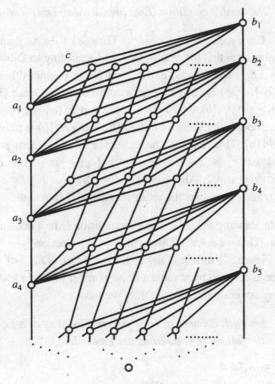

Figure 3.7

implication (Sm)$_d$ $\Rightarrow$ (Dil) is left as an exercise. The lattice of Figure 3.7 (cf. Dilworth [1941b], Figure 2) shows that the preceding implication is nonreversible; note that this lattice satisfies the ACC.

In the lattice L of Figure 3.7 all elements distinct from 0 form a dual ideal **a**, which is generated by $a_1, a_2, a_3, \ldots$. The elements $b_1, b_2, b_3, \ldots$ form a dual ideal **b**, and **b** $>$ **a**. Take an element $c \notin \mathbf{b}$, and consider the dual ideal **c** generated by c and **b**. Then **b** $>$ **c** $\geq$ **a**. Let $y \in \mathbf{c}$, $y \notin \mathbf{b}$. Then there exists an element b_i such that $b_i \succ x \geq y$, $x \notin \mathbf{b}$. By the method of construction there exists an integer j such that $x \wedge b_k = a_k$ for all $k \geq j$. Hence $\mathbf{a} \geq [x) \wedge \mathbf{b} \geq [x) \wedge \mathbf{b} \geq \mathbf{c} \wedge \mathbf{b} = \mathbf{c}$, and therefore $\mathbf{c} = \mathbf{a}$ and **b** $\succ$ **a**. Clearly $\mathbf{b} \wedge [a_1) = \mathbf{a}$ and $\mathbf{b} \vee [a_1) = [b_1)$. But then $\mathbf{b} \succ [a_1) \wedge \mathbf{b}$, while $[a_1) \vee b$ is not an upper cover of $[a_1)$. It follows that (Sm) does not hold in $\mathbf{D}(L)$, that is, L does not satisfy (Sm)$_d$. On the other hand, L satisfies (Dil), since **a** is the only nonprincipal dual ideal that covers a principal ideal dual (namely [0)) and every element distinct from 0 is contained in **a**.

The following result is due to Dilworth [1941b], Theorem 6.1.

Theorem 3.2.2 *Dilworth's condition* (Dil) *implies Mac Lane's condition* (Mac).

Sketch of proof. Let $a \wedge c < b < a < a \vee c$. Then $[c) > [a \wedge c)$, and hence there exists a dual ideal **p** such that $[c) \geq \mathbf{p} \succ [a \wedge c)$ (according to Dilworth [1941b], Theorem 2.1, covering ideals always exist). Now $[b) \not\geq \mathbf{p}$, since otherwise $[b) \geq \mathbf{p}$ and $[c) \geq \mathbf{p}$ imply $[b \wedge c) = [a \wedge c) \geq \mathbf{p}$, a contradiction. Hence $[b) \vee \mathbf{p} \succ [b)$. But then $[b) \vee \mathbf{p} \geq [a) \wedge ([b) \vee \mathbf{p}) \geq [b)$. If $[b) \vee \mathbf{p} = [a) \wedge ([b) \vee \mathbf{p})$, we have $[a) \geq \mathbf{p}$, which is impossible. Thus we obtain $[b) = [a) \wedge ([b) \vee \mathbf{p})$, which implies (cf. Dilworth [1941b], Theorem 2.2) the existence of an element $p \in \mathbf{p}$ such that $b = a \wedge (b \vee p)$. Let $c_1 = c \wedge p$. Then $[c) \geq [c_1) \geq p \succ [a \wedge c)$, and hence $c \geq c_1 \geq a \wedge c$, $c_1 \neq a \wedge c$. Also $b = a \wedge (b \vee p) \geq a \wedge (b \vee c_1) \geq b$. Thus $b = a \wedge (b \vee c_1)$, and c_1 satisfies the requirements of (Mac). ∎

For an example showing that Mac Lane's condition (Mac) does not imply Dilworth's condition (Dil) we refer to Dilworth [1941b], Section 8.

The lattice in Figure 3.7 satisfies Dilworth's condition (Dil), but *not* every element of the lattice is covered by at most a finite number of dual ideals. If there are only finitely many covering dual ideals, then we have

Theorem 3.2.3 *Let each element of L be covered by at most a finite number of dual ideals. Then the following conditions are equivalent:*

 (i) *L satisfies* $(\mathrm{Sm})_d$;
 (ii) *L satisfies Dilworth's condition* (Dil);
 (iii) *L satisfies Mac Lane's condition* (Mac).

For a proof we refer to Dilworth [1941b], Theorem 6.7.

Figure 3.8 shows the interrelationships between conditions related to semimodularity. (Recall that by a *condition related to semimodularity* we mean a condition that is equivalent to semimodularity for lattices of finite length.) Figure 3.8 is a

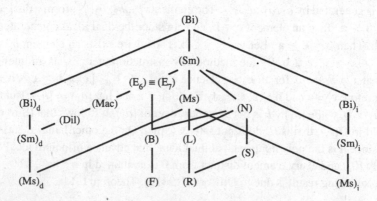

Figure 3.8

refinement of Figure 3.5 (Section 3.1). Note, however, that for the sake of simplicity we have omitted arrowheads and modularity (which implies all other conditions). In what follows we give some more explanations and comments concerning the abbreviations used in Figure 3.8:

(Bi)   *Birkhoff's condition*, also denoted as *weak (upper) semimodularity* or *double covering property*. Birkhoff called it condition (ξ'), and Croisot called it condition (1).

(Sm)   *(Upper) semimodularity*, also called the *neighborhood condition*. It is Mac Lane's condition (E_4) and Croisot's condition (2).

(Ms)   *M-symmetry*, also called *Wilcox's condition*; it is Croisot's condition (α).

(Mac)   *Mac Lane's condition*; Mac Lane himself denoted it by (E_5).

(E_6), (E_7) Equivalent conditions introduced by Mac Lane.

(Dil)   *Dilworth's condition*.

$(Bi)_i$ Birkhoff's condition for the lattice of ideals; Croisot's condition (1_i).

$(Sm)_i$ (Upper) semimodularity for the lattice of ideals; Croisot's condition (2_i).

$(Ms)_i$ *M*-symmetry for the lattice of ideals; Croisot's condition (α_i).

$(Bi)_d$ Birkhoff's condition for the lattice of dual ideals; Croisot's condition (1^i).

$(Sm)_d$ (Upper) semimodularity for the lattice of dual ideals; Croisot's condition (2^i).

$(Ms)_d$ *M*-symmetry for the lattice of dual ideals; Croisot's condition (α^i).

(F), (B)   Conditions involving maximal chains (see Section 3.1, Notes).

(L), (N)   Conditions involving maximal chains (see Notes and Exercises below).

(R), (S)   Conditions involving maximal chains (see Notes and Exercises below).

The abbreviations (F), (B), (L), (R), (S) are those of Croisot [1951]. However, Croisot's original abbreviation (M) (which we use for modularity) is changed here into (N).

Notes

It seems to be an open question whether the implications $(Ms)_d \Rightarrow (Sm)_d$ and $(Ms)_i \Rightarrow (Bi)_i$ are reversible (see Croisot [1951]).

There exist a great number of other conditions equivalent to semimodularity in lattices of finite length, for example, the following properties (L), (N), (R), (S) (see also Exercises below):

A pair (x, y) of lattice elements is said to possess the property (l) if, for each maximal chain $\{x_k\}$ in $[x \wedge y, x]$ having the form $x_k = y_k \wedge x$ with $\{y_k\}$ being a chain in $[y, x \vee y]$, the chain $\{y_k\}$ is necessarily maximal. A lattice is said to have the property (L) if each pair (x, y) of lattice elements has the property (l).

A pair (x, y) of lattice elements is said to possess the property (n) if, for each maximal chain $\{x_k\}$ in $[x \wedge y, x]$ having the form $x_k = y_k \wedge x$ with $\{y_k\}$ being a chain in $[y, x \vee y]$, the chain $\{y_k\}$ can be chosen maximal. A lattice is said to have the property (N) if each pair (x, y) of lattice elements has the property (n).

A pair (x, y) of lattice elements is said to possess the property (r) if, for each maximal chain $\{z_k\}$ in $[x \wedge y, x \vee y]$ having the form $z_k = x_k \wedge y_k$ with $\{(x_k, y_k)\}$ being a chain of the cardinal product $[x, x \vee y] \times [y, x \vee y]$, the chain $\{(x_k, y_k)\}$ is necessarily maximal. A lattice is said to have the property (R) if each pair (x, y) of lattice elements has the property (r).

A pair (x, y) of lattice elements is said to possess the property (s) if, for each maximal chain $\{z_k\}$ in $[x \wedge y, x \vee y]$ having the form $z_k = x_k \wedge y_k$ with $\{(x_k, y_k)\}$ being a chain of the cardinal product $[x, x \vee y] \times [y, x \vee y]$, the chain $\{(x_k, y_k)\}$ can be chosen maximal. A lattice is said to have the property (S) if each pair (x, y) of lattice elements has the property (s).

Exercises

1. Prove the implications $(Ms)_i \Rightarrow (Ms)$ and $(Ms)_d \Rightarrow (Ms)$ (Croisot [1951]).
2. Prove the implication $(Sm)_i \Rightarrow (Sm)$, and show by example that the converse does not hold (Croisot [1951]).
3. Concerning the conditions (L), (N), (R), and (S) defined in the preceding Notes, show that :
   (a) (R) implies (L) and (S);
   (b) (L) and (S) imply (N);
   (c) (N) implies (Sm);
   (d) neither of the conditions (L) and (S) implies the other one.
   Give examples showing that none of the preceding implications is reversible.
4. Show that condition (B) (cf. Section 3.1, Notes) implies (N), but not conversely.
5. Show that none of the conditions $(Sm)_i$, $(Bi)_i$, $(Sm)_d$, $(Bi)_d$ together with the corresponding dual condition implies modularity (even if the lattice satisfies a chain condition) (Croisot [1951]).

References

Bogart, K. P., Freese, R., and Kung, J. P. S. (eds.) [1990] *The Dilworth Theorems. Selected Papers of Robert P. Dilworth*, Birkhäuser, Boston.

Croisot, R. [1951] Contribution à l'étude des treillis semi-modulaires de longueur infinie, Ann. Sci. Ecole Norm. Sup. (3) 68, 203–265.

Dilworth, R. P. [1940] Lattices with unique irreducible decompositions, Ann. of Math. 41, 771–777.

Dilworth, R. P. [1941a] The arithmetical theory of Birkhoff lattices, Duke Math. J. 8, 286–299.

Dilworth, R. P. [1941b] Ideals in Birkhoff lattices, Trans. Amer. Math. Soc. 49, 325–353.

Dilworth, R. P. [1961] Structure and decomposition theory of lattices, in *Lattice Theory*, Proceedings of Symposia in Pure Mathematics, Vol. 2 (ed. R. P. Dilworth), Amer. Math. Soc., Providence, R. I., pp. 3–16.

Dubreil-Jacotin, M. L., L. Lesieur, and R. Croisot [1953] *Leçons sur la théorie des treillis, des structures algébriques ordonnées et des treillis géométriques*, Gauthier-Villards, Paris.

Jónsson, B. [1990] Dilworth's work on decompositions in semimodular lattices, in: Bogart et al. [1990], pp. 187–191.

Klein-Barmen, F. [1937] Birkhoffsche und harmonische Verbände, Math. Z. 42, 58–81.

Mac Lane, S. [1938] A lattice formulation for transcendence degrees and p-bases, Duke Math. J. 4, 455–468.

3.3 Interrelationships in Lattices with a Chain Condition

Summary. Several concepts occurring in Figure 3.8 (Section 3.2) coincide if we restrict ourselves to lattices satisfying a chain condition. Here we first have a look at lattices satisfying the ascending chain condition (ACC) and then at lattices satisfying the descending chain condition (DCC). Finally we consider lattices of finite length. In this case all notions of Figure 3.8 coincide, that is, each of these condition characterizes semimodularity in the class of lattices of finite length.

We begin with lattices satisfying the ascending chain condition.

Theorem 3.3.1 *In a lattice satisfying the ascending chain condition one has the interrelationships visualized in Figure 3.9.*

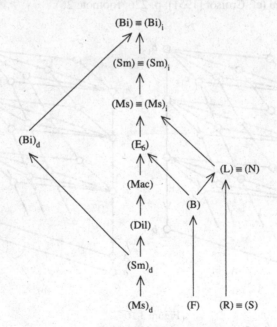

Figure 3.9

For the abbreviations see the comments after Figure 3.8 (Section 3.2). In Figure 3.9 an arrow indicates logical implication, while the symbol $\equiv$ indicates logical equivalence.

The implications (Ms) $\Rightarrow$ (Ms)$_i$, (Sm) $\Rightarrow$ (Sm)$_i$, and (Bi) $\Rightarrow$ (Bi)$_i$ follow from the fact that, for a lattice L satisfying the ACC one has $\mathbf{I}(L) \cong L$. For the implication (E$_6$) $\Rightarrow$ (Ms) we refer to Croisot [1951], pp. 233–234. In order to see that (Bi)$_d$ does not imply (Sm), consider Figure 3.10(a), (b) (from Croisot [1951], Example B16, pp. 248–249).

The elements of the lattice L in Figure 3.10(a) are the two chains $\{a_n\}$, $\{b_n\}$ (with n running through the positive integers), the elements c_{ij} where i and j are arbitrary positive integers, and the elements a and b. It is easy to see that the partial order defined by the diagram is a lattice satisfying the ACC. The lattice of dual ideals $\mathbf{D}(L)$ is obtained from L by inserting the two nonprincipal dual ideals A and B [see Fig. 3.10(b)]. The lattice $\mathbf{D}(L)$ satisfies (Bi), and thus the lattice L satisfies (Bi)$_d$. On the other hand, the element b covers $a = a_1 \wedge b$ in L, whereas $b_1 = a_1 \vee b$ does not cover a_1, that is, L does not satisfy (Sm).

For a lattice showing that (Mac) does not imply (Bi)$_d$ see Croisot [1951], Example B14, pp. 249–250. Ramalho [1994], Example 1, gives an upper semimodular lower continuous lattice with ACC but not satisfying Mac Lane's condition, that is, (Sm) does not imply (Mac). We have already stated (cf. Figure 3.7, Section 3.2) that, even in a lattice satisfying the ACC, Dilworth's condition (Dil) does not imply (Sm)$_d$. The problem whether (Sm)$_d$ implies (Ms)$_d$ in a lattice satisfying the ACC seems to be open (cf. Croisot [1951], p. 226, footnote 26).

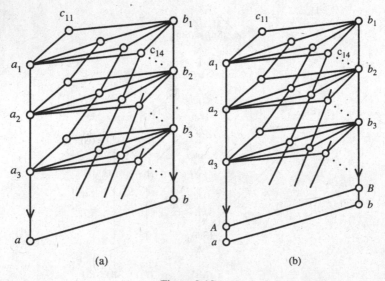

(a)   (b)

Figure 3.10

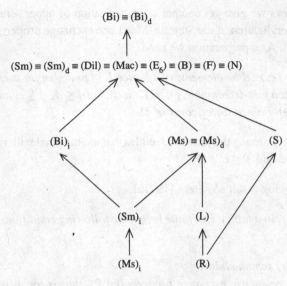

Figure 3.11

For lattices satisfying the descending chain condition we have the following result (cf. Croisot [1951], Théorème 3, p. 234)

Theorem 3.3.2   *In a lattice satisfying the descending chain condition one has the interrelationships visualized in Figure 3.11.*

For the abbreviations see the comments after Figure 3.8 (Section 3.2). Again, in Figure 3.11 an arrow indicates logical implication, while the symbol $\equiv$ indicates logical equivalence.

The implications $(Ms) \Rightarrow (Ms)_d$, $(Sm) \Rightarrow (Sm)_d$, and $(Bi) \Rightarrow (Bi)_d$ follow from the fact that, for a lattice L satisfying the DCC, one has $\mathbf{D}(L) \cong L$. The implications $(Bi)_i \Rightarrow (Sm)$ and $(Sm)_i \Rightarrow (Ms)$ are left as exercises. The following question seems to be open (see Croisot [1951], p. 235, footnote 31): Does $(Bi)_i$ imply $(Ms)_i$ in a lattice satisfying the DCC? For examples showing that the remaining implications cannot be reversed see Croisot [1951].

A lattice of finite length satisfies both the ascending and the descending chain condition. Thus Theorem 3.3.1 and Theorem 3.3.2 imply

Corollary 3.3.3   *In a lattice of finite length, the conditions* (Ms), (Sm), (Bi), $(Ms)_i$, $(Sm)_i$, $(Bi)_i$, $(Ms)_d$, $(Sm)_d$, $(Bi)_d$, (Mac), (Dil), (F), (B), (L), (N), (R), (S), and (E_6) are all equivalent. Hence each of these conditions characterizes upper semimodularity in lattices of finite length.*

In what follows we give yet another characterization of upper semimodularity involving a generalization of the Steinitz–Mac Lane exchange property (for which see Section 1.7). As a preparation we need

Lemma 3.3.4 *Let c, d be elements of a lattice L of finite length such that $c \prec d$. Then there exists a join-irreducible $j \in J(L)$ such that $j \leq d$, $j \nleq c$, and $j \wedge c = j'$ (the uniquely determined lower cover of j).*

Proof. Choose j among the join-irreducibles that are minimal with respect to the properties $j \leq d$ and $j \nleq c$. ∎

For the following result see Stern [1991a].

Theorem 3.3.5 *In a lattice L of finite length the following conditions are equivalent:*

 (i) *L is (upper) semimodular;*
 (ii) *L has the geometric exchange property (GEP), that is for $u, v \in J(L)$ and $b \in L$ the relations $v \leq b \vee u$ and $v \nleq b \vee u'$ imply $u \leq b \vee v \vee u'$;*
 (iii) *$u \wedge b = u' \prec u$ implies $b \prec u \vee b$ for any $u \in L(J)$ and $b \in L$;*
 (iv) *$u \wedge b = u' \prec u$ implies $u \, M^* \, b$.*

Proof. (i) $\Rightarrow$ (ii): Let L be a semimodular lattice of finite length, and assume that for $u, v \in J(L)$ and $b \in L$ we have $v \leq b \vee u$ but $v \nleq b \vee u'$. Then $u \wedge (b \vee u') = u' \prec u$, which implies (by semimodularity) $b \vee u' \prec b \vee u$. From this and from $v \nleq b \vee u'$ we get $v \vee b \vee u' = b \vee u$ implying $u \leq b \vee v \vee u'$.

(ii) $\Rightarrow$ (iii): Assume that (iii) does not hold. Then there exists an element $c \in L$ such that $u \vee b > c \succ b$ and an element $v \in J(L)$ such that $v \leq c$ but $v \nleq b$. It follows that $v \leq b \vee u$ and $v \nleq b = b \vee u'$. On the other hand, we have $c = b \vee u$ and therefore $c = b \vee u \vee u'$. This means that $u \nleq b \vee v \vee u'$, that is, (ii) does not holds.

(iii) $\Rightarrow$ (iv): Let condition (iii) be satisfied and assume that $u \wedge b = u' \prec u$ holds for $u \in J(L)$ and $b \in L$. From Maeda & Maeda [1970], Theorem 7.5.2, it follows that $u \, M^* \, b$.

(iv) $\Rightarrow$ (i): Assume that condition (iv) holds, and let $a, b \in L$ be incomparable elements for which $a \wedge b \prec a$. By Lemma 3.3.4 there exists a join-irreducible element $u \in J(L)$ such that $u \wedge b = u'$ and $a = (a \wedge b) \vee u$. Now $u \wedge b = u'$ implies by (iv) that $u \, M^* \, b$, and this in turn yields $b \prec u \vee b$ by Maeda & Maeda [1970], Theorem 7.5.4. Since $a \vee b = (a \wedge b) \vee u \vee b = u \vee b$, we obtain $b \prec a \vee b$. ∎

Condition (ii) of the preceding result is closely related to the exchange property $\Delta 5$ of Finkbeiner [1951] (cf. also Dilworth [1944] and Faigle [1980a]). In fact, Finkbeiner considers lattices in which the poset of join-irreducibles satisfies the DCC. Condition (ii) characterizes semimodularity in lattices of finite length via an exchange property of the associated closure operator on the set of join-irreducible elements. For a more general setting see Faigle [1980c], [1986b] (cf. also Section 6.3). For the equivalence of conditions (i) and (iii) see also Teo [1988].

If each join-irreducible is an atom, then the geometric exchange property (GEP) reduces to the Steinitz–Mac Lane exchange property (EP) and we get Theorem 1.7.2 (in the finite-length case).

Notes

In accordance with our previously adopted terminology we may also say that the conditions of Corollary 3.3.3 are conditions related to semimodularity. Some of the equivalences of this corollary had already been shown earlier, for example, (Ms) $\Leftrightarrow$ (Sm) $\Leftrightarrow$ (Bi) (cf. Section 1.9). In fact, several of these equivalences have been proved for classes of lattices properly containing the class of lattices of finite length. For example, it was shown (Theorem 3.1.7) that (Ms), (Sm), (Bi), and (Mac) are equivalent in upper continuous strongly atomic lattices.

We have also considered other conditions characterizing upper semimodularity in lattices of finite length, for example, a condition involving the submodularity of the rank function (cf. Corollary 1.9.10). A number of further conditions equivalent to upper semimodularity in lattices of finite length has not been treated here. Most of the older conditions are discussed in Croisot [1951].

Leclerc [1990] gives a characterization of upper semimodularity in finite lattices using the concepts of medians and majority rule. For background concerning these concepts see also Barthélemy et al. [1986].

Exercises

1. Show that in a lattice satisfying the ACC, (Ms)$_i$ and (Ms)$_d$ together imply modularity (Croisot).
2. Verify that the implication (Bi)$_i$ $\Rightarrow$ (Sm) holds in a lattice satisfying the DCC (Croisot [1951], pp. 237–238).
3. Verify that the implication (Sm)$_i$ $\Rightarrow$ (Ms) holds in a lattice satisfying the DCC (Croisot [1951], pp. 237–238).

References

Barthélemy, J. P., B. Leclerc, and B. Monjardet [1986] On the use of ordered sets in problems of comparison and consensus of classification, J. Classification 3, 187–224.

Croisot, R. [1951] Contribution à l'étude des treillis semi-modulaires de longueur infinie, Ann. Sci. Ecole Norm. Sup. (3) 68, 203–265.

Dilworth, R. P. [1941b] Ideals in Birkhoff lattices, Trans. Amer. Math. Soc. 49, 325–353.

Dilworth, R. P. [1944] Dependence relations in a semimodular lattice, Duke Math. J. 11, 575–587.

Faigle, U. [1980a] Geometries on partially ordered sets, J. Combin. Theory Ser. B 28, 26–51.

Faigle, U. [1980c] Extensions and duality of finite geometric closure operators, J. Geometry 14, 23–34.

Faigle, U. [1986b] Exchange properties of combinatorial closure spaces, Discrete Applied Math. 15, 249–260.

Finkbeiner, D. T. [1951] A general dependence relation for lattices, Proc. Amer. Math. Soc. 2, 756–759.

Leclerc, B. [1990] Medians and majorities in semimodular lattices, SIAM J. Discrete Math. 3, 266–276.

Maeda, F. and S. Maeda [1970] *Theory of Symmetric Lattices*, Springer-Verlag, Berlin.

Ramalho, M. [1994] On upper continuous and semimodular lattices, Algebra Universalis 32, 330–340.

Stern, M. [1991a] *Semimodular Lattices*, Teubner, Stuttgart.

Teo, K. L. [1988] Diagrammatic characterizations of semimodular lattices of finite length, Southeast Asian Bull. Math. 12, 135–140.

3.4 0-Conditions and Disjointness Properties

Summary. Distributivity, modularity, and the conditions equivalent to semimodularity in lattices of finite length might be called global conditions. We shall relate some of them now to the corresponding local conditions at 0 (0-conditions) as well as to the so-called disjointness properties. These local conditions at 0 are, of course, not to be confused with local distributivity and local modularity in the sense of Dilworth.

We shall restrict our attention to distributivity [condition (D) of Section 1.2], modularity [condition (M) of Section 1.2], Mac Lane's condition (Mac) [see Section 3.1], M-symmetry [condition (Ms)] and upper semimodularity [condition (Sm)]. Distributivity leads to the local condition (D_0) known as 0-distributivity, while modularity leads to the local condition (M_0), which is also known as 0-modularity. Mac Lane's condition (Mac) leads to a local condition denoted by (Mac_0), which is the same as 0-semimodularity in the sense of Salii [1980]. We have already seen that the counterpart of M-symmetry at 0 is nothing else than $\perp$-symmetry in the sense of Maeda & Maeda [1970]. This condition was denoted by (Ms_0) (see Section 2.3). The local property (Sm_0) corresponding to upper semimodularity is just the atomic covering property (C). The local properties (M_0), (Mac_0), and (C) are in turn related to certain *disjointness properties*, namely to what we call the *general disjointness property* (GD), the *weak disjointness property* (WD), and the *atomic disjointness property* (AD).

Before giving the corresponding definitions and before going into any details, we give Figure 3.12 for a first orientation (with the question mark indicating that no suitable concept has yet been found to be inserted at this place). By $(D_0 M_0)$ we mean the logical conjunction of (D_0) and (M_0).

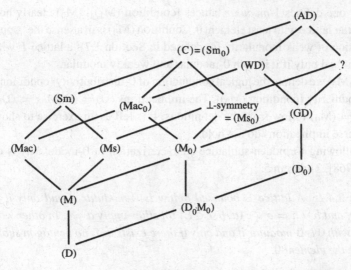

Figure 3.12

Let us now consider the 0-conditions (D_0), (M_0), and (Mac_0) in more detail. A lattice L with least element 0 is said to be 0-*distributive* if, for $a, b, c \in L$,

(D_0) $a \wedge b = 0$ and $a \wedge c = 0$ imply $a \wedge (b \vee c) = 0$.

Dually we define 1-*distributive* lattices [condition (D_1)]. It is clear that 0-distributivity generalizes pseudocomplementation (for which see Section 1.4), since condition (D_0) can be reformulated in the following way: For any element a of the lattice, the subset of all elements disjoint from a forms an ideal. In the case of pseudocomplementation the subset of all elements disjoint from any given lattice element is required to form a principal ideal. While a pseudocomplemented lattice always has a greatest element, a 0-distributive lattice is not necessarily bounded from above.

It is well known that the ideal lattice $\mathbf{I}(L)$ of a pseudocomplemented lattice L or of a distributive lattice L is pseudocomplemented. The converse is not true, that is, $\mathbf{I}(L)$ can be pseudocomplemented with L being neither pseudocomplemented nor distributive: for example, take for L the lattice dual to that of Figure 1.21. Since L is 0-distributive, we also see that (D_0) does not imply pseudocomplementation. In fact, it can be shown that a lattice with 0 is 0-distributive if and only if its ideal lattice is pseudocomplemented (Varlet [1968], Theorem 1). The "only if" part is a special case of the following general assertion: An algebraic 0-distributive lattice is pseudocomplemented (Varlet [1968], Theorem 2). We leave both assertions as exercises.

A lattice L with least element 0 is said to be 0-*modular* if, for $a, b, c \in L$,

(M_0) $a \leq c$ and $b \wedge c = 0$ imply $(a \vee b) \wedge c = a$.

Dually one defines 1-*modular* lattices [condition (M_1)]. (M_0) clearly holds in any modular lattice with least element 0. Condition (M_0) is in a sense the "opposite" to the notion of weak modularity (as defined in Section 1.7): a lattice L with 0 is modular if and only if it is both 0-modular and weakly modular.

By (D_0M_0) we denote the logical conjunction of 0-distributivity [condition (D_0)] and 0-modularity [condition (M_0)]. The implications $(D) \Rightarrow (D_0M_0) \Rightarrow (D_0)$ and $(D_0M_0) \Rightarrow (M_0)$ follow from the definitions. It is left as an exercise to show that the converse implications do not hold.

The following forbidden-sublattice characterization of 0-modularity is due to Varlet [1968], Theorem 5.

Theorem 3.4.1   *A lattice L bounded below is 0-modular if and only if $a \le c$, $b \wedge c = 0$, and $b \vee a = b \vee c$ $(a, b, c \in L)$ together imply $a = c$. In other words, a lattice L with 0 is 0-modular if and only if there exists in L no pentagon sublattice including the element 0.*

The diamond M_3 is 0-modular but not 0-distributive, while the pentagon N_5 is 0-distributive but not 0-modular. Hence the notions of 0-modularity and 0-distributivity are independent of each other, that is, neither of them implies the other one.

While 0-modularity can be characterized by means of certain forbidden sublattices, no such characterization is known for 0-distributivity. Under the additional assumption that the lattice is modular, the following characterization was given by Grillet (see Varlet [1968]): A modular lattice with 0 is 0-distributive if and only if it contains no sublattice isomorphic to the lattice of Figure 3.13(a) or to the lattice of Figure 3.13(b) .

In lattices with 0, modularity is generalized by 0-modularity. Similarly, upper semimodularity [condition (Sm)], Mac Lane's condition (Mac), and M-symmetry [condition (Ms)] also have their counterparts at 0. In fact, the counterpart of (Sm)

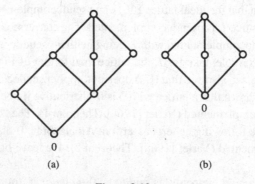

Figure 3.13

at 0, which could be denoted by (Sm_0), is just the atomic covering property (C) (see Section 1.7). Using the notion of a modular pair, 0-modularity can be reformulated in the following way:

(M_0) $b \wedge c = 0$ implies $b \, M \, c$ for all elements b, c of a lattice with 0,

that is, every disjoint pair is a modular pair. Recall that by $\perp$-symmetry (see Section 2.3) we mean that

(Ms_0) $b \, M \, c$ and $b \wedge c = 0$ together imply $c \, M \, b$ for all elements b, c of a lattice with 0,

that is, modularity of disjoint pairs is symmetric. It is clear that 0-modularity implies $\perp$-symmetry, while the reverse implication does not hold in general (example?). Similarly to the proof that M-symmetry $\Rightarrow$ (Sm) (see Lemma 1.9.17) one shows that $\perp$-symmetry $\Rightarrow$ (C); it is also easy to provide an example showing that the converse implication does not hold.

A lattice L with 0 satisfies the *local Mac Lane condition* (Mac_0) if, for any $x, y, z \in L$ such that $y \wedge z = 0$ and $0 < x < z < y \vee x$, there exists an element $t \in L$ such that $0 < t \leq y$ and $x = (x \vee t) \wedge z$. In other words, we obtain (Mac_0) if we replace $y \wedge z$ in the definition of (Mac_1) (one of the equivalent forms of Mac Lane's condition – see Section 3.1) by $y \wedge z = 0$.

The implications $(M) \Rightarrow (M_0)$, $(Ms) \Rightarrow \perp$-symmetry, $(Mac) \Rightarrow (Mac_0)$, and $(Sm) \Rightarrow (C)$ (indicated by ascending lines in Figure 3.12) clearly hold in lattices with 0. The corresponding reverse implications do not hold in general (examples?).

The implications $(M_0) \Rightarrow (Mac_0) \Rightarrow (C)$, $(M_0) \Rightarrow \perp$-symmetry $\Rightarrow (C)$ are shown similarly to the corresponding conditions "without 0." It is also not difficult to see that the converse implications do not hold in general (examples?) and that neither of the conditions (Mac_0) and $\perp$-symmetry implies the other one.

Let us finally turn to the disjointness properties (GD) (general disjointness property), (WD) (weak disjointness property), and (AD) (atomic disjointness property). These disjointness properties are formulated for lattices with least element 0 and with $a, b, c, x, y, z, t, \dots$ denoting arbitrary lattice elements.

We say that a lattice with 0 has the *general disjointness property* if

(GD) $a \wedge b = 0$ and $(a \vee b) \wedge c = 0$ imply $a \wedge (b \vee c) = 0$.

We say that a lattice with 0 has the *weak disjointness property* if

(WD) $x \wedge y = (x \vee y) \wedge z = 0$ and $z > 0$ imply the existence of an element t such that $0 < t \leq z$ and $x \wedge (y \vee t) = 0$.

Finally we say that a lattice with 0 has the *atomic disjointness property* if

(AD) $a \wedge b = 0$, $(a \vee b) \wedge p = 0$, and $p \succ 0$ imply $a \wedge (b \vee p) = 0$.

No suitable disjointness property has yet been found that fits in the part indicated by a question mark in Figure 3.12.

The implications $(D_0) \Rightarrow (GD)$, $(Mac_0) \Rightarrow (WD)$, $(C) \Rightarrow (AD)$, and $(GD) \Rightarrow (WD) \Rightarrow (AD)$ are well known or obvious.

In what follows we establish some more of the implications indicated in Figure 3.12. First we note that (besides distributive and pseudocomplemented lattices) the class of lattices with the general disjointness property (GD) also includes modular lattices and even 0-modular lattices (Saarimäki [1992]):

Lemma 3.4.2   *In a lattice bounded below, 0-modularity implies the general disjointness property* (GD).

Proof.   Let L be a 0-modular lattice and $a, b, c \in L$ such that $a \wedge b = 0$ and $(a \vee b) \wedge c = 0$. Putting $d = (a \vee b) \wedge (b \vee c)$, we have $b \leq d \leq b \vee c$ and thus $d \vee c = b \vee c$. On the other hand, $d \leq a \vee b$ and hence $d \wedge c = 0$. By assumption we have $c \, M \, d$ and therefore $b = (b \vee c) \wedge d = d$. Thus we get $a \wedge (b \vee c) = a \wedge d = a \wedge b = 0$. ∎

An atomistic lattice need not satisfy (GD) if it satisfies (AD): to see this consider, for example, the lattice of Figure 1.24(b) (Section 1.7), which is even semimodular. On the other hand, the example of Figure 3.14 (due to Saarimäki) satisfies (GD) but is not semimodular. Hence the properties (GD) and (C) are stronger than (AD), but they are independent of each other even in atomistic lattices.

The following characterization of the general disjointness property is due to Saarimäki (personal communication).

Theorem 3.4.3   *A lattice L satisfying the* DCC *has the general disjointness property* (GD) *if and only if it has no sublattice isomorphic to the lattice shown in Figure 3.15(a) or to the lattice shown in Figure 3.15(b) (with 0 denoting the least element of L).*

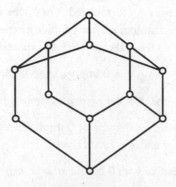

Figure 3.14

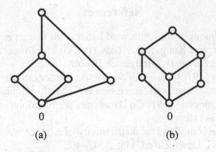

0   0

(a)   (b)

Figure 3.15

Note that the lattice of Figure 3.15(a) is a special type of the lattice L_4 shown in Figure 1.6 (Section 1.3), namely an L_4 including the least element of the lattice. The lattice of Figure 3.15(b) is a special type of the lattice S_7^* (the dual of the centered hexagon S_7), namely an S_7^* including the least element of the lattice.

Notes

The concepts of 0-distributivity [condition (D_0)] and 0-modularity [condition (M_0)] and their duals were introduced by Varlet [1968]. We have drawn here on this paper and on Varlet [1963].

Condition (Mac_0) was first considered by Salii [1980], who called it "0-semi-modularity" (see also Salii [1988], p. 50, and Adams [1990]). Vernikov [1992] uses the term 0-*semimodularity* to denote the atomic covering property (C).

Saarimäki & Sorjonen [1991] introduced (GD) and (AD) in order to investigate the existence of antitonic complement functions in upper continuous lattices. (WD) is due to Saarimäki (personal communication). The general disjointness property (GD) should be compared with property ($\perp$4) after Corollary 3.5.25.

Under certain additional assumptions (e.g. if each element has a section semi-complement – see Section 1.5), some of the implications in Figure 3.12 can be reversed. For example, Varlet [1968] showed that a 0-distributive lattice that is section-semicomplemented is distributive. For other reversions we refer to the following section.

Exercises

1. Prove that a lattice with 0 is 0-distributive if and only if its ideal lattice is pseudocomplemented (Varlet [1968], Theorem 1).
2. Prove that an algebraic 0-distributive lattice is pseudocomplemented. (This assertion implies the "only if" in the preceding exercise.) (Varlet [1968], Theorem 2.)
3. Give an example of a nonmodular bounded lattice satisfying both (M_0) and (M_1).

References

Adams, M. E. [1990] Uniquely complemented lattices, in: Bogart et al. [1990], pp. 79–84.
Bogart, K. P., Freese, R., and Kung, J. P. S. (eds.) [1990] *The Dilworth Theorems. Selected Papers of Robert P. Dilworth*, Birkhäuser, Boston.
Maeda, F. and S. Maeda [1970] *Theory of Symmetric Lattices*, Springer-Verlag, Berlin.
Saarimäki, M. [1992] Disjointness of lattice elements, Math. Nachr. 159, 169–774.
Saarimäki, M. and P. Sorjonen [1991] On Banaschewski functions in lattices, Algebra Universalis 28, 103–118.
Salii, V. N. [1980] Some conditions for distributivity of a lattice with unique complements (Russian), Izv. Vysš. Uceb. Zaved. Mat. 5, 47–49.
Salii, V. N. [1988] *Lattices with Unique Complements*, Translations of Mathematical Monographs, Vol. 69, Amer. Math. Soc, Providence, R.I. (translation of the 1984 Russian edition, Nauka, Moscow.)
Varlet, J. C. [1968] A generalization of the notion of pseudo-complementedness, Bull. Soc. Roy. Sci. Liège 36, 149–158.
Vernikov, B. M. [1992] Semicomplements in lattices of varieties, Algebra Universalis 29, 227–231.

3.5 Interrelationships in Lattices with Complementation

Summary.   In this section we have a look at the reversibility of the implications in Figure 3.12 (Section 3.4) in case the lattices carry some kind of complementation. We consider lattices with unique complements, section-semicomplemented lattices, and section-complemented lattices. Then we investigate the question: When is a $\perp$-symmetric lattice M-symmetric? Finally we cast a glance at two problems of Grillet & Varlet [1967].

If a lattice is distributive, complements are unique. It was conjectured long ago that the converse also holds, that is, a uniquely complemented lattice is distributive and hence a Boolean lattice. This conjecture has become known as Huntington's problem (cf. Huntington [1904]). It could indeed be proved that uniquely complemented lattices satisfying certain additional assumptions (modularity, atomicity, De Morgan's laws, etc.) are distributive. One of the first results in this direction was

Theorem 3.5.1   *A modular uniquely complemented lattice is distributive.*

For the various sources of this result see Salii [1988], p. 41. The first proof was given in Birkhoff [1948] (Exercise to §X.13 and Theorem VIII.10), who also refers to von Neumann. For this reason the preceding result is also known as the Birkhoff–von Neumann theorem.

Successively it was also established that a uniquely complemented lattice is distributive if it is orthocomplemented or atomic or relatively complemented or if it satisfies De Morgan's laws. For instance, by McLaughlin's theorem (Theorem 1.2.3) a uniquely complemented atomic lattice is modular, and hence it is distributive by Theorem 3.5.1. This was proved by Birkhoff & Ward [1939] under the additional assumption that the lattice is complete. It came as a surprise when

Dilworth [1945] proved that any lattice can be embedded in a uniquely complemented lattice. For more background and details on Huntington's problem and Dilworth's proof see Salii [1988] and Adams [1990].

Following Salii's approach (see Salii [1988]), we outline a distributivity criterion for uniquely complemented lattices. As corollaries to Salii's main results (Theorem 3.5.4 and Theorem 3.5.10 below) one can derive the distributivity of certain special classes of uniquely complemented lattices.

In this section we denote by $\bar{x}$ the complement of x in a uniquely complemented lattice. We shall use the following obvious

Lemma 3.5.2   *If $x < y$ in a uniquely complemented lattice, then $\bar{x} \wedge y \neq 0$.*

An element b of a uniquely complemented lattice L is called *regular* if, for any $x, y \in L$, $b \wedge x = b \wedge y = 0$ implies $b \wedge (x \vee y) = 0$. (Note the relationship between regular elements and dually distributive elements as defined in Section 2.2.) An element is called *irregular* if it is not regular. Clearly, 0 and 1 are regular elements. The proof of the next result is left as an exercise.

Lemma 3.5.3   *In a uniquely complemented lattice each atom is a regular element.*

The following characterization is due to Salii [1988].

Theorem 3.5.4   *A uniquely complemented lattice is distributive if and only if each nonzero element contains a nonzero regular element.*

Proof.   Necessity is obvious, since in a distributive lattice all elements are regular. Suppose now that x, y, z are arbitrary elements of a uniquely complemented lattice in which each nonzero element contains a nonzero regular element. Assume that $u = (x \wedge y) \vee (x \wedge z) \neq x \wedge (y \vee z) = v$, that is, $u < v$. Then Lemma 3.5.2 implies $t = \bar{u} \wedge v > 0$. Also $t \wedge y = (\bar{u} \wedge v) \wedge y \leq (\bar{u} \wedge x) \wedge y = \bar{u} \wedge (x \wedge y) \leq \bar{u} \wedge u = 0$, and similarly $t \wedge z = 0$. If now a is a regular element contained in t, then $a \wedge (y \vee z) = 0$. But $a \leq t \leq v \leq y \vee z$ and hence $a = 0$. Thus the nonzero element t contains no nonzero regular element, which contradicts our hypothesis. ∎

As a consequence of Theorem 3.5.4 and Lemma 3.5.3 we obtain

Corollary 3.5.5   *An atomic uniquely complemented lattice is distributive.*

This result was proved by Birkhoff & Ward [1939] under the additional assumption of completeness. Ogasawara & Sasaki [1949] provided a short proof in the general case. Another related result is

Corollary 3.5.6   *If the greatest element of a uniquely complemented lattice is the least upper bound of atoms, then the lattice is distributive.*

Theorem 3.5.4 emphasizes the role of irregular elements in nondistributive uniquely complemented lattices. The most important property of irregular elements is given in

Theorem 3.5.7 *An element x of a uniquely complemented lattice L is irregular if and only if there exists a nonzero element $y \in L$ such that $x \wedge y = \bar{x} \wedge y = 0$.*

Proof. If x is an irregular element, then there exists at least one element $z \in L$ such that $x \wedge z = 0$ but z is not $\leq \bar{z}$. Then $\bar{x} \wedge z < z$ and (by Lemma 3.5.2) $y = \overline{(\bar{x} \wedge z)} \wedge z > 0$. This y is the element we are looking for, since $x \wedge y \leq x \wedge z = 0$ and

$$\bar{x} \wedge y = \bar{x} \wedge (\overline{(\bar{x} \wedge z)} \wedge z) = (\bar{x} \wedge z) \wedge \overline{(\bar{x} \wedge z)} = 0.$$

On the other hand, if $y \neq 0$ and $\bar{x} \wedge y = 0$, then $x \wedge (\bar{x} \vee y) \neq 0$ (otherwise x would have two distinct complements: $\bar{x}$ and $\bar{x} \vee y$). But this means that in case $x \wedge y = 0$ the element x would not be regular. ∎

At this point the following result of Grillet & Varlet [1967] can be derived:

Theorem 3.5.8 *Any section-complemented uniquely complemented lattice is distributive.*

Proof. Suppose a uniquely complemented lattice L is section-complemented but not distributive. Then it contains at least one irregular element x. By Theorem 3.5.7 there exists an element $y > 0$ such that $x \wedge y = \bar{x} \wedge y = 0$. It is not possible that both x and $\bar{x}$ are $\leq \bar{y}$ (for otherwise we should have $\bar{y} = 1$ and hence $y = 0$). Without loss of generality we may assume that $\bar{x} \not\leq \bar{y}$. Then $x \vee \bar{y} \neq 1$, since otherwise a section complement z of $x \wedge \bar{y}$ in the interval $[0, \bar{y}]$ would be a complement of x different from $\bar{x} : x \vee z = (x \vee (x \wedge \bar{y})) \vee z = x \vee ((x \wedge \bar{y}) \vee z) = x \vee \bar{y} = 1$. It follows that a section complement u of $(x \vee \bar{y}) \wedge \bar{x}$ in the interval $[0, \bar{x}]$ and a section complement v of $(x \vee \bar{y}) \wedge y$ in $[0, y]$ are distinct (since $u \wedge v \leq \bar{x} \wedge y = 0$) complements of $x \vee \bar{y}$. Indeed, we have

$$(x \vee \bar{y}) \vee u = x \vee ((x \vee \bar{y}) \vee u) \geq x \vee (((x \vee \bar{y}) \wedge \bar{x}) \vee u) = x \vee \bar{x} = 1,$$
$$(x \vee \bar{y}) \vee v = \bar{y} \vee ((x \vee \bar{y}) \vee v) \geq \bar{y} \vee (((x \vee \bar{y}) \wedge y) \vee v) = \bar{y} \vee y = 1$$

and

$$(x \vee \bar{y}) \wedge u = (x \vee \bar{y}) \wedge (u \wedge \bar{x}) = ((x \vee \bar{y}) \wedge \bar{x}) \wedge u = 0,$$
$$(x \vee \bar{y}) \wedge v = (x \vee \bar{y}) \wedge (v \wedge \bar{x}) = ((x \vee \bar{y}) \wedge y) \wedge v = 0.$$

Thus we have arrived at a contradiction and conclude that the lattice does not contain irregular elements. From Theorem 3.5.4 it follows that the lattice is distributive. ∎

As a consequence we get a result of Szász [1958]:

Corollary 3.5.9 *A relatively complemented uniquely complemented lattice is distributive.*

The following criterion due to Salii [1980] for the distributivity of a uniquely complemented lattice will be useful.

Theorem 3.5.10 *A uniquely complemented lattice is distributive if and only if* $(x \vee ((x \vee y) \wedge \bar{x})) \wedge y > 0$ *holds for any x and for any $y > 0$.*

Proof. Let us first remark that if $x \wedge y \neq 0$ or $\bar{x} \wedge y \neq 0$, then the above inequality becomes trivial, and thus it is significant only if $x \wedge y = \bar{x} \wedge y = 0$, that is, if x is an irregular element. The "only if" part is clear, since in a Boolean lattice the left-hand side of the inequality is y. Concerning the "if" part we show that the lattice is section-complemented, whence distributivity follows from Theorem 3.5.8. Assume on the contrary that an element $x \in [0, z]$ has no complement in this interval. Then $x \vee (\bar{x} \wedge z) < z$ (note that $\bar{x} \wedge z > 0$ by Lemma 3.5.2). The element $y = x \vee (\bar{x} \wedge z) \wedge z$ is likewise different from 0, and we have $x \vee y < z$. Hence $(x \vee y) \wedge \bar{x} \leq \bar{x} \wedge z$ and therefore $x \vee ((x \vee y) \wedge \bar{x}) \leq x \vee (\bar{x} \wedge z)$ (which is, in fact, an equality). But then $(x \vee ((x \vee y) \wedge \bar{x})) \wedge y \leq (x \vee (\bar{x} \wedge z)) \wedge \overline{(x \vee (\bar{x} \wedge z))} = 0$ contradicting the hypothesis. ∎

As corollaries we get some results mentioned in the beginning of this section.

Corollary 3.5.11 *In a uniquely complemented lattice, the condition $x \leq y \Rightarrow \bar{x} \geq \bar{y}$ implies distributivity.*

Corollary 3.5.12 *In a uniquely complemented lattice, each of the De Morgan laws implies distributivity.*

In fact, distributivity in uniquely complemented lattices is even guaranteed by weak forms of the De Morgan laws (see Szász [1978]).

Corollary 3.5.13 *A uniquely complemented ortholattice is distributive.*

Corollary 3.5.14 *A uniquely complemented orthomodular lattice is distributive.*

For a direct proof of this corollary see Kalmbach [1983]. We have already stated the Birkhoff–von Neumann theorem asserting that in the class of modular lattices all uniquely complemented lattices are distributive (Theorem 3.5.1). Salii [1980] proved that they are distributive in a much wider class:

Theorem 3.5.15 *A uniquely complemented lattice satisfying the local Mac Lane condition* (Mac$_0$) *is distributive.*

Proof. Assume there exists a nondistributive uniquely complemented lattice satisfying (Mac$_0$). By Theorem 3.5.10 there exist elements $x \neq 0$, $y \neq 0$ such that $x \wedge y = \bar{x} \wedge y = 0$ and the meet of $\dot{z} = x \vee ((x \vee y) \wedge \bar{z}$ and y is 0. Now by (Mac$_0$) there exists an element t such that $0 < t \leq y$ and $(x \vee t) \wedge z = x$. Thus we have

$$(x \vee t) \wedge \bar{x} = ((x \vee t) \wedge (x \vee y)) \wedge \bar{x} = (x \vee t) \wedge ((x \vee y) \wedge \bar{x})$$
$$= ((x \vee t) \wedge z) \wedge ((x \vee y) \wedge \bar{x}) = x \wedge ((x \vee y) \wedge \bar{x}) \leq x \wedge \bar{x} = 0.$$

This means that $x \vee t$ is a complement of $\bar{x}$ distinct from x, contradicting the assumption of unique complementedness. ∎

Corollary 3.5.16 *A uniquely complemented lattice satisfying Mac Lane's condition* (Mac) *is distributive.*

A further consequence is a result of Grillet & Varlet [1967]:

Corollary 3.5.17 *A uniquely complemented 0-modular lattice is distributive.*

Since any modular lattice with 0 satisfies both Mac Lane's condition (Mac) and 0-modularity, we obtain that Corollary 3.5.16 or Corollary 3.5.17 implies the Birkhoff–von Neumann theorem (Theorem 3.5.1).

Let us now turn to disjointness in section-semicomplemented lattices. For the notion of section semicomplementedness see Section 1.5. In section-semicomplemented lattices one can likewise show the reversibility of some implications in Figure 3.12 (Section 3.4). For example, we have at the "lower level" of Figure 3.12 the following result due to Varlet [1968]: If a 0-distributive lattice is section-semicomplemented, then it is distributive. At the "upper level" of Figure 3.12 we note that the atomic covering property (C) in conjunction with "atomistic" (which implies section semicomplementedness) yields upper semimodularity [condition (Sm)]. Concerning the "upper level" we have the following result of Saarimäki [1992]:

Lemma 3.5.18 *If a lattice having the atomic disjointness property* (AD) *is section-semicomplemented, then it has the atomic covering property* (C).

Proof. Let p be an atom and $a \wedge p = 0$. We show that $a \prec a \vee p$. Suppose there is an element b such that $a < b < a \vee p$. As L is section-semicomplemented, there is a nonzero element $c \leq b$ such that $c \wedge a = 0$. Then $(c \vee a) \wedge p \leq b \wedge p = 0$, and (AD) implies that $c \wedge (a \vee p) = 0$, which in turn yields $c = 0$, a contradiction. ∎

Let us now indicate some results concerning the "intermediate levels" of Figure 3.13. First we mention another result of Saarimäki [1992]:

Theorem 3.5.19 *If a lattice having the general disjointness property* (GD) *is section-semicomplemented, then it is 0-modular and semimodular.*

Proof. Let L be an section-semicomplemented lattice with (GD), and let $a \wedge b = 0$ and $c \leq b$. As in the proof of Theorem 3.5.18, it can be shown that the element c cannot have any nonzero semicomplement in $(a \vee c) \wedge b$. Hence $c = (a \vee c) \wedge b$, and therefore we have $a \, M \, b$.

To show that L is upper semimodular let $d = a \wedge b \prec b$. As L is section-semicomplemented, there is a nonzero element c such that $d \wedge c = 0$, $c \leq b$. Since b covers d, it follows that $b = d \vee c$. By 0-modularity we have $c \, M \, d$ and thus (by Maeda & Maeda [1970], (7.5.3)) the element c is an atom. Since $a \vee b = a \vee c$, the assertion $a \prec a \vee b$ follows from the atomic covering property (C), which in turn is a consequence of 0-modularity. ∎

Corollary 3.5.20 *A complemented 0-modular lattice is semimodular.*

Proof. It is sufficient to note that a complemented 0-modular lattice is section-semicomplemented (cf. Varlet [1968]). ∎

In the preceding corollary semimodularity can be replaced by the stronger Mac Lane condition (Mac). This is a consequence of the following result, whose proof is left as an exercise:

Lemma 3.5.21 *If a lattice satisfying the local Mac Lane condition (Mac$_0$) is section-semicomplemented, then it satisfies the Mac Lane condition (Mac).*

Corollary 3.5.22 *A complemented 0-modular lattice satisfies Mac Lane's condition (Mac).*

Proof. A complemented 0-modular lattice is section-semicomplemented (cf. Varlet [1968]). A 0-modular lattice satisfies the local Mac Lane condition (Mac$_0$). (Mac$_0$) together with section semicomplementedness yields the Mac Lane condition (Mac) (by Lemma 3.5.21). ∎

We recall that $\perp$-symmetry together with complementedness does not imply M-symmetry (see Figure 2.3, Section 2.3).

It seems to be an open question whether a section-semicomplemented lattice having the weak disjointness property (WD) is 0-modular. If the lattice satisfies the DCC, this question has an affirmative answer (see Exercise 4).

We now consider the reversibility of the implications indicated in Figure 3.12 (Section 3.4) in section-complemented lattices. Loosely speaking, it is not in general possible to jump in Figure 3.12 from a "higher level" to a "lower level."

For example, the diamond M_3 is 0-modular, and thus it has the general disjointness property (GD); it is section-complemented but not 0-distributive. The lattice of all subsets, except the three-element subsets, of a four-element set is semimodular (cf. Grätzer [1978], Figure 3, p. 174), and thus it has the atomic disjointness property (AD); it is section-complemented but does not have the general disjointness property (GD).

On the other hand, we mentioned above that distributivity and 0-distributivity are equivalent already in section-semicomplemented lattices (Varlet [1968]). Also, upper semimodularity, the atomic covering property, and the atomic disjointness property are equivalent in atomistic lattices. A further result concerning the reversibility of implications indicated in Figure 3.12 is due to Saarimäki & Sorjonen [1991]:

Theorem 3.5.23   *If a lattice having the general disjointness property* (GD) *is section-complemented, then it is modular.*

Proof.   Let L be a section-complemented lattice having (GD), and let $a, b \in L$. We show that $a \, M \, b$, that is, $c \leq b \Rightarrow (c \vee a) \wedge b = c \vee (a \wedge b)$.

Assume first that $a \wedge b = 0$. If $c \leq b$ then $(c \vee a) \wedge b \geq c \vee (a \wedge b) = c$. Since the lattice is section-complemented, there exists an element $\bar{c}$ such that $c \wedge \bar{c} = 0$ and $c \vee \bar{c} = (c \vee a) \wedge b$. Moreover, $(c \vee \bar{c}) \wedge a \leq b \wedge a = 0$. Applying (GD) to the triple $(\bar{c}, c, a)$, we get $\bar{c} \wedge (c \vee a) = 0$. Because $\bar{c} \leq (c \vee a) \wedge b \leq c \vee a$, this implies $\bar{c} = 0$. Thus $c = (c \vee a) \wedge b$, and we have shown that $a \, M \, b$ holds in this special case.

Let now $a, b \in L$ be arbitrary elements and $c \leq b$. Then the element $d = a \wedge b$ has a complement $\bar{d}$ in $[0, a]$. Since $\bar{d} \wedge b = \bar{d} \wedge a \wedge b = \bar{d} \wedge d = 0$, the preceding considerations show that we have $\bar{d} \, M \, b$. Using this and $c \vee d \leq b$, we obtain

$$(c \vee a) \wedge b = (c \vee d \vee \bar{d}) \wedge b = (c \vee d) \wedge (\bar{d} \wedge b) = c \vee (a \wedge b),$$

that is, $a \, M \, b$.   ∎

Corollary 3.5.24   *A geometric lattice is modular if and only if it has the general disjointness property* (GD). *Similarly, an orthomodular lattice is modular if and only if it has* (GD).

Corollary 3.5.25   *A 0-modular relatively complemented lattice is modular.*

We now ask a question concerning the reversibility of one more implication in Figure 3.12 (Section 3.4): When is a $\perp$-symmetric lattice M-symmetric? To give an answer we need some more concepts. Maeda & Maeda [1970] call a lattice L with 0 a *semiortholattice* if there exists a binary relation $\perp$ (*semiorthogonality*) on L having the following four properties:

(i) $a \perp a \Rightarrow a = 0$;
(ii) $a \perp b \Rightarrow b \perp a$;
(iii) $a \perp b, \; c \leq a \Rightarrow c \perp b$;
(iv) $a \perp b, \; a \vee b \perp c \Rightarrow a \perp b \vee c$.

A bounded semiortholattice L is called *semiorthocomplemented* if, for every $a \in L$, there exists an element $b \in L$ with $a \vee b = 1$ and $a \perp b$. The element b is said to be a *semiorthocomplement* of a. Note that $a \perp b$ implies $a \wedge b = 0$. Hence a semiorthocomplement of an element a is a complement of a.

If a lattice L with 0 has the general disjointness property (GD) then the statement $a \wedge b = 0$ conversely defines a semiorthogonality relation on L. Hence complemented lattices with (GD) are semiorthocomplemented with $a \perp b$ defined as $a \wedge b = 0$. Of course, orthocomplemented lattices are also semiorthocomplemented with $a \perp b$ defined as $a \leq b^{\perp}$. A semiortholattice is called *relatively semiorthocomplemented* if for all $a \leq b$ there exists an element c such that $b = a \vee c$ and $a \perp c$. The element c is called a *relative semiorthocomplement* of a in b.

A relatively semiorthocomplemented lattice is relatively complemented. For this and other details on semiorthogonality in lattices see Maeda & Maeda [1970].

Here we focus on the following problem (Maeda & Maeda [1970], Problem 1, p. 29):

> Is a $\perp$-symmetric lattice M-symmetric if it is relatively semiorthocomplemented?

This problem was answered in the affirmative by Constantin [1973] and Padmanabhan [1974]. Here we follow the more general approach of Padmanabhan [1974].

Theorem 3.5.26 *For a relatively semiorthocomplemented lattice L the following statements are equivalent:*

(i) *L is M-symmetric.*
(ii) *L is $\perp$-symmetric.*
(iii) *If $a\,M\,b$, then a has a left complement within b.*

Sketch of proof.   (i) $\Rightarrow$ (ii): Clear.

(ii) $\Rightarrow$ (iii): (For the notion of left complement see Section 2.3.) Let $a\,M\,b$. If $a \wedge b = 0$, then $b\,M\,a$ and b itself is a left complement of a. If $a \wedge b > 0$, then choose a relative semiorthocomplement c of $a \wedge b$ in b. We have $(a \wedge b) \vee c = b$ and $(a \wedge b) \perp c$. Thus $a \wedge b \wedge c = 0$ and $(a \wedge b)\,M\,c$ (cf. Maeda & Maeda [1970], §2). It follows that $a \vee c = a \vee (a \wedge b) \vee c = a \vee b$ and $a \wedge c = 0$.

Since $a\,M\,b$ and $(a \wedge b)\,M\,c$, Lemma 1.9.14 yields that the maps $\varphi_b : [a, a \vee b] \to [a \wedge b, b]$ and $\varphi_c : [a \wedge b, b] \to [0, c]$ are onto. Thus $x\varphi_b\varphi_c = x\varphi_c$ is onto, and hence the map $\varphi_c : [a, a \vee c] \to [a \wedge c, c]$ is onto. Lemma 1.9.14 implies $a\,M\,c$. Since $a \wedge c = 0$ and L is $\perp$-symmetric, we get $c\,M\,a$. It follows that c is a left complement of a within b.

(iii) $\Rightarrow$ (i): Let $a\,M\,b$. By (iii) we choose a left complement c of a within b. Let $y \in [a \wedge b, a] \subseteq [a \wedge c, a]$. By $c\,M\,a$ and Lemma 1.9.14, there exists an $x \in [c, a \vee c]$ such that $x\varphi_a = y$. Hence $x \geq y \geq a \wedge b$, and therefore $x \vee c \geq$

$(a \wedge b) \vee c = b$, that is, $x \geq b$. Thus x belongs to the interval $[b, a \vee b]$, and hence the mapping $\varphi_a : [b, a \vee b] \to [a \wedge b, a]$ is onto. Therefore we have $b \, M \, a$.   ∎

As a corollary we obtain Theorem 2.3.1, since any lattice satisfying the assumptions of this theorem is relatively semiorthocomplemented by Maeda & Maeda [1970], Lemma 3.6.

We close this section with results concerning two problems of Grillet & Varlet [1967]. Recall first that a 0-modular uniquely complemented lattice is distributive [viz., a complemented 0-modular lattice satisfies Mac Lane's condition (Mac) by Corollary 3.5.22, (Mac) implies the local condition (Mac$_0$), and distributivity follows from Theorem 3.5.15].

If instead of unique complements we only require the existence of complements, then it is not even sure whether the lattice is modular (cf. Grillet & Varlet [1967]):

Is a 0-modular complemented lattice modular ?

If the answer is no, that is, if there exists a 0-modular complemented lattice that is not modular, then this lattice must have infinite length: This follows from McLaughlin's theorem (Theorem 1.2.3) and from the characterization of 0-modularity by means of forbidden special pentagons including the least element of the lattice (Theorem 3.4.1). The preceding observation is contrasted with the following result of Saarimäki [1998].

Theorem 3.5.27   *If a complete 0-modular semicomplemented lattice is upper continuous, then it is modular and complemented.*

Grillet & Varlet [1967] also posed the following problem:

Is a 0-modular section semicomplemented or orthocomplemented lattice modular?

It was noticed by Savel'zon [1978] that the answer is affirmative in the orthocomplemented case. To see this we note that a 0-modular lattice is $\perp$-symmetric and that a 0-modular orthocomplemented lattice L is also 1-modular. Thus the orthocomplement $a^\perp$ of an arbitrary element $a \in L$ will satisfy $a \, M \, a^\perp$ and $a^\perp \, M^* \, a$ (and the lattice is M-symmetric by Theorem 2.3.1). L is relatively semiorthocomplemented, and thus it is relatively complemented (Maeda & Maeda [1970], Lemma 3.6 and Corollary 2.10). Finally, a relatively complemented 0-modular lattice is modular (by Corollary 3.5.25).

Exercises

1. Prove that in a uniquely complemented lattice each atom is a regular element (Salii [1988], Lemma 3, p. 43).

2. Use the preceding result to show that if the greatest element of a uniquely complemented lattice is the least upper bound of atoms, then the lattice is distributive.
3. If a lattice satisfying the local Mac Lane condition (Mac_0) is section-semicomplemented, then it satisfies Mac Lane's condition (Mac) (Lemma 3.5.21).
4. Show that if a section-semicomplemented lattice has the weak disjointness property and satisfies the DCC, then it also satisfies the local Mac Lane condition (Mac_0). This implication even holds when DCC is replaced by "strongly atomic." (Saarimäki.)
5. If a complemented lattice is both 0-distributive and 0-modular, then it is distributive (Saarimäki [1992]).

References

Adams, M. E. [1990] Uniquely complemented lattices, in: Bogart et al. [1990], pp. 79–84.
Birkhoff, G. and M. Ward [1939] A characterization of Boolean algebras, Ann. of Math. 40, 609–610.
Birkhoff, G. [1948] *Lattice Theory* (2nd ed.), Amer. Math. Soc. Colloquium Publications, Vol. 25, New York.
Bogart, K. P., R. Freese, and J. P. S. Kung (eds.) [1990] *The Dilworth Theorems. Selected Papers of Robert P. Dilworth*, Birkhäuser, Boston.
Constantin, J. [1973] Note sur un problème de Maeda, Canad. Math. Bull. 16, 193.
Dilworth, R. P. [1945] Lattices with unique complements, Trans. Amer. Math. Soc. 57, 123–154.
Grätzer, G. [1978] *General Lattice Theory*, Birkhäuser, Basel.
Grillet, P. A. and J. C. Varlet [1967] Complementedness conditions in lattices, Bull. Soc. Roy. Sci. Liège 36, 628–642.
Huntington, E. V. [1904] Sets of independent postulates for the algebra of logic, Trans. Amer. Math. Soc. 5, 288–309.
Kalmbach, G. [1983] *Orthomodular Lattices*, Academic Press, London.
Maeda, F. and S. Maeda [1970] *Theory of Symmetric Lattices*, Springer-Verlag, Berlin.
McLaughlin, J. E. [1956] Atomic lattices with unique comparable complements, Proc. Amer. Math. Soc. 7, 864–866.
Ogasawara, T. and U. Sasaki [1949] On a theorem in lattice theory, J. Sci. Hiroshima Univ. Ser. A 14, 13.
Padmanabhan, R. [1974] On M-symmetric lattices, Canad. Math. Bull. 17, 85–86.
Saarimäki, M. [1992] Disjointness of lattice elements, Math. Nachrichten 159, 169–774.
Saarimäki, M. [1998] Disjointness and complementedness in upper continuous lattices, Report 78, University of Jyväskylä.
Saarimäki, M. and P. Sorjonen [1991] On Banaschewski functions in lattices, Algebra Universalis 28, 103–118.
Salii, V. N. [1980] Some conditions for distributivity of a lattice with unique complements (Russian), Izv. Vysš. Uceb. Zaved. Mat. 5, 47–49.
Salii, V. N. [1988] *Lattices with Unique Complements*, Translations of Mathematical Monographs, Vol. 69, Amer. Math. Soc., Providence, R. I. (translation of the 1984 Russian edition, Nauka, Moscow).
Savel'zon, O. I. [1978] 0-modular lattices (Russian), Uporyad. Mnozestva i Reshotky 5, 97–107.

Szász, G. [1958] On complemented lattices, Acta Sci. Math. 19, 77–81.

Szász, G. [1978] On the De Morgan formulae and the antitony of complements in lattices, Czech. Math. J. 28, 400–406.

Varlet, J. C. [1968] A generalization of the notion of pseudo-complementedness, Bull. Soc. Roy. Sci. Liège 36, 149–158.

4

Supersolvable and Admissible Lattices; Consistent and Strong Lattices

4.1 The Möbius Function

Summary. The Möbius function of a poset is an important object in enumerative combinatorics. Thus it is of interest to find out how to compute the Möbius function for particular posets and lattices. We give a definition of the Möbius function μ as the inverse of the zeta function ζ in the algebra of incidence functions of a poset P over a field F. We shall see that the Möbius function of a finite semimodular lattice alternates in sign and deal with other results on the Möbius function for finite semimodular lattices.

Let P be a locally finite poset, and F be a field. A function $f : P \times P \to F$ is said to be an *incidence function* on P with values in F if $f(x, y) = 0$ whenever $x \not\leq y$. Important examples of incidence functions are

The *Kronecker delta*: $\delta(x, y) = 1$ if and only if $x = y$.
The *zeta function*: $\zeta(x, y) = 1$ if and only if $x \leq y$.
The *strict zeta function*: $n(x, y) = 1$ if and only if $x < y$.
The Möbius function: $\mu(x, y)$ (to be defined below).

The set A of all incidence functions of P over F is endowed in a natural way with the structure of an algebra: addition and scalar multiplication are defined as usual, and the product h of $f, g \in A$ is defined to be the *convolution*, that is,

$$h(x, y) = \sum_z f(x, z)g(z, y).$$

This algebra is the *incidence algebra* of P over F. It is an associative algebra and is noncommutative in general. The term "algebra" is used here in the sense of "bilinear algebra over a field." The notion of an algebra as used here is not synonymous with the use of this word in the sense of universal algebra as we apply it in Sections 1.2, 1.3 and in Chapter 9.

Concerning the above-defined functions we have that $\zeta = n + \delta$ and δ is an identity in A. A function f is *invertible* in A if and only if $f(x, x) \neq 0$ for all $x \in P$; moreover, left inverses and right inverses coincide and are unique (see Hall

149

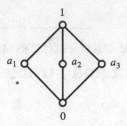

Figure 4.1

[1986] or Bogart [1990]). Thus ζ is invertible in A, and we define the *Möbius function* μ of P to be the inverse of ζ. Alternatively, the Möbius function can be defined inductively by setting

$$\mu(x, x) = 1 \quad \text{and} \quad \mu(x, y) = - \sum_{x < z \leq y} \mu(z, y).$$

In particular, if y covers x then $\mu(x, y) = -1$. For the diamond M_3 with the elements designated as in Figure 4.1 we have for $i = 1, 2, 3$

$$\mu(0, 0) = \mu(1, 1) = \mu(a_i, a_i) = 1, \qquad \mu(0, a_i) = \mu(a_i, 1) = -1,$$

$$\text{and} \quad \mu(0, 1) = 2.$$

The proofs of the following two properties are left as exercises:

(1) If μ^* is the Möbius function of the dual partially ordered set, then $\mu^*(y, x) = \mu(x, y)$.
(2) If μ_P and μ_Q are the Möbius functions of the posets P and Q, respectively, then the Möbius function of $P \times Q$ is given by $\mu_{P \times Q}((x, y), (u, v)) = \mu_P(x, u)\mu_Q(y, v)$.

Let us now give some more examples.

If $a_0 \leq a_1 \leq \cdots$ is a locally finite chain, then for any $i \geq 0$ we have $\mu(a_i, a_i) = 1$, $\mu(a_i, a_{i+1}) = -1$, and $\mu(a_i, a_{i+r}) = 0$ $(r \geq 2)$.

The classical Möbius function μ' (see e.g. Hardy & Wright [1954], pp. 232–258) is obtained by taking P to be the locally finite distributive lattice of the positive integers partially ordered by divisibility. To see this, observe that P is the direct product of the chains $1 < p < p^2 < \cdots$ for every prime p. Thus if x divides y and $x = \prod p_i^{\alpha_i}$, $y = \prod p_i^{\beta_i}$, then $\mu_D(x, y) = \prod_p \mu_P(p^{\alpha_i}, p^{\beta_i})$.

Using the preceding example it is easy to see that if x divides y, then $\mu_D(x, y) = \mu'(y/x)$. In connection with this example one can show that the Möbius function of any locally finite distributive lattice assumes only the values 0, 1, and -1 (see Exercise 1).

Let now B be the Boolean lattice of all subsets of a finite set S. Then if $S = \{a_1, \ldots, a_n\}$, we can describe the subsets T of S by means of n-tuples $(\delta_1, \ldots, \delta_n)$ with $\delta_i = 1$ if a_i is in T, and 0 if not. B is isomorphic to the direct product of n copies of $\{0, 1\}$, and we can therefore use the above property (2) and our example involving one locally finite chain to describe the Möbius function of B, namely: If $T_1 \subseteq T_2$, then $\mu(T_1, T_2) = (-1)^{|T_2| - |T_1|}$.

The incidence algebra can be viewed as a *matrix algebra*, consisting of all $|P| \times |P|$ matrices over F in which entries corresponding to pairs $x \not\leq y$ are required to be 0. In this way we get, for example, the *zeta matrix* of a finite poset P. More precisely, let P be a finite poset whose elements are given a listing $x_1, x_2, \ldots, x_n$. We define elements z_{ij} by stipulating that

$$z_{ij} = \begin{cases} 1 & \text{if } x_i \leq x_j, \\ 0 & \text{otherwise,} \end{cases}$$

and call $Z_{ij} = (z_{ij})$ the *zeta matrix* of P [this is the matrix version of the zeta function $\zeta(x, y)$ defined above]. For example, let the elements of the centered hexagon S_7 be listed as in Figure 4.2. The zeta matrix associated with the lattice of Figure 4.2 is shown in the following table:

	x_1	x_2	x_3	x_4	x_5	x_6	x_7
x_1	1	1	1	1	1	1	1
x_2	0	1	0	1	1	0	1
x_3	0	0	1	0	1	1	1
x_4	0	0	0	1	0	0	1
x_5	0	0	0	0	1	0	1
x_6	0	0	0	0	0	1	1
x_7	0	0	0	0	0	0	1

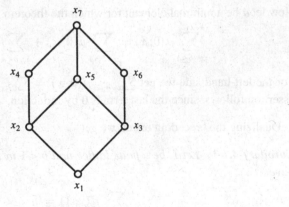

Figure 4.2

Since a partial ordering P of a set $X = \{x_1, x_2, \ldots, x_n\}$ is an oriented graph on X, the zeta matrix of P is the *adjacency matrix* of this graph. The zeta matrix of S_7 as given in the preceding table is an *upper triangular* matrix. The elements of any finite poset can be numbered so that its zeta matrix will be upper triangular. This reflects once more the invertibility of the function ζ. In accordance with the terminology introduced above, the inverse of the zeta matrix of a poset P will be called the *Möbius matrix* of P.

The following result is known as *Philip Hall's theorem*.

Theorem 4.1.1   *Let P be a finite poset, and let c_k denote the number of chains of k elements whose endpoints are x and y. Then $\mu(x, y) = c_1 - c_2 + c_3 - c_4 + \cdots$.*

For a proof (resting on incidence-algebra methods) see Hall [1932] or Rota [1964]. By $\Delta(P)$ we denote the *order complex* of P, that is, the simplicial complex whose points are the elements of P and whose simplices are the chains of P. Then $\chi(x, y) = 1 + \mu(x, y)$, where $\chi(x, y)$ denotes the *Euler characteristic* of the order complex of the open interval (x, y) in P. The value $\mu(x, y)$ is also called the *reduced Euler characteristic* of the open interval (x, y).

The next result is due to Weisner [1935] (for the presentation given here see Stanley [1986], Corollary 3.9.4).

Theorem 4.1.2   *Let L be a finite lattice with at least two elements and $b > 0$ in L. Then for any $a \in L$ we have*

$$\sum_{x \vee b = a} \mu(0, x) = 0.$$

Proof.   If $a = b$, then

$$\sum_{x \vee b = b} \mu(0, x) = \sum_{0 \le x \le b} \mu(0, x) = 0.$$

Now let a be a minimal element for which the theorem is not yet proved. Then

$$\sum_{x \vee b \le a} \mu(0, x) = \sum_{x \vee b = a} \mu(0, x) + \sum_{x \vee b < a} \mu(0, x).$$

For the left-hand side we get $\sum_{x \vee b \le a} \mu(0, x) = \sum_{0 \le x \le a} \mu(0, x) = 0$. Now the assertion follows, since the last term is 0 by induction. ∎

Dualizing the preceding result, we get

Corollary 4.1.3   *Let L be a finite lattice and $b < 1$ in L. Then for any $a \in L$ we have*

$$\sum_{x \wedge b = a} \mu(x, 1) = 0.$$

This corollary can be used to compute, for example, the Möbius function of the lattice of subspaces of a finite vector space or the Möbius function of a finite partition lattice (cf. Stanley [1986]). For the latter we also refer to Exercise 2.

For the following result see Stanley [1986], Proposition 3.10.1.

Theorem 4.1.4 *The Möbius function of a finite semimodular lattice alternates in sign.*

Proof. Take $a = 1$, and let b be an atom such that $x \vee b = 1$. If also $b \leq x$, then $x = 1$. Thus we have either $x = 1$ or $b \wedge x = 0$. Hence from $r(x) + r(b) \geq r(x \wedge b) + r(x \vee b)$ it follows that either $x = 1$ or $r(x) + 1 \geq 0 + n$ (with n denoting the rank of the lattice), that is, either $x = 1$ or x is a coatom. Therefore Theorem 4.1.2 (for $a = 1$) implies

$$\mu(0, 1) = - \sum_{\substack{\text{coatoms } x: \\ x \wedge b = 0}} \mu(0, x).$$

The assertion follows now by induction on n, since every interval of a semimodular lattice is again semimodular. ∎

In the geometric case we have the following result due to Rota [1964]:

Theorem 4.1.5 *Let L be a finite geometric lattice, and let $x \leq y$ in L. Then $\mu(x, y) \neq 0$ and has sign $(-1)^{r(y)-r(x)}$.*

This result is false for semimodular lattices in general, as the lattice S_7 shows: with the notation of Figure 4.2(a) we have

$$\mu(x_1, x_4) = - \sum_{x_1 \leq z < x_4} \mu(x_1, z) = -[\mu(x_1, x_1) + \mu(x_1, x_2)] = -[1 - 1] = 0.$$

Below (see Lemma 4.1.9) we shall treat a substitute for Theorem 4.1.5 for semi-modular lattices. According to Rota [1964], we define a *cross-cut* C of a finite lattice L as a subset of L having the following three properties:

(i) $0 \notin C$ and $1 \notin C$, that is, $C \subseteq L - \{0, 1\}$;
(ii) $x, y \in C$ and $x \leq y$ imply $x = y$;
(iii) any maximal chain from 0 to 1 intersects C.

Thus a cross-cut is an antichain that meets every maximal chain. *A spanning subset* of a finite lattice L is a subset S whose join is 1 and whose meet is 0. *Rota's cross-cut theorem* is

Theorem 4.1.6 *If C is a cross-cut of a nontrivial finite lattice and if q_k denotes the number of spanning subsets of C with k elements $(k \geq 2)$, then $\mu(0, 1) = \sum_k (-1)^k q_k$.*

For a proof see Rota [1964] or Aigner [1979]. The result can be used to show that the Euler characteristic $E(C)$ of the cross-cut C is in fact independent of the cross-cut. We shall need the following special case of Rota's theorem:

Corollary 4.1.7 *Let L be a finite lattice. Then $\mu(0, 1) = \sum_k (-1)^k q_k$, where q_k is the number of k-element subsets of atoms whose join is 1.*

An important consequence is that in a finite lattice L the value of $\mu(0, 1)$ depends only on the join sublattice spanned by the atoms of L or, dually, on the meet sublattice spanned by the coatoms of L. Another consequence is the joins-of-atoms theorem due to Hall [1936]:

Corollary 4.1.8 *If, in a finite lattice, 1 is not a join of atoms or 0 is not a meet of coatoms, then $\mu(0, 1) = 0$.*

This result is also an immediate consequence of Crapo's complementation theorem, for which see below (Theorem 4.1.10). Using Corollary 4.1.7, we can now prove the following substitute for Theorem 4.1.5 for arbitrary finite semimodular lattices (cf. Kung [1986b], Proposition 2.4.3).

Lemma 4.1.9 *Let $x \leq y$ in a finite semimodular lattice L. Then y is a join of atoms in the upper interval $[x, 1]$ if and only if $\mu(x, y) \neq 0$.*

Proof. If y is a join of atoms in $[x, 1]$, then by Corollary 4.1.7, the Möbius function $\mu(x, y)$ in $[x, 1]$ equals the Möbius function $\mu(x, y)$ in L^x, the sublattice in L generated by the atoms in $[x, 1]$. Since L^x is a geometric lattice, $\mu(x, y) \neq 0$. Conversely, if y is not a join of atoms in $[x, 1]$, the sum in Corollary 4.1.7 is empty and $\mu(x, y) = 0$. ∎

Crapo [1968] proved the following complementation theorem.

Theorem 4.1.10 *Let L be a finite lattice, and let $a \in L - \{0, 1\}$. Then*

$$\mu(0, 1) = \sum_{x \leq y} \mu(0, x)\mu(y, 1),$$

where the sum ranges over all complements x and y of a.

Corollary 4.1.11 *Let L be a finite lattice in which some element x fails to have a complement. Then $\mu(0, 1) = 0$.*

Suppose now that L is not complemented. Then $\mu(0, 1) = 0$, and this implies $\chi(\Delta(\bar{L})) = 1$, where $\bar{L} = L - \{0, 1\}$ is the so-called *proper part* of L, $\Delta(\bar{L})$ denotes the order complex of the poset $\bar{L}$, and χ is the Euler characteristic. Thus, if a finite lattice L is not complemented, then $\Delta(\bar{L})$ has the Euler characteristic of a point. In fact, much more can be shown: Baclawski & Björner [1979] proved that if a finite lattice L is not complemented, then $\bar{L}$ is contractible (see Section 4.2).

Notes

For detailed information concerning incidence functions on posets in general and the Möbius function in particular we refer to the books Aigner [1979], Bogart [1990], Hall [1986], Stanley [1986] and to the articles Rota [1964], Barnabei et al. [1982], Greene [1982], Zaslavsky [1987].

Greene [1982] gives an expository account (with complete proofs) of some basic results in the theory of the Möbius functions on a partially ordered set and traces the historical development. The paper has an extensive bibliography.

The modern era of the Möbius function begins with Rota [1964]. Rota's paper marks the time at which the Möbius function emerged as a fundamental invariant, which unifies both enumerative and structural aspects of the theory of partially ordered sets. Rota's results were extended by Crapo [1966], [1968].

Exercises

1. Show that for a locally finite distributive lattice, the Möbius function assumes only the values 0, 1, and -1.

2. Let Π_n be the lattice of partitions of a set of n elements. The partition containing only one block is denoted by 1, and that containing n blocks with one element each is denoted by 0. Show that for the Möbius function of Π_n one has $\mu_n(0, 1) = -(n-1)\mu_{n-1}(0, 1) = (-1)^{n-1}(n-1)!$ (Hint: Apply Corollary 4.1.3 with b chosen to be a dual atom.)

3. Let L be a finite lattice, and denote by $A(L)$ the set of atoms of L. If $A(L)$ is independent, then the Möbius values of L are all 0 or ± 1. Specifically, if $x \in L$, then $\mu(x) = (-1)^{|B|}$ if $x = \bigvee B$ for some $B \subseteq A(L)$, and $\mu(x) = 0$ otherwise. (Hint: The assertion follows from Rota's cross-cut theorem. The corollary also follows from Theorem 1.2 of Sagan [1995].)

4. Show that for the Tamari lattice T_n (see Section 1.4) one has $\mu(T_n) = (-1)^{n-1}$ (Sagan [1995], Proposition 2.4). (Hint: Note that T_n has $n-1$ atoms and that the atom set is independent (see Huang & Tamari [1972]), and apply the preceding exercise.)

References

Aigner, M. [1979] *Combinatorial Theory*, Springer-Verlag, Berlin.

Baclawski, K. and A. Björner [1979] Fixed points in partially ordered sets, Adv. in Math. 31, 263–287.

Barnabei, M., A. Brini, and G.-C. Rota [1982] Un'introduzione alla teoria delle funzioni di Möbius, in: *Matroid Theory and Its Applications* (ed. A. Barlotti), Liguori Editore, Napoli, pp. 7–109.

Bogart, K. P. [1990] *Introductory Combinatorics* (2nd edition), Harcourt Brace Jovanovich, San Diego (1st edition, Pitman, Marshfield, Mass., 1983).

Crapo, H. H. [1966] The Möbius function of a lattice, J. Combin. Theory 1, 126–131.

Crapo, H. H. [1968] Möbius inversion in lattices, Archiv d. Math. 19, 595–607.

Greene, C. [1982] The Möbius function of a partially ordered set, in: Rival [1982a], pp. 555–581.

Hall, M. [1986] *Combinatorial Theory* (2nd edition), Wiley, New York.

Hall, P. [1932] A contribution to the theory of groups of prime power order, Proc. London Math. Soc. 36, 39–95.

Hall, P. [1936] The Eulerian functions of a group, Quart. J. Math. (Oxford), 134–151.

Hardy, G. H. and E. M. Wright [1954] *An Introduction to the Theory of Numbers*, Oxford.

Huang, S. and D. Tamari [1972] Problems of associativity: a simple proof for the lattice property of systems ordered by a semi-associative law, J. Combin. Theory Ser. A 13, 7–13.

Kung, J. P. S. [1986b] Radon transforms in combinatorics and lattice theory, in: *Combinatorics and Ordered Sets*, Contemporary Mathematics, Vol. 57 (ed. I. Rival), pp. 33–74.

Rota, G.-C. [1964] On the foundations of combinatorial theory. I. Theory of Möbius functions, Z. Wahrsch. u. Verw. Gebiete 2, 340–368.

Sagan, B. E. [1995] A generalization of Rota's NBC theorem, Adv. in Math. 111, 195–207.

Stanley, R. [1986] *Enumerative Combinatorics* I, Wadsworth & Brooks/Cole, Monterey, Calif.

Weisner, L. [1935] Abstract theory of inversion of finite series, Trans. Amer. Math. Soc. 38, 474–484.

White, N. (ed.) [1986] *Combinatorial Geometries*, Cambridge Univ. Press, New York.

Zaslavsky, T. [1987] The Möbius function and the characteristic polynomial, in: White [1986], pp. 114–138.

4.2 Complements and Fixed Points

Summary. In the preceding section we noted that if a finite lattice of L is not complemented, then the reduced lattice $\bar{L}$ is contractible. If L is semimodular, then the converse is also true. This and other conditions characterizing the fixed-point property for $\bar{L}$ (where L is a semimodular lattice of finite length) follow from results of Baclawski & Björner [1979] and Björner & Rival [1980].

Let P and Q be posets. A map $f : P \to Q$ is *order-preserving* if $x \leq y$ implies $f(x) \leq f(y)$ for all $x, y \in P$. A *self-map* is a function $f : P \to P$. The *fixed-point set* of f is the subposet

$$P^f = \{x \in P \; : \; x = f(x)\}.$$

Discrete fixed-point theory deals among other things with properties of the set P^f. The simplest property is whether or not P^f is nonempty. A poset P such that the fixed point set P^f is nonempty for any order-preserving self-map f is said to have the *fixed-point property*. By convention, the empty set does not have the fixed-point property. For surveys and results on discrete fixed-point theory we refer to Rival [1980], Baclawski [1982], and Baclawski & Björner [1979].

Tarski [1955] proved that if P is a complete lattice and if f is an order-preserving self-map (of P), then P^f is nonempty and forms a complete lattice under the induced order. In a set-theoretic setting this result was already given by Knaster

[1928], which explains the name "Knaster–Tarski theorem." Davis [1955] showed that if every order-preserving map of a lattice to itself has a fixed point, then the lattice is complete. Thus a lattice has the fixed-point property if and only if it is complete. This result, which characterizes the fixed-point property for lattices, is also referred to as the Tarski–Davis theorem.

In particular, every lattice of finite length has the fixed-point property. It is therefore more interesting to consider the proper part $\bar{L}$ of a given lattice L of finite length. A poset of the form $\bar{L}$ is also called a *reduced lattice*. The problem is then to find necessary and sufficient conditions for $\bar{L}$ to have the fixed-point property. The problem of characterizing general posets having the fixed-point property is unsolved. However, some sufficient conditions and some necessary conditions have been proved for this property to hold (cf. Rival [1980] for a survey).

Following the presentation of Baclawski & Björner [1979] and Björner & Rival [1980], we first recall some general notions and then give their result concerning semimodular lattices of finite length. By $\Delta(P)$ we denoted the order complex of a poset P. The partial order on the vertices of $\Delta(P)$ defines an orientation on each simplex of $\Delta(P)$. As a simplicial complex, the order complex $\Delta(P)$ gives rise to a topological space $|\Delta(P)|$, the *geometric realization* of P, that is, the polyhedron associated to $\Delta(P)$.

For instance, Figure 4.3(b) shows a finite poset [the face poset of the regular cell decomposition of the 2-sphere shown in Figure 4.3(a)] and Figure 4.3(c) its geometric realization.

Topological properties of posets (that is, of the topological spaces attached to them) are of great importance for many algebraic and combinatorial considerations. For instance, the investigations of Baclawski, Björner, Stanley, and others have shown that such central combinatorial results as Rota's cross-cut theorem (Theorem 4.1.6) and Crapo's complementation theorem (Theorem 4.1.10) rest on

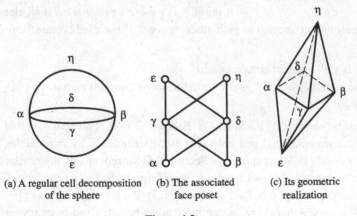

| (a) A regular cell decomposition | (b) The associated | (c) Its geometric |
| of the sphere | face poset | realization |

Figure 4.3

topological foundations (cf. e.g. Budach et al. [1988]). Let us give some more notions and results.

Let K be a field, and denote by $C_*(P, K)$ the algebraic chain complex over $\Delta(P)$ with coefficients in K. We write $d_n : C_n(P, K) \to C_{n-1}(P, K)$ for the differential (boundary map) of the complex and $C_*(P, K)$, and $B_n = \mathrm{Im}(d_{n+1})$, $Z_n = \mathrm{Ker}(d_n)$, and $H_n(P, K) = Z_n/B_n$. The K-vector spaces $H_n(P, K)$ are the *homology groups* of P *with coefficients in* K. A poset P for which $H_i(P, K) = 0$ for $i \neq 0$ and $H_0(P, K) = K$ is said to be *acyclic over the field* K, or K-*acyclic*. Thus, P is $\mathbb{Q}$-acyclic if all the reduced simplicial homology groups of $\Delta(P)$ with rational coefficients vanish.

Baclawski & Björner [1979], Theorem 2.1, proved that if a finite poset P is $\mathbb{Q}$-acyclic, then P has a strong fixed-point property (which will not be defined here).

If the poset is $\mathbb{Q}$-acyclic but infinite, then it does not have the fixed-point property in general. For example, take the set of integers ordered in the usual way. This poset is acyclic, but it has a fixed-point-free automorphism. The poset of the preceding example is even contractible in the following sense.

A poset P for which the geometric realization $|\Delta(P)|$ has the homotopy type of a point is said to be *contractible*. That is, P is contractible if the geometric realization $|\Delta(P)|$ is contractible as a topological space. Baclawski & Björner [1979] introduced the class of dismantlable posets; this is a class of contractible posets that contains the complete lattices and whose members have the fixed-point property. To define dismantlable posets, we need some more notions.

A chain is *(dually) well-ordered* if every nonempty subset of it has a minimal (maximal) element. We say that a poset P is *well-ordered complete* if every nonempty well-ordered chain of P has a join in P and every nonempty dually well-ordered chain of P has a meet in P. If a and b are elements of a poset P, we say that a is *connected* to b if there exist elements $c_1, c_2, \ldots, c_n$ of P such that $a \leq c_1 \geq c_2 \leq \cdots \geq c_n \leq b$. A poset P is said to be *connected* if all elements of P are pairwise connected to each other. A poset P is called *dismantlable* if

  (i) P is well-ordered complete, and
 (ii) the identity map of P is connected to some constant map in P^P.

It can be shown (cf. Baclawski & Björner [1979], Theorem 4.1) that a finite poset is dismantlable if and only if it is "dismantlable by irreducibles" in the sense of Rival [1974] as applied in Section 5.6. Moreover, any dismantlable poset is contractible; this follows from the preceding condition (ii) and from Quillen's "homotopy property" (see Quillen [1978], 1.3). Baclawski & Björner [1979], Theorem 4.2, proved that every dismantlable poset has the fixed-point property.

Finally we need the concept of a retract. Let P and Q be posets. Then Q is a *retract* of P if there are order-preserving maps f of Q to P and g of P to Q such that $g \circ f$ is the identity map of Q (cf. Duffus et al. [1980]).

Using the above-mentioned concepts and results, Björner & Rival [1980] proved

Theorem 4.2.1   *For a semimodular lattice L of finite length the following conditions are equivalent:*

(a) $\bar{L}$ *has the fixed-point property;*
(b) L *is not complemented;*
(c) $\sup\{p : p \text{ is an atom of } L\} < 1;$
(d) $\bar{L}$ *is dismantlable;*
(e) $\bar{2^n}$ *is not a retract of $\bar{L}$, where n is the rank of L;*
(f) $\bar{L}$ *is contractible;*
(g) $\bar{L}$ *is $\mathbb{Q}$-acyclic.*

Notes

The existence of a connection between fixed points and complements in lattices was first established by Baclawski & Björner [1979].

Exercises

1. Give an example of a lattice satisfying condition (d) but not (b) in Theorem 4.2.1 (Björner & Rival [1980]).
2. Give an example of a lattice satisfying condition (b) but not (d) in Theorem 4.2.1 (Björner & Rival [1980]).
3. Give an example of a lattice satisfying condition (f) but not (a) in Theorem 4.2.1 (Björner & Rival [1980]).
4. Give an example of a finite noncomplemented lattice with the property that each of its elements is the join of atoms and the meet of coatoms (Björner & Rival [1980]).
5. Following Björner [1981], we call an interval $[x, y]$ of a lattice L of finite length $\vee$-*regular* if y is the join of the atoms of $[x, y]$. Dually, $[x, y]$ will be called $\wedge$-*regular* if x is the meet of the coatoms of $[x, y]$. An interval $[x, y]$ is called *upper* if $y = 1$ and *lower* if $x = 0$.
   (a) Give an example showing that $[0,1]$ may be the only interval of a complemented lattice that is $\vee$-regular.
   (b) Let L be a lattice of finite length such that all upper intervals are $\vee$-regular. Then L is complemented. (Hint: Use induction on the length of L.) (Björner [1981], Theorem 1.)

(c) Give an example showing that the family of all upper intervals cannot be replaced by the family of all lower intervals in (b).

6. Let L be a lattice of finite length. Then the following conditions are equivalent:

   (i) L is relatively complemented;
   (ii) all intervals are $\vee$-regular;
   (iii) all intervals are $\wedge$-regular;
   (iv) L has no three-element interval.

(Björner [1981], Theorem 2.)

References

Baclawski, K. [1982] Combinatorics: trends and examples, in: *New Directions in Applied Mathematics* (ed. P. J. Hilton and G. S. Young), Springer-Verlag, pp. 1–10.

Baclawski, K. and A. Björner [1979] Fixed points in partially ordered sets, Adv. in Math. 31, 263–287.

Björner, A. [1981] On complements in lattices of finite length, Discrete Math. 36, 325–326.

Björner, A. and I. Rival [1980] A note on fixed points in semimodular lattices, Discrete Math. 29, 245–250.

Björner, A., A. Garsia, and R. Stanley [1982] An introduction to Cohen–Macaulay partially ordered sets, in: Rival [1982a], pp. 583–615.

Budach, L., B. Graw, C. Meinel, and S. Waack [1988] *Algebraic and Topological Properties of Finite Partially Ordered Sets*, Teubner-Texte zur Mathematik, Vol. 109, Teubner-Verlag, Leipzig.

Davis, A. C. [1955] A characterization of complete lattices, Pacific J. Math. 5, 311–319.

Duffus, D., W. Poguntke, and J. Rival [1980] Retracts and the fixed point problem for finite partially ordered sets, Canad. Math. Bull. 23, 231–236.

Knaster, B. [1928] Un théorème sur les fonctions d'ensembles, Ann. Soc. Polon. Math. 6, 133–134.

Quillen, D. [1978] Homotopy properties of the poset of non-trivial p-subgroups of a group, Advances in Math. 28, 101–128.

Rival, J. [1974] Lattices with doubly irreducible elements, Canad. Math. Bull. 17, 91–95.

Rival, I. [1980] The problem of fixed points in ordered sets, Ann. Discrete Math. 8, 283–292.

Rival, I. (Ed.) [1982a] *Ordered Sets. Proc. NATO Advanced Study Inst., Conf. Held at Banff, Canada, Aug 28–Sep 12, 1981.*

Tarski, A. [1955] A lattice-theoretical fixpoint theorem and its applications, Pacific J. Math. 5, 285–309.

4.3 Supersolvable Lattices

Summary. We recall the notion and some properties of supersolvable lattices as introduced by Stanley [1972] (see also Stanley [1986]). Examples include the subgroup lattices of finite supersolvable groups (which gave rise to the name) and partition lattices. Supersolvable lattices may or may not be semimodular. Every finite modular lattice is supersolvable, and every supersolvable lattice satisfies the Jordan–Dedekind chain condition.

A finite lattice L is *supersolvable* (SS, for short) if it possesses a maximal chain D, called a *modular chain* or *M-chain*, with the property that the sublattice of L generated by D and any other chain of L is distributive. If L is an SS lattice whose M-chain D has length n (or cardinality $n + 1$), then every maximal chain K of L has length n, since all maximal chains of the distributive lattice generated by D and K have the same length. Hence any SS lattice satisfies the Jordan–Dedekind chain condition. The following result (Stanley [1972], Proposition 2.1) relates M-chains to modular elements and thereby gives rise to several important examples of SS lattices.

Lemma 4.3.1 *Let L be a finite lattice, and D a maximal chain with the property that, for all $x \in L$ and for all $d \in D$, both xMd and dMx hold. Then D is an M-chain of L, that is, the sublattice of L generated by D and any other chain of L is distributive.*

Sketch of proof. The conditions xMd and dMx for all $x \in L$ mean that d is a left and right modular element. The proof is essentially the same as Birkhoff's proof of the less general result that the free modular lattice generated by two finite chains is a finite distributive lattice (cf. Birkhoff [1967], p. 65–66). Birkhoff uses modularity only for showing the identities

(i) $(a_1 \wedge b_1) \vee \cdots \vee (a_r \wedge b_r) = a_1 \wedge (b_1 \vee a_2) \wedge \cdots \wedge (b_{r-1} \vee a_r) \wedge b_r$ and

(ii) $(b_1 \vee a_1) \wedge \cdots \wedge (b_r \vee a_r) = b_1 \vee (a_1 \wedge b_2) \vee \cdots \vee (a_{r-1} \wedge b_r) \vee a_r$

when $a_i \geq a_{i+1}$ and $b_i \leq b_{i+1}$ in a modular lattice. Stanley [1972] showed that these identities still hold if one only assumes that the a_i's are modular elements. (Note that it is not possible to dualize (ii) in order to prove (i), since the property of being a modular element is not self-dual.) ∎

The following example shows that the converse to Lemma 4.3.1 is false in general: Let L be the lattice of subsets of $\{a, b, c\}$ ordered by inclusion except that the relation $\{a\} \subset \{a, b\}$ is excluded. Let D be the chain $\emptyset < \{c\} < \{b, c\} < \{a, b, c\}$. Then D is an M-chain but $\{a\} M \{b, c\}$ does not hold and hence $\{b, c\}$ is neither a left nor a right modular element. However, for SS lattices Stanley [1972], Proposition 2.2, provided the following partial converse to Lemma 4.3.1.

Lemma 4.3.2 *Let L be an SS lattice with an M-chain D. If $d \in D$ and $y \in L$, then dMy.*

Proof. Let $z \leq y$. Since the sublattice generated by D, y, and z is distributive, so is the sublattice generated by d, y, and z. Therefore dMy. ∎

The converse to the preceding proposition is also false, as the pentagon shows: if D denotes the longer of the two maximal chains in a pentagon, then we have xMp

for all $x \in D$ with p denoting the middle element of the shorter maximal chain. However, a pentagon is not an SS lattice (such lattices satisfy the Jordan–Dedekind chain condition). Since a finite upper semimodular lattice is M-symmetric (see Section 1.9), Lemma 4.3.1 and Lemma 4.3.2 yield

Corollary 4.3.3   *Let D be a maximal chain of a finite upper semimodular lattice. Then D is an* M*-chain if and only if every element of D is both left and right modular.*

Let us now consider some examples of supersolvable lattices.

Let G be a finite supersolvable group (see Section 8.3 for a definition), and $L(G)$ its lattice of subgroups. Every normal subgroup of a group G is a modular element in $L(G)$ (see Birkhoff [1967], p. 172). Since G is supersolvable, $L(G)$ contains a maximal chain of normal subgroups (corresponding to a chief series of G). Thus $L(G)$ is an SS lattice, and every chief series of G is an M-chain (there may be other M-chains). This example also explains Stanley's terminology "supersolvable" lattice, since it is the supersolvability of G that implies the existence of an M-chain in $L(G)$. It was already noted in Section 1.9 that a finite supersolvable group G is a J-group, that is, its subgroup lattice $L(G)$ satisfies Jordan–Dedekind chain condition.

Every finite modular lattice is supersolvable, since any maximal chain is an M-chain.

Consider now the lattice Π_n of partitions of an n-element set S. A partition π of S is a modular element of Π_n if and only if at most one block of π has more than one element. From this it follows that Π_n is an SS lattice with exactly $n/2$ M-chains ($n > 1$). Stanley [1972] gives two examples of SS lattices generalizing the preceding example.

In accordance with the above terminology a finite geometric lattice is called supersolvable if it contains a maximal chain of modular elements, and a matroid is called supersolvable if its lattice of flats is supersolvable (Stanley [1971a], [1972]). We leave it as an exercise to give an example of a geometric lattice that is not supersolvable. For an investigation of supersolvable geometric lattices and supersolvable matroids see Wanner & Ziegler [1991].

Let $\mathbf{N}$ denote the set of all partial orderings P on the set $\{1, 2, \ldots, n\}$ such that $i < j$ in P implies $i < j$ as integers. Define $P \le Q$ in $\mathbf{N}$ if $i \le j$ in P implies $i \le j$ in Q. Dean & Keller [1968] proved that this order relation makes $\mathbf{N}$ into a lower semimodular lattice of rank $\binom{n}{2}$. In fact, Dean & Keller [1968] and Avann [1972] showed that this "lattice of natural partial orders" is even lower locally distributive (meet-distributive). Stanley [1972] observed that $\mathbf{N}$ is an SS lattice with an M-chain given by $0 = P_0 < P_1 < \cdots < P_N = 1$, where the P_i are generated by the first i terms of the sequence $1 < n, 1 < n-1, 2 < n-2, \ldots, 1 < 2, 2 < n, 2 < n-1, \ldots, 2 < 3, 3 < n, 3 < n-1, \ldots, 3 < 4, \ldots, n-1 < n$.

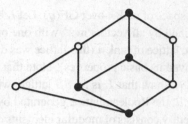

Figure 4.4

The preceding example is also an instance of a lattice that is supersolvable but not upper semimodular. Another example of such a lattice is given in Figure 4.4 (cf. Stanley [1986], p. 134) with an M-chain denoted by solid dots.

Notes

Stanley [1972] gives further examples of SS lattices. He shows that SS lattices have a number of interesting combinatorial properties connected with the counting of chains, which can be formulated in terms of Möbius functions. SS lattices that are also upper semimodular (USS lattices, for short) enjoy a number of properties not shared by general SS lattices. For example, recall that the degree of a projective geometry V, deg V, is one less than the number of points on a line. In particular, if V is coordinatized by the field GF(q), then deg $V = q$. If L is the lattice of subspaces of a projective geometry of degree q, then the integer q enters into many of the combinatorial properties of L. Stanley [1972], Section 6, defines a general class of upper semimodular lattices L based on the integer q, and, in the SS case ("q-USS lattices") he shows how q enters into the combinatorial properties of L.

We have already indicated that the set of closure operators on a finite poset P forms an upper semimodular lattice (in fact, an upper locally distributive or join-distributive lattice) when partially ordered by setting $H \leq K$ if $H(x) \leq K(x)$ for all $x \in P$, with H and K denoting closure operators (see Notes to Section 1.7). Hawrylycz & Reiner [1993] proved that this lattice is supersolvable (as suggested by Rota).

For detailed information on supersolvable lattices see Björner et al. [1982] and Stanley [1986].

Exercises

1. Let L be a finite upper semimodular lattice. Suppose $x \leq y$ in L, y is a modular element of L, and x is a modular element of the interval $[0, y]$. Then x is a modular element of L. In particular, a maximal chain $0 = x_0 \prec x_1 \prec \cdots \prec x_n = 1$ of L is an M-chain if each x_{i-1} is a modular element of the interval $[0, x_i]$. (Stanley [1971b], Lemma 4.3.)

2. Let V be a projective space of rank n over GF(q). Let L be the lattice of flats of the geometry determined by all vectors in V with one or two nonzero entries. Then L is a geometric lattice of rank n (this lattice was discovered by Dowling [1973], who also derived its basic properties). Note that for $q = 2$ one gets the partition lattices Π_{n+1}. Show that L is an SS lattice, with $n!$ M-chains when $q > 2$. (Hint: For any q, the Boolean lattice generated by the vectors in V that contain one nonzero entry consists of modular elements of L.) (Stanley [1972], Example 2.9.)

3. Give an example of a geometric lattice that is not supersolvable.

References

Avann, S. P. [1972] The lattice of natural partial orders, Aequationes Math. 8, 95–102.

Birkhoff, G. [1967] *Lattice Theory*, Amer. Math. Soc. Colloquium Publications, Vol. 25, Providence, R.I.

Björner, A., A. Garsia, and R. Stanley [1982] An introduction to Cohen–Macaulay partially ordered sets, in: Rival [1982a] pp. 583–615.

Dean, R. A. and G. Keller [1968] Natural partial orders, Canad. J. Math. 20, 535–554.

Dowling, T. A. [1973] A q-analog of the partition lattice, in: *A Survey of Combinatorial Theory, Proc. Int. Symp. on Combinatorial Mathematics and Its Applications, Colorado State Univ., Fort Collins, September 9–11, 1971* (ed. J. N. Srivastava), North-Holland, Amsterdam, pp. 101–115.

Hawrylycz, M. and V. Reiner [1993] The lattice of closure relations on a poset, Algebra Universalis 30, 301–310.

Rival, I. (Ed.) [1982a] *Ordered Sets, Proc. NATO Advanced Study Inst., Conf. Held at Banff, Canada, Aug. 28–Sep. 12, 1981*.

Stanley, R. [1971a] Modular elements in geometric lattices, Algebra Universalis 1, 214–217.

Stanley, R. [1971b] Supersolvable semimodular lattices, in: *Möbius Algebras (Proc. Conf., Univ. of Waterloo, Waterloo, Ont., 1971)*, Univ. of Waterloo, Waterloo, Ont., pp. 80–142.

Stanley, R. [1972] Supersolvable lattices, Algebra Universalis 2, 197–217.

Stanley, R. [1974] Finite lattices and Jordan–Hölder sets, Algebra Universalis 4, 361–371.

Stanley, R. [1986] *Enumerative Combinatorics I*, Wadsworth & Brooks/Cole, Monterey, Calif.

Wanner, T. and G. M. Ziegler [1991] Supersolvable and modularly complemented matroid extensions, Eur. J. Combin. 12, 341–360.

4.4 Admissible Lattices and Cohen–Macaulay Posets

Summary. The theory of Cohen–Macaulay posets due to Baclawski, Hochster, Stanley, Reisner, Björner, and Garsia is on the borderline of combinatorics, algebra, and topology and has many applications. Here we indicate how finite semimodular lattices fit into this theory. We mainly draw on Baclawski [1980], Björner [1980], and the comprehensive survey of Björner et al. [1982]. For further information we also refer to the more recent monograph of Hibi [1992].

A *shelling* of a simplicial complex Σ is a linear order on its maximal faces, which allows Σ to be built up inductively in a relatively simple fashion. The notion of a shelling is of recent origin (see Bruggesser & Mani [1971]); it has been used extensively in studying the topology and combinatorics of simplicial complexes. The notion of shellability, which originated in polyhedral theory, is a useful concept also in combinatorics with applications in matroid theory and order theory.

A finite graded poset P is *shellable* if its order complex $\Delta(P)$ is shellable. A finite simplicial complex Δ is said to be *shellable* if its maximal faces $F_1, F_2, \ldots, F_n$ can be ordered in such a way that $F_k \cap (\bigcup_{i=1}^{k-1} F_i)$ is a nonempty union of maximal proper faces of F_k for $k = 2, 3, \ldots, n$. A finite graded poset P is called *Cohen–Macaulay* (briefly: *CM poset*) if its order complex $\Delta(P)$ is a Cohen–Macaulay complex. It is known that a shellable complex Δ is Cohen–Macaulay, that is, a certain commutative ring associated with Δ is a Cohen–Macaulay ring. We refer to Stanley [1983] for an introduction to the combinatorics of order complexes and their homology.

For a finite poset P we denote by $C_r(P)$ its covering relation,

$$C_r(P) = \{(x, y) \in P \times P : x \prec y\}.$$

An *edge labeling* of P is a map $\lambda : C_r(P) \to \Lambda$, where Λ is some poset. An edge labeling thus corresponds to an assignment of elements of Λ to the edges of the Hasse diagram of P. An unrefinable chain $x_0 \prec x_1 \prec x_2 \prec \cdots \prec x_n$ in a poset with an edge labeling λ will be called *rising* if $\lambda(x_0, x_1) \leq \lambda(x_1, x_2) \leq \cdots \leq \lambda(x_{n-1}, x_n)$.

Let now $\lambda : C_r(P) \to \Lambda$ be an edge labeling of a graded poset P. Then λ is said to be an *R labeling* if in every interval $[x, y]$ of P there is a unique rising unrefinable chain $x = x_0 \prec x_1 \prec \cdots \prec x_n = y$. λ is said to be an *edgewise lexicographical labeling* (*EL labeling* for short), if

 (i) λ is an R labeling and
(ii) for every interval $[x, y]$ of P if $x = x_0 \prec x_1 \prec \cdots \prec x_n = y$ is the unique rising unrefinable chain and $x \prec z \leq y$, $z \neq x_1$, then $\lambda(x, x_1) < \lambda(x, z)$.

A poset is said to be *edgewise lexicographically shellable* (*EL-shellable*, for short) if it is graded and admits an EL labeling. For example, the edge labelings of the poset given in Figure 4.5(a) (the poset of the partitions of the integer 6 ordered by refinement) and in Figure 4.5(b) (the face lattice of a square) are EL labelings. An example of a poset P with an edge-labeling that is not an EL labeling is shown in Figure 4.6. In the Examples of Figure 4.5(a), (b) and of Figure 4.6 we have $\Lambda = \mathbb{Z}$.

The following fundamental result is due to Björner [1980], Theorem 2.3, to which we refer for a proof.

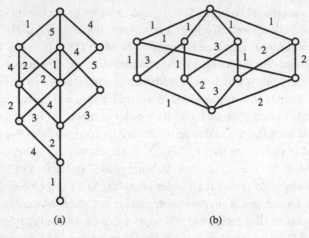

(a)   (b)

Figure 4.5

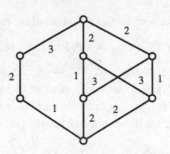

Figure 4.6

Theorem 4.4.1   *Let P be a edgewise lexicographically shellable poset. Then P is shellable.*

We leave it as an exercise to give an example of shellable poset that is not edgewise lexicographically shellable (cf. Vincze & Wachs [1985]). Note also that by Theorem 4.4.1 there does not exist any EL labeling for the poset of Figure 4.6, since this poset P is obviously not shellable.

An important subclass of lexicographically shellable posets are the admissible lattices introduced by Stanley [1974]. Let L be a finite lattice and $\omega : J(L) \to \mathbb{P}$ a map from the set $J(L)$ of join-irreducibles ($\neq 0$) of L to the set $\mathbb{P}$ of positive integers. Such a map induces an edge labeling $\gamma : C_r(P) \to \mathbb{P}$ of L by the rule

$$\gamma(x \prec y) = \min\{\omega(j) \: : \: j \in J(L), \: x < x \vee z = y\}.$$

Thus $\gamma(x \prec y)$ is the least label of a join-irreducible that is less than or equal to y but not less than or equal to x. The value $\gamma(x \prec y)$ is always defined, since y

is a join of join-irreducibles. If γ is an R-labeling, then ω is called an *admissible map*. A finite lattice L is said to be *admissible* if it is graded and there exists an admissible map $\omega : J(L) \to \mathbb{P}$.

Stanley [1974] proved that all supersolvable and all finite upper semimodular lattices are admissible (see below for examples illustrating these results). There are other classes of admissible lattices, but these seem to be of lesser interest. Björner [1980], Theorem 3.1, proved that admissible lattices are lexicographically shellable. From Theorem 4.4.1 it follows that admissible lattices are shellable and hence Cohen–Macaulay.

Let us first consider finite distributive lattices. Although finite distributive lattices are upper semimodular, they are dealt with separately here, since the method of labeling applied to them can be directly generalized to supersolvable lattices.

Let P be a finite poset with $d + 1$ elements. Recall that by ord(P) we denote the set of all order ideals of P, ordered by set inclusion. Then ord(P) is a distributive lattice (see Corollary 1.3.2) of rank $d + 1$. Let $[d + 1]$ denote the set $\{1, 2, \ldots, d, d + 1\}$, and let $\omega : P \to [d + 1]$ be an order-preserving bijection. If $I \prec I'$ is an edge in ord(P), then $I' - I = \{x\}$ and we can label $\lambda_\omega(I \prec I') = \omega(x)$. It can be shown that this induces an edgewise lexicographical labeling.

Consider for example the poset P given in Figure 4.7(a) with the map ω shown in Figure 4.7(b). The lattice ord(P) is shown in Figure 4.7(c), and the induced edge labeling in Figure 4.7(d).

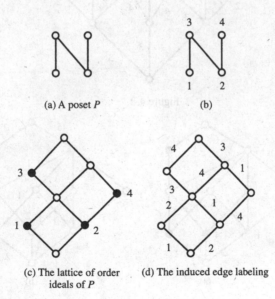

(a) A poset P (b)

(c) The lattice of order   (d) The induced edge labeling
ideals of P

Figure 4.7

Provan [1977] had already shown that finite distributive lattices are shellable.

Let now L be an supersolvable lattice. Any M-chain D of L generates a distributive lattice. Each such distributive lattice can be labeled by the method described in the preceding example. It turns out that this will assign a unique label to each edge of L and that this induces an admissible map. Figure 4.8 shows a supersolvable lattice with an induced edge labeling.

Let now L be a finite upper semimodular lattice. Give the join-irreducibles of L a linear order $j_1, j_2, \ldots, j_s$ that extends their partial order, that is, $j_m < j_n$ implies $m < n$. Then $\lambda(x \prec y) = \min\{m : x < x \vee j_m = y\}$ defines an edgewise lexicographical labeling of L. For example, Figure 4.9(a) shows a semimodular lattice with a linear order on the join-irreducibles extending their partial order. Figure 4.9(b) shows the induced edge labeling.

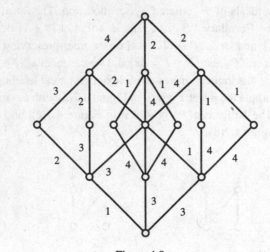

Figure 4.8

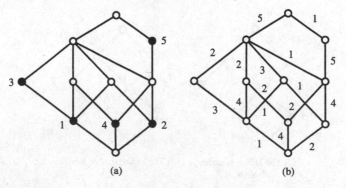

(a)   (b)

Figure 4.9

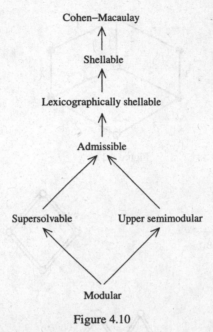

Figure 4.10

It was previously known that finite geometric lattices are Cohen–Macaulay (Folkman [1966]). Figure 4.10 visualizes the interrelationships between some of the concepts considered in Sections 4.3 and 4.4.

Notes

Some of the important examples of lexicographically shellable posets admit EL labelings possessing a stronger property. This stronger property is described by the concept of *SL* labeling (strongly lexicographical labeling):

An *SL labeling* λ of a graded poset P is an EL labeling $\lambda : C_r(P) \to \Lambda$ such that if $x = x_0 \prec x_1 \prec \cdots \prec x_n = y$ is the unique rising unrefinable chain in the interval $[x, y]$, then $\lambda(x_{n-1}, y) > \lambda(z, y)$ for all $z \neq x_{n-1}$ such that $x \le z \prec y$. A poset is said to be *strongly lexicographically shellable* (*SL-shellable*, for short) if it is graded and admits an SL labeling (cf. Björner [1980], Definition 3.4).

Figure 4.5(a) above shows an SL labeling, and Figure 4.5(b) shows an EL labeling that is not an SL labeling. The induced edge labeling of an admissible map need not be an SL labeling, as the example of Figure 4.11 shows.

It is not known whether all admissible lattices are SL-shellable, but Björner and Stanley have shown that the important examples are: finite upper semimodular, finite lower semimodular, and supersolvable lattices are SL-shellable (cf. Björner [1980], Theorem 3.7).

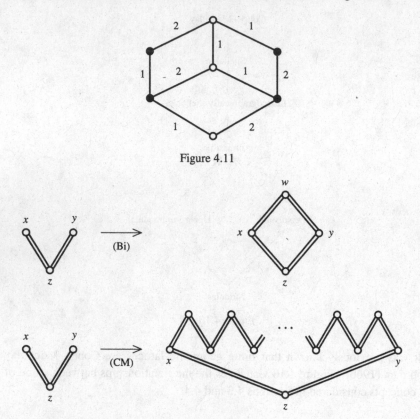

Figure 4.11

Figure 4.12

Extending Stanley's above-mentioned result that any finite upper semimodular is admissible, Rival [1976a], Theorem 2, proved that a finite lattice L is upper semimodular if and only if every injective order-preserving map $\omega : J(L) \to \mathbb{P}$ is admissible .

For an elementary proof that shellable posets are Cohen–Macaulay we refer to Garsia [1980]. In particular, Garsia derives the Cohen–Macaulayness of distributive and of upper semimodular lattices.

Abels [1991] investigates the geometry of the so-called chamber system of a semimodular lattice.

Figure 4.10 indicates that CM posets may be viewed as far-reaching generalizations of finite upper semimodular lattices. In fact, we may interpret the homotopy theorem for CM posets (cf. Baclawski [1980], Theorem 8.1) as implying that these posets obey a "very" weak semimodularity law [in fact, a weakening of Birkhoff's condition (Bi)]: If x and y cover z, then there exists a path of coverings from x to y and all these paths are "combinatorially" homotopic (cf. Figure 4.12).

CM posets have been further generalized to Buchsbaum posets (see Stückrad & Vogel [1986]).

Exercises

1. The notion of semimodularity can be generalized from lattices to posets in many ways. One possibility is to define a finite poset P as (upper) semimodular if whenever x and y cover z in P, then there exists an element w in P that covers both x and y. Give an example of a finite poset that is semimodular in this sense but not Cohen–Macaulay (Baclawski [1980]). Note that this is in contrast to the situation in which the poset is a lattice: in the lattice case semimodularity is inherited by every closed interval, which is not so in the case of posets.

2. A finite poset P is said to be locally semimodular if every closed interval of P is a semimodular poset (as defined in Exercise 1). Show that if P is locally semimodular poset, then $\hat{P}$ (i.e. the poset obtained by adjoining new elements $\hat{0}$ and $\hat{1}$ such that $\hat{0} < x < \hat{1}$ for all $x \in P$) is shellable and hence CM (Baclawski [1980], Proposition 4.3).

3. Give an example of shellable poset that is not edgewise lexicographically shellable (Vincze & Wachs [1985]).

References

Abels, H. [1991] The geometry of the chamber system of a semimodular lattice, Order 8, 143–158.

Baclawski, K. [1980] Cohen–Macaulay ordered sets, J. Algebra 63, 226–258.

Björner, A. [1980] Shellable and Cohen–Macaulay partially ordered sets, Trans. Amer. Math. Soc. 260, 159–183.

Björner, A. [1992] Homology and shellability of matroids and geometric lattices, in: White [1992], pp. 226–283.

Björner, A., A. Garsia, and R. Stanley [1982] An introduction to Cohen–Macaulay partially ordered sets, in: Rival [1982a], pp. 583–615.

Bruggesser, H. and P. Mani [1971] Shellable decompositions of cells and spheres, Math. Scand. 29, 197–205.

Folkman, J. [1966] The homology groups of a lattice, J. Math. Mech. 15, 631–636.

Garsia, A. [1980] Combinatorial methods in the theory of Cohen–Macaulay rings, Adv. in Math. 38, 229–266.

Hibi, T. [1992] *Algebraic Combinatorics on Convex Polytopes*, Carslaw, Glebe, Australia.

Provan, S. [1977] Decompositions, shellings, and diameters of simplicial complexes and convex polyhedra, Thesis, Cornell Univ., Ithaca, N.Y.

Reisner, G. [1976] Cohen–Macaulay quotients of polynomial rings, Adv. in Math. 21, 30–49.

Rival, I. [1976a] A note on linear extensions of irreducible elements in a finite lattice, Algebra Universalis 6, 99–103.

Stanley, R. [1972] Supersolvable lattices, Algebra Universalis 2, 197–217.

Stanley, R. [1974] Finite lattices and Jordan–Hölder sets, Algebra Universalis 4, 361–371.

Stanley, R. [1977] Cohen–Macaulay complexes, in: *Higher Combinatorics* (ed. M. Aigner), Reidel, Dordrecht, pp. 55–62.

Stanley, R. [1983] *Combinatorics and Commutative Algebra*, Progress in Mathematics, Vol. 41, Birkhäuser, Basel.

Stückrad, J. and W. Vogel [1986] *Buchsbaum Rings and Applications*, Springer-Verlag, Berlin.

Vincze, A. and M. Wachs [1985] A shellable poset that is not lexicographically shellable, Combinatorica 5, 257–260.

4.5 Consistent Lattices

Summary. Following Kung [1985], we consider here the concept of consistency. This property is a direct consequence of Dedekind's isomorphism theorem, and it generalizes modularity in a different direction than semimodularity or supersolvability. We give several examples and some ways of obtaining new consistent lattices from old ones. Finally we consider the relationship of consistency with the Kurosh–Ore replacement property and a property due to Crawley. This relationship will play an important role in Chapter 6 (Sections 6.1 and 6.5) and Chapter 8.

Let L be a lattice of finite length. A join-irreducible j of L is said to be *consistent* if, for every element x in L, the element $x \vee j$ is a join-irreducible in the upper interval $[x, 1]$. A lattice of finite length is said to be *consistent* if every of its join-irreducible elements is consistent. A lattice L of finite length is said to be *dually consistent* if its dual L^* is consistent. More generally, Gragg & Kung [1992] define consistency relative to a property (P) that may or may not be satisfied by a lattice element (e.g. the property of covering at most k elements): An element j of a lattice of finite length L is said to be *consistent relative to the property* (P) if j satisfies (P) and, for every element x in L, $x \vee j$ satisfies (P) in the upper interval $[x, 1]$.

It is clear that modular lattices of finite length and geometric lattices are consistent upper semimodular lattices. The centered hexagon S_7 (cf. Figure 4.13) is an upper semimodular lattice that is not consistent: the element d is a join-irreducible, but not a consistent join-irreducible, since $b \vee d (= 1)$ is not join-irreducible in the upper interval $[b, 1]$. S_7^*, the dual of S_7, is consistent and lower semimodular. Recall that S_7^* is a lower locally distributive (meet-distributive) lattice. It is easy to see that any finite lower locally distributive lattice is consistent. A direct proof

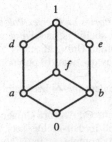

Figure 4.13

of this is left as an exercise. We also refer to Corollary 4.6.3, where this result will follow in a broader context.

Let us introduce some additional notation. Let L be a lattice of finite length. By $J(1, L)$ we denote the *set of all join-irreducibles of L*, that is, the set of all elements of L covering at most one element. Similarly we denote by $M(1, L)$ *the set of all meet-irreducibles of L*, that is, the set of all elements of L covered by at most one element. Thus we have $J(1, L) = J(L) \cup \{0\}$ and $M(1, L) = M(L) \cup \{1\}$. By $J_c(1, L)$ we denote the set of all consistent join-irreducibles of L. With this notation, we have for the lattice S_7 of Figure 4.13

$$J_c(1, S_7) = \{0, a, b\}, \qquad J(1, S_7) = \{0, a, b, d, e\}, \qquad M(1, S_7) = \{f, d, e, 1\}.$$

Note that $|J_c(1, S_7)| \le |M(1, S_7)|$. A similar inequality holds in arbitrary finite lattices, that is, if L is a finite lattice then $|J_c(1, L)| \le |M(1, L)|$. This statement is part of a general theory developed by Kung in a series of papers (see Kung [1985], [1986b], [1987]). Parts of this theory will be considered in more detail in Chapter 6 (Sections 6.1 and 6.5) and Chapter 8 (Section 8.4).

Figure 4.14 shows a consistent lower semimodular lattice. We have already noted (see Section 1.7) that the lattice of Figure 4.14 (which is the dual of the lattice shown in Figure 1.27, Section 1.7) is isomorphic to the lattice of subgroups of the group D_4, the group of symmetries of the square. Since every subgroup is a subnormal subgroup, this lattice is also isomorphic with the lattice of subnormal subgroups of D_4. We refer to Section 8.3 for a formal definition of subnormal subgroups and for more details on the lattice of subnormal subgroups of a group.

The pentagon N_5 and the lattice $L(A_4)$ of subgroups of the alternating group A_4 (cf. Fig. 4.15) are consistent. The lattices N_5 and $L(A_4)$ are neither upper nor lower semimodular; in fact, they do not even satisfy the Jordan–Dedekind chain condition. In particular, $L(A_4)$ is not a supersolvable lattice (the alternating group A_4 is solvable but not supersolvable). N_5 shows that consistency together with dual consistency does not imply supersolvability. We also note that supersolvability does not imply consistency (see Figure 4.4, Section 4.3).

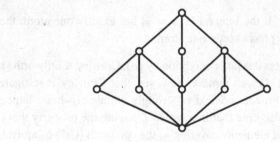

Figure 4.14

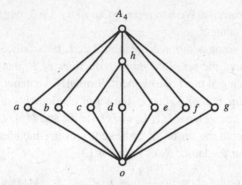

Figure 4.15

There are several ways of making new consistent lattices from old ones. For example, the direct product of two consistent lattices is consistent. Let K and L be disjoint lattices. The *concatenation* $K * L$ is the lattice obtained by identifying 1_K and 0_L and putting $x \geq y$ for all $x \in L$ and $y \in K$. If K and L are consistent, then so is $K * L$. Let $L_1, L_2, \ldots, L_k$ be pairwise disjoint lattices. The *parallel union* $L_1 \uparrow L_2 \uparrow \cdots \uparrow L_k$ is obtained by identifying the least elements of these lattices and adding a new greatest element $\tilde{1}$ such that $\tilde{1} \geq x$ for all elements $x \in L_i$ (and no other new order relations). For example, each of the lattices N_5 and $L(A_4)$ is a parallel union of modular lattices. The parallel union of finitely many pairwise disjoint consistent lattices is consistent.

Reuter [1989] provided a construction principle for finite consistent upper semi-modular lattices by gluing finite geometric lattices. We shall have a look at Reuter's construction in the following section.

In a lattice L of finite length, consistency is equivalent to the following condition (Cr), which was first investigated by Crawley [1961]:

(Cr) For all $x, y \in L$, if the interval $[x \wedge y, y]$ has exactly one dual atom, then the interval $[x, x \vee y]$ has exactly one dual atom.

In fact, Crawley formulates the dual condition

(Cr*) For all $x, y \in L$, if the interval $[x, x \vee y]$ has exactly one atom, then the interval $[x \wedge y, y]$ has exactly one atom.

Crawley [1961] denotes this latter condition by (ρ); Crawley & Dilworth [1973], p. 53, denote it by (*). Let us remark, however, that Crawley investigates the condition (Cr*) in the broader context of strongly atomic algebraic lattices (cf. also Section 8.1). We also note that in a lattice L having the property that for all $y > x$ in L, y is a join of elements covering x, the condition (Cr*) is equivalent to lower semimodularity. In particular, a geometric lattice satisfying (Cr*) is modular.

Crawley used his condition (Cr*) in order to characterize the Kurosh–Ore replacement property for meet decompositions, ∧-KORP (see Section 1.6). Although their origins are very different, dual consistency and the ∧-KORP are equivalent:

Theorem 4.5.1 *Let L be a lattice of finite length. Then the following three conditions are equivalent:*

(i) *L has the* ∧-KORP;
(ii) *L satisfies* (Cr*);
(iii) *L is dually consistent.*

The equivalence of (i) and (ii) is due to Crawley [1961], who proved this more generally for strongly atomic algebraic lattices (see also Crawley & Dilworth [1973], Theorem 5.5). The equivalence of (ii) and (iii) is implicit in Crawley's results. For lattices in which every element has a finite join decomposition, the equivalence of (iii) and (i) was explicitly observed by Reuter [1989]. Dualizing the preceding theorem, we get that, in lattices of finite length, consistency, (Cr), and the ∨-KORP are equivalent.

Let us make now some remarks on the ∨-KORP from the viewpoint of closure structures. For a finite set X, let (X, cl) denote a closure structure on X. We consider the following exchange property: Let $\mathrm{cl}(A) = \mathrm{cl}(B)$. Then for $a \in A$ there exists $b \in B$ such that $\mathrm{cl}(A) = \mathrm{cl}((A - \{a\}) \cup \{b\})$. This exchange property, which may or may not hold for the closure operator cl, is also called the *Kurosh–Ore exchange property* or *basis exchange property* (cf. Kung [1986c]). A closure structure has

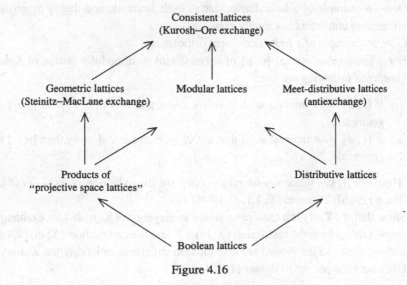

Figure 4.16

the Kurosh–Ore exchange property if and only if the lattice of its closed subsets has the $\vee$-KORP, that is, if and only if this lattice is consistent.

In Section 1.7 we indicated the connection between geometric lattices and the Steinitz–Mac Lane exchange property. The so-called antiexchange property was mentioned in the Notes to Section 2.5. The antiexchange property is related to lower locally distributive (meet-distributive) lattices (for more details see Section 7.3). The interrelationships between some classes of lattices mentioned before are visualized in Figure 4.16 (cf. Reuter [1989]).

Notes

Reuter [1989] has shown that the construction of minors, as done in matroid theory (cf. Brylawski [1986]), also works in the case of closure structures having the Kurosh–Ore exchange property: Let (X, cl) be a closure structure and $Y \subseteq X$. The set Y together with the operator $\text{cl}_Y : A \to \text{cl}(A) \cap Y$ for $A \subseteq Y$ is called the *restriction* of (X, cl) to Y and is denoted by $(X, \text{cl}) \mid Y$. The set $X - Y$ together with the operator $\text{cl}_{X/Y} : A \to \text{cl}(A \cup Y) \backslash Y$ for $A \subseteq X \backslash Y$ is called the *contraction* of (X, cl) through Y, denoted by $(X, \text{cl})/Y$. Now if (X, cl) is a closure structure satisfying the Kurosh–Ore exchange property then both the restriction $(X, \text{cl}) \mid Y$ and the contraction $(X, \text{cl})/Y$ are also closure structures satisfying the Kurosh–Ore exchange property (Reuter [1989], Theorem 2). The proofs are left as exercises.

Exercises

1. Give an example of a finite lattice that is both atomistic and dually atomistic but neither consistent nor dually consistent.
2. Give an example of a finite consistent ortholattice.
3. For a geometric interval $[a, b]$ of a consistent semimodular lattice of finite length the following are true:

   (i) If $[c, d]$ is an interval such that $b \vee c = d$ and $b \wedge c = a$, then $[c, d]$ is geometric.
   (ii) If $[c, d]$ is an interval such that $a \wedge d = c$ and $a \vee d = b$, then $[c, d]$ is geometric.

   [Hint: For (i) use induction on $r(c) - r(a)$, for (ii) induction on $r(b) - r(d)$.] (Race [1986], Theorem IV. 3.1, p. 48–49.)
4. Show that if (X, cl) is a closure structure satisfying the Kurosh–Ore exchange property then both the restriction $(X, \text{cl}) \mid Y$ and the contraction $(X, \text{cl})/Y$ (as defined above in the Notes) are also closure structures satisfying the Kurosh–Ore exchange property (Reuter [1989]).

References

Brylawski, T. [1986] Constructions, in: White [1986], pp. 127–223.

Crawley, P. [1961] Decomposition theory for non-semimodular lattices, Trans. Amer. Math. Soc. 99, 246–254.

Crawley, P. and R. P. Dilworth [1973] *Algebraic Theory of Lattices*, Prentice-Hall, Englewood Cliffs, N.J.

Gragg, K. and J. P. S. Kung [1992] Consistent dually semimodular lattices, J. Combin. Theory Ser. A 60, 246–263.

Kung, J. P. S. [1985] Matchings and Radon transforms in lattices I. Consistent lattices, Order 2, 105–112.

Kung, J. P. S. [1986b] Radon transforms in combinatorics and lattice theory, in: *Combinatorics and Ordered Sets*, Contemporary Mathematics, Vol. 63 (ed. I. Rival), pp. 33–74.

Kung, J. P. S. [1986c] Basis-exchange properties, in: White [1986], pp. 62–75.

Kung, J. P. S. [1987] Matchings and Radon transforms in lattices II. Concordant sets, Math. Proc. Camb. Phil. Soc. 101, 221–231.

Race, D. M. [1986] Consistency in Lattices, Ph.D. thesis, North Texas State Univ., Denton, Tex.

Reuter, K. [1989] The Kurosh–Ore exchange property, Acta Math. Hungar. 53, 119–127.

White, N. L. (ed.) [1986] *Theory of Matroids*, Encyclopedia of Mathematics and its Applications, Vol. 26, Cambridge Univ. Press, Cambridge.

4.6 Strong Lattices and Balanced Lattices

Summary. The concept of strongness is due to Faigle [1980a]. We give an equivalent definition and some examples. Then we characterize strongness by means of special forbidden pentagons. Neither of the properties "strong" and "consistent" implies the other one. Similarly, neither of the properties "strong" and "supersolvable" implies the other one. On the other hand, it will be shown that a finite semimodular lattice is strong if and only if it is consistent. This result follows from Reuter's method of gluing finite geometric lattices (Reuter [1989]).

Let L be a lattice of finite length. A join-irreducible $j \, (\neq 0)$ is called *strong* if, for all $x \in L$

$$j \leq x \vee j' \;\Rightarrow\; j \leq x.$$

Recall that j' denotes the uniquely determined lower cover of j. A lattice of finite length is called *strong* if each of its join-irreducibles $(\neq 0)$ is strong. A lattice L of finite length is said to be *dually strong* if the order dual L^* of L is strong.

It is obvious that any atomistic lattice of finite length – in particular, any geometric lattice – is strong. It is also readily checked that any modular lattice of finite length is strong. The pentagon N_5 is consistent and dually consistent, but neither strong nor dually strong. The lattice in Figure 4.17 (cf. Reuter [1989]) is both strong and dually strong, but neither consistent nor dually consistent; we also note that this lattice is neither upper nor lower semimodular (in fact, it does not even satisfy the Jordan–Dedekind chain condition).

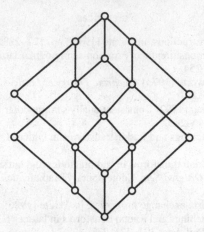

Figure 4.17

An upper semimodular lattice of finite length may be strong (as in the case of geometric lattices or modular lattices) or may not be strong. For example, the centered hexagon S_7 is an upper semimodular lattice that is not strong. We already noted in Section 4.5 that S_7 is not consistent. In fact, it will be shown (see Theorem 4.6.8 below) that a semimodular lattice of finite length is strong if and only if it is consistent.

Let us also note that strongness and supersolvability are unrelated, that is, neither of these two concepts implies the other one. The lattice of Figure 4.4 (Section 4.3) is supersolvable but not strong. On the other hand, we shall see that any lower semimodular lattice is strong (cf. Corollary 4.6.3). Since there are lower semimodular lattices that are not supersolvable (it is left as an exercise to give an example), we see that strongness does not imply supersolvability.

The following characterization of strongness by means of forbidden special pentagons was given in Richter & Stern [1984]. The proof is left as an exercise.

Lemma 4.6.1 *A lattice L of finite length is strong if and only if it does not contain a special pentagon sublattice of the form represented in Figure 4.18 with $j \in J(L)$.*

Let now L be a lattice of finite length, and let $a \in L$, $a \neq 0$. We have already associated with a the element $a_+ = \bigwedge (c \in L : c \prec a)$, that is, the meet of all lower covers of a in L. In case $j \in J(L)$ (that is, j is a join-irreducible $\neq 0$), we used the notation j' for the uniquely determined lower cover of j. Sometimes j' is also called the *derivation* of j. Trivially we have the equality $j' = j_+$ for any join irreducible j ($\neq 0$) in a lattice of finite length. Our aim is to extend the operation of derivation from join irreducibles ($\neq 0$) to arbitrary elements a ($\neq 0$) in such a way that the equality $a' = a_+$ holds for consistent semimodular lattices (see Section 6.5).

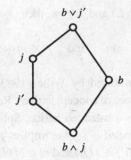

Figure 4.18

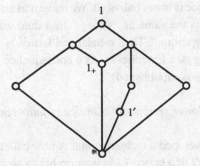

Figure 4.19

With this in mind we define the *derivation* of an element a ($\neq 0$) by

$$a' = \bigvee (j' \; : \; j \in J(L), \; j \leq a).$$

This concept of derivation has its roots in lattice-theoretic investigations of the subgroup lattice of abelian groups. For some general properties of the derivation and for further references see for example Richter [1983].

In arbitrary lattices of finite length, the elements a' and a_+ need not be comparable, as Figure 4.19 shows. On the other hand we have

Lemma 4.6.2   *A lattice of finite length is strong if and only if $a' \leq a_+$ holds for all a ($\neq 0$).*

The proof is left as an exercise (see Exercise 3). In view of Exercise 4, the preceding result generalizes Reuter [1989], Lemma 6. In Chapter 6 (Corollary 6.5.3) we shall see that, in a semimodular lattice of finite length, the reverse inequality $a_+ \leq a'$ also holds for all a ($\neq 0$).

Recall that m^* denotes the uniquely determined upper cover of $m \in M(L)$. For a lattice L of finite length let us now define the *arrow relations* between $J(L)$ and

$M(L)$: If $j \in J(L)$ and $m \in M(L)$ and $j \not\leq m$, then

$$j \nearrow m \Leftrightarrow j \vee m = m^* \quad \text{and} \quad j \swarrow m \Leftrightarrow j \wedge m = j'.$$

The arrow relations were introduced by Wille [1983] in his investigations on subdirect-product decompositions of concept lattices. Reuter [1989] used the arrow relation to define what he calls a balanced lattice. Splitting up Reuter's concept, we call the lattice L *lower balanced* if $j \nearrow m$ implies $j \swarrow m$ and *upper balanced* if $j \swarrow m$ implies $j \nearrow m$ for all $j \in J(L)$ and $m \in M(L)$. A lattice of finite length is said to be *balanced* if it is both upper and lower balanced.

It is obvious that an upper semimodular lattice is upper balanced. Similarly, a lower semimodular lattice is lower balanced. We leave it as an exercise to show that "lower balanced" means the same as "strong." In a dual way, "upper balanced" is equivalent to "dually strong." Thus a balanced lattice is a lattice that is both strong and dually strong (see Exercise 4). As a consequence we get the following observation due to Faigle (unpublished).

Corollary 4.6.3 *Any lower semimodular lattice of finite length is strong.*

Dilworth [1941a] developed a technique that is now called *Dilworth gluing*: If a nonempty dual ideal D of a lattice L_0 is isomorphic to an ideal I of a lattice L_1, let L be the union of L_0 and L_1 with the elements of D and I identified via the isomorphism. L can be ordered with the transitive closure of the union of the orders on L_0 and L_1. Moreover, L is a lattice with respect to this order. Dilworth showed that L is modular provided both L_0 and L_1 are modular. Hall & Dilworth [1944] further developed the gluing construction to show that there are finite modular lattices that cannot be embedded into a complemented modular lattice.

An extension of Dilworth's gluing is due to Herrmann [1973]. Herrmann's method (S-*glued sums of lattices*) provided a tool for examining the structure of modular lattices of finite length as well as for the construction of examples and counterexamples. For a survey we refer to Day & Freese [1990]. Herrmann's method has in turn been generalized in several ways. Here we consider a generalization due to Reuter [1989] whose approach will lead to a structural description of finite consistent semimodular lattices as glued by finite geometric lattices.

Herrmann's S-glued sums of lattices can be treated within the framework of tolerance relations, which will be done here. The connection between gluing and tolerance relations was studied by Wille [1983]. Let us first recall some facts. For more details on tolerances in general see Chajda [1991]; for tolerances on lattices see Chajda & Zelinka [1974], Bandelt [1981], and Czédli [1982].

A binary relation Θ on a lattice L is called a *tolerance relation* if Θ is reflexive, symmetric, and compatible with the lattice operations, that is, $x_1 \Theta y_1$ and $x_2 \Theta y_2$

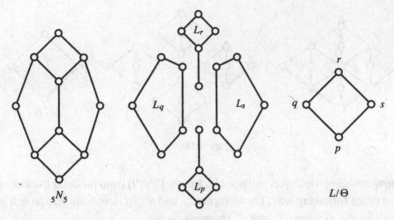

Figure 4.20

imply $(x_1 \wedge x_2) \Theta (y_1 \wedge y_2)$ and $(x_1 \vee x_2) \Theta (y_1 \vee y_2)$. We define

$$a_\Theta = \bigwedge(x \in L : a \Theta x) \quad \text{and} \quad a^\Theta = \bigvee(x \in L : a \Theta x)$$

for $a \in L$ and call the intervals $[a]_\Theta = [a_\Theta, (a_\Theta)^\Theta]$ the *blocks* of Θ. The compatibility property of Θ makes it possible to build a factor lattice L/Θ whose elements are the blocks of Θ (see Czédli [1982]). For a given lattice $Q \cong L/\Theta$ we describe the blocks of Θ by indexing L with the elements of Q, that is, by L_q ($q \in Q$).

It is clear from the definition that congruence relations are special tolerance relations. While a congruence relation induces a partition, we shall be concerned here with overlapping subsets induced by so-called glued tolerance relations: A tolerance relation Θ is called a *glued tolerance relation* if $L_p \cap L_q \neq \emptyset$ for all $p, q \in Q$ with $p \prec q$. The above described procedure is illustrated in Figure 4.20 by means of the lattice $_5N_5$.

Next we recall that a glued tolerance relation Θ of a finite lattice L contains all covering pairs of L. Moreover this property characterizes glued tolerance relations (cf. Wille [1985], Proposition 9). From this characterization it follows that there exists a smallest glued tolerance relation of L, denoted by $\Sigma(L)$ and called the *skeleton* of L. How do we find the skeleton relation $\Sigma(L)$ of a given finite lattice L?

If L is modular, an answer was given by Herrmann [1973]: the *blocks* of $\Sigma(L)$ are the maximal atomistic intervals of L. Namely, we know that $(x, x^+) \in \Sigma(L)$ and $(x_+, x) \in \Sigma(L)$ for each $x \in L$, because $\Sigma(L)$ contains all covering pairs of L and is compatible with $\wedge$ and $\vee$. Hence the intervals $[x_+, (x_+)^+]$ are contained in the blocks of $\Sigma(L)$. Now if L is modular, it can be shown that these intervals are exactly the blocks of L (Herrmann [1973], Lemma 6.1). We illustrate this situation in Figure 4.21.

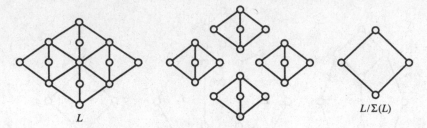

Figure 4.21

Herrmann's approach was extended by Reuter [1989] from modular to balanced lattices in the following way. The elements x_+ and x^+ in Herrmann's approach are replaced by the elements x' and x^*, respectively:

$$x' = \bigvee (j' : j \in J(L), \ j \leq x) \quad \text{and} \quad x^* = \bigwedge (m^* : m \in M(L), \ x \leq m).$$

Following Reuter [1989], we show that in a finite balanced lattice the intervals $[x', (x')^*]$ are exactly the blocks of $\Sigma(L)$. As a preparation we relate the mappings $\sigma : x \to x'$ and $\pi : x \to x^*$ to the property of being balanced:

Lemma 4.6.4 *A finite lattice L is balanced if and only if $a' \leq b \Leftrightarrow a \leq b^*$ holds for all $a, b \in L$, that is, (σ, π) is a mixed Galois connection.*

We leave the proof as an exercise. For the notion of a *mixed Galois connection (Galois pair)* see for example Aigner [1979].

Corollary 4.6.5 *Let L be a finite balanced lattice. Then the following statements and their duals hold for all $a, b \in L$:*

  (i) *$a' \leq a$.*
 (ii) *$a \leq (a')^*$.*
(iii) *$a \leq b$ implies $a' \leq b'$.*
 (iv) *$(a \vee b)' \leq a' \vee b'$.*

The first statement is obvious from the definition of a'. The other assertions are standard results of Galois connections.

For the following result see Reuter [1989], Theorem 3.

Theorem 4.6.6 *Let L be a finite balanced lattice. Then the intervals $[a', (a')^*]$ $(a \in L)$ are the blocks of the skeleton relation $\Sigma(L)$.*

Proof. Let $a' \leq x_1, x_2 \leq (a')^*$ and $b' \leq y_1, y_2 \leq (b')^*$. Then $(a \vee b)' = a' \vee b' \leq x_1 \vee y_1$ [or else $x_2 \vee y_2 \leq (a')^* \vee (b')^* \leq (a' \vee b')^* = ((a \vee b)')^*$] and $((a' \wedge b')^*)' \leq a' \wedge b' \leq x_1 \wedge y_1$ [or else $x_2 \wedge y_2 \leq (a')^* \wedge (b')^* = (a' \wedge b')^*$]. Hence the intervals $[a', (a')^*]$ are the blocks of a tolerance relation Θ. We proceed to show that Θ is a glued tolerance relation. For $j \in J(L)$ we clearly have $(j', j) \in \Theta$. For a covering pair $x \prec y$ there exists a $j \in J(L)$ such that $j' \leq x$ and $j \vee x = y$. Thus $(j', j) \in \Theta$ implies $(j \vee x, j' \vee x) = (x, y) \in \Theta$ and Θ must be the skeleton relation. ∎

We also note (cf. Reuter [1989], Lemma 5):

Lemma 4.6.7 *Let L be a finite lattice such that all blocks of $\Sigma(L)$ are both atomistic and dually atomistic. Then L is a balanced lattice.*

Proof. We show that $j \nearrow m$ implies $j \swarrow m$ for $j \in J(L)$ and $m \in M(L)$, that is, L is upper balanced. Let $j \in J(L)$ and $m \in M(L)$ such that $j \nearrow m$, that is, $j \vee m = m^*, j \nleq m$.

From $(m, m^*) \in \Sigma(L)$ it follows that $(m \wedge j, j) \in \Sigma(L)$, which means that there exists an atomistic block containing $j \wedge m$ and j. Since j is join-irreducible, it must be an atom of this block. Hence $j \wedge m = j'$. ∎

The lattice of Figure 4.17 shows that the converse to this lemma does not hold.

The method of gluing developed above will now be applied to finite geometric lattices. A finite lattice L is said to be *glued by geometric lattices* if all blocks of $\Sigma(L)$ are geometric lattices. These lattices are characterized by the following result due to Faigle and Reuter, which also shows that, in finite semimodular lattices, consistency, strongness, and the property of being balanced have the same meaning.

Theorem 4.6.8 *For a finite lattice L the following conditions are equivalent:*

  (i) *L is semimodular and has the $\vee$-KORP;*
 (ii) *L is glued by geometric lattices;*
(iii) *L is a semimodular and balanced;*
 (iv) *L is semimodular and consistent;*
  (v) *L is semimodular and strong.*

Remark. The equivalence of conditions (i), (iii), (iv), and (v) is also valid for lattices of finite length (and for even more general lattices – see the Notes below).

Sketch of proof. (i) $\Leftrightarrow$ (iv): Follows from the dual of Crawley's result characterizing the Kurosh–Ore replacement property for meet decompositions (Theorem 4.5.1).

(ii) $\Rightarrow$ (iii): By Lemma 4.6.7, L is a balanced lattice. To see that L is semimodular, let for $x \wedge y \prec x, y$ for $x, y \in L$. Since $\Sigma(L)$ contains all covering pairs, there exists a block that contains and $x \wedge y, x, y$, and $x \vee y$. Thus $x \vee y \succ x, y$ follows from the semimodularity of this block.

(iii) $\Rightarrow$ (ii): The blocks of $\Sigma(L)$ are of the form $[a', (a')^*]$ or, equivalently, $[(a^*)', a^*]$. They are semimodular, since they are intervals of the semimodular

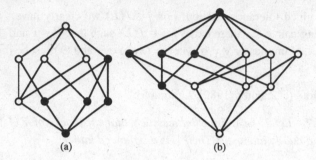

Figure 4.22

lattice L. It remains to show that they are atomistic. Set $b = a^*$, and let $x \in [b', b]$. Then $x = \bigvee\{j \in J(L) : j \leq x\} = \bigvee\{j \vee b' : j \in J(L), \ j \leq x\}$. The elements $j \vee b'$ that are not b' itself cover b', by the semimodularity of L. Thus x is b' or x is the join of atoms of $[b', b]$.

(iii) $\Rightarrow$ (iv): Suppose L is not consistent. Then there are elements $j \in J(L)$ and $x \in L$ such that $x \vee j \notin J([x, 1])$. By semimodularity $x \vee j$ covers $x \vee j'$. Because $x \vee j \notin J([x, 1])$, there exists besides $x \vee j'$ another lower cover, say c, of $x \vee j$ with $c \geq x$. From $x \vee j' \not\leq c$ we get $j' \not\leq c$. Choose a maximal element m such that $m \geq c$ and $m \not\geq x \vee j$. It follows that m is meet irreducible. Moreover, $m \not\geq j'$ (since otherwise m and $x \vee j$ do not have a unique greatest lower bound). Thus $j \vee m = m^*$ but $j \wedge m \neq j'$.

(iv) $\Rightarrow$ (v): Suppose L is semimodular and consistent but not strong. Let $x \in L$ be of minimal height such that there are $j_1, j_2 \in J(L)$ with $j_1 < j_2$ and $j_1, j_2 \not\leq x$ and $x \vee j_1 = x \vee j_2$. Then $x \neq 0$ and if y is covered by x, then $y \vee j_1 < y \vee j_2$. Now $y \vee j_1 \neq x \vee j_1$, since otherwise $y \vee j_1 = y \vee j_2$. By semimodularity $y \vee j_1$ is covered by $x \vee j_1$. It follows that $x \vee j_2 = x \vee j_1 = y \vee j_2$. Now $y \vee j_2 \notin J([y, 1])$ because of $y \vee j_2 = x \vee (y \vee j_1)$, a contradiction.

(v) $\Rightarrow$ (iii): By semimodularity $j \wedge m = j'$ implies $j \vee m = m^*$ for $j \in J(L), m \in M(L)$. Suppose L is a strong semimodular lattice, but $j \vee m = m^*$ and $j' \not\leq m$ for some $j \in J(L), m \in M(L)$. There exists a $j_1 \in J(L)$ such that $j_1 \leq j'$ and $j' \not\leq m$. Now $j_1 \vee m \geq j$ and $m \not\geq j$ yield a contradiction. ∎

The equivalence of (i), (iv), and (v) was shown by Faigle [1980a]. The preceding proof is due to Reuter [1989], Theorem 3.

As an illustration consider the geometric lattices L_1 and L_2 of Figure 4.22(a) and (b), respectively. These lattices appear already in Section 1.7 (Figure 1.23 and Figure 1.24, respectively), but here we denote by solid dots the ideal of L_1 and the dual ideal of L_2 that are identified by the gluing procedure. As a result of the gluing procedure we get the lattice shown in Figure 4.23.

For further properties of consistent semimodular lattices see Section 6.5. In Figure 4.24 we visualize the interrelationships between several concepts of

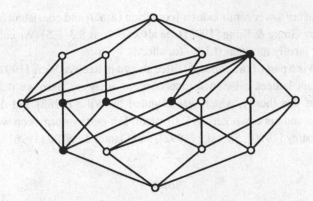

Figure 4.23

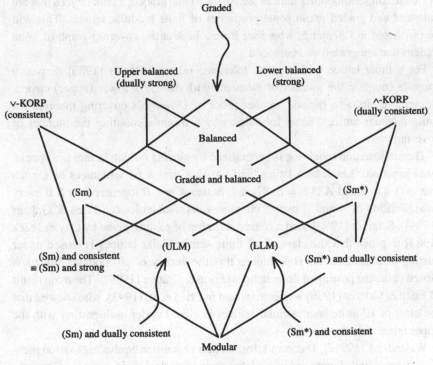

Figure 4.24

Sections 4.5 and 4.6 and some concepts mentioned earlier. Let us give some more comments concerning Figure 4.24.

Lattices that are upper semimodular [condition (Sm)] and dually consistent are locally modular in the sense of Crawley & Dilworth [1973] (cf. Section 8.2). More precisely, we called them upper locally modular (ULM, for short).

Lattices that are lower semimodular [condition (Sm*)] and consistent have been investigated by Gragg & Kung [1992] (see also Sections 8.3–8.5). We called these lattices lower locally modular (LLM, for short).

Figure 4.23 is a part of the concept lattice shown in Reeg & Weiß [1991], Figure 6.10.4. Concept lattices arise in formal concept analysis. Here we don't make explicit use of this theory, which was founded by Wille [1982] and developed mainly by him and his coworkers in Darmstadt. For more information we refer to Davey & Priestley [1990], Skorsky [1992], and Ganter & Wille [1996].

Notes

For a brief survey of the impact of Dilworth's work on the Kurosh–Ore theorem in semimodular lattices see Stern [1990b].

Consistent semimodular lattices are balanced and graded. Finite lattices that are balanced and graded retain some properties of finite modular lattices. This will be illustrated in Chapter 5, where we have a look at the covering graph of finite lattices that are graded and balanced.

For a finite lattice with a glued tolerance relation, Reuter [1985] derived a formula counting the number of elements with exactly k lower (upper) covers. Reuter's results also include another proof of Dilworth's covering theorem for finite modular lattices, sieve formulas, and a formula counting the number of covering pairs.

The construction by gluing is generalized by *pasting* (which in turn is a special amalgamation): Let L be a lattice, and let A, B, and S be sublattices of L such that $A \cap B = S$ and $A \cup B = L$. Then L pastes A and B together over S if every amalgamation of A and B over S contains L as a sublattice (cf. Fried & Grätzer [1989]). Schmidt [1990] used a characterization of pasting given by Day & Ježek [1983] to prove that the class of all finite semimodular lattices is closed under pasting of finite lattices. This implies that the variety of all modular lattices is closed under the pasting of finite lattices (Fried & Grätzer [1989]). The main result of Fried & Grätzer [1989] was generalized in Fried et al. [1993], who showed that the class of all finite semimodular lattices is closed under multipasting with the upper interpolation property.

Walendziak [1994c], Theorem 1, formulated a condition equivalent to strongness for lattices of finite length and proved that a semimodular lower continuous strongly coatomic lattice is consistent if and only if it is strong. (Note that for lower continuous strongly coatomic lattices consistency can be defined in the same way as for lattices of finite length.) The results of Walendziak [1994c], Theorem 1, and Richter [1982], Theorem 11, imply that conditions (i), (iv), and (v) in Theorem 4.6.8 remain equivalent in case "finite" is replaced by "lower continuous + strongly coatomic."

We also note the following characterization of upper semimodularity due to Faigle [1980a] and Faigle & Herrmann [1981]. This characterization involves only join-irreducibles and meet-irreducibles: A lattice L of finite length is upper semimodular if and only if for all $m_1, m_2 \in M(L)$ and for all $g, h \in J(L)$ the relations $g \swarrow m_1, g \swarrow m_2, h \leq m_1, h \nleq m_2$ imply the existence of an $m \in M(L)$ such that $h \nleq m, g \leq m$, and $m_1 \wedge m_2 \leq m$. This condition turns out to be particularly suitable for testing upper semimodularity in a finite lattice (cf. Skorsky [1992].)

Exercises

1. Give an example of a lower semimodular (and hence strong) lattice that is not supersolvable.
2. Prove the characterization of strongness by forbidden special pentagons (Lemma 4.6.1)
3. Prove that a lattice of finite length is strong if and only if $a' \leq a_+$ holds for all a ($\neq 0$) (Lemma 4.6.2).
4. Prove that a lattice of finite length is balanced if and only if it is both strong and dually strong.
5. A finite lattice L is balanced if and only if $a' \nleq b \Leftrightarrow a \leq b^*$ holds for all $a, b \in L$, that is, (σ, π) is a mixed Galois connection (Galois pair) (Lemma 4.6.4).
6. If L is a balanced lattice of finite length, then $l(M(L)) = l(J(L))$ with l denoting the length of the corresponding poset. [For modular lattices this is derived in Ganter & Rival [1975] from the inequality $l(M(L)) \leq l(J(L))$, which they show to hold in a semimodular lattice L of finite length.]

References

Aigner, M. [1979] *Combinatorial Theory*, Springer-Verlag, Berlin.

Bandelt, H.-J. [1981] Tolerance relations on lattices, Bull. Austral. Math. Soc. 23, 367–381.

Bogart, K. P., R. Freese, and J. P. S., Kung (eds.) [1990] *The Dilworth Theorems. Selected Papers of Robert P. Dilworth*, Birkhäuser, Boston.

Chajda, I. [1991] *Algebraic Theory of Tolerance Relations*, Univ. Palackeho. Olomous.

Chajda, I. and B. Zelinka [1974] Tolerance relations on lattices, Časop. Pešt. Mat. 99, 394–399.

Crawley, P. and R. P. Dilworth [1973] *Algebraic Theory of Lattices*, Prentice-Hall, Englewood Cliffs, N.J.

Czédli, G. [1982] Factor lattices by tolerances, Acta Sci. Math. (Szeged) 44, 35–42.

Davey, B. A. and H. A. Priestley [1990] *Introduction to Lattices and Order*, Cambridge Univ. Press, Cambridge.

Day, A. and R. Freese [1990] The role of gluing constructions in modular lattice theory, in: Bogart et al. [1990], pp. 251–260.

Day, A. and J. Ježek [1983] The amalgamation property for varieties of lattices, Mathematics Report No. 1, Lakehead Univ.

Dilworth, R. P. [1941a] The arithmetical theory of Birkhoff lattices, Duke Math. J. 8, 286–299.

Dilworth, R. P. [1990b] Background to Chapter 4, in: Bogart et al. [1990], pp. 205–209.

Faigle, U. [1980a] Geometries on partially ordered sets, J. Combin. Theory Ser. B 28, 26–51.

Fried, E. and G. Grätzer [1989] Pasting and modular lattices, Proc. Amer. Math. Soc. 106, 885–890.

Fried, E., G. Grätzer, and E. T. Schmidt [1993] Multipasting of lattices Algebra Universalis 30, 241–261.

Ganter, B. and I. Rival [1975] An arithmetical theorem for modular lattices, Algebra Universalis 5, 395–396.

Ganter, B. and R. Wille [1996] *Formale Begriffsanalyse: Mathematische Grundlagen*, Springer-Verlag, Heidelberg.

Gragg, K. and J. P. S. Kung [1992] Consistent dually semimodular lattices, J. Combin. Theory Ser. A 60, 246–263.

Hall, M. and R. P. Dilworth [1944] The imbedding problem for modular lattices, Ann. Math. 45, 450–456.

Herrmann, C. [1973] S-verklebte Summen von Verbänden, Math. Z. 130, 255–274.

Reeg, S. and W. Weiß [1991] Properties of finite lattices, Diplomarbeit, FB Mathematik, TH Darmstadt.

Reuter, K. [1985] Counting formulas for glued lattices, Order 1, 265–276.

Reuter, K. [1989] The Kurosh–Ore exchange property, Acta Math. Hungar. 53, 119–127.

Richter, G. [1982] The Kuroš–Ore theorem, finite and infinite decompositions, *Studia Sci. Math. Hungar.* 17, 243–250.

Richter, G. [1983] Applications of some lattice theoretic results on group theory, in: *Proc. Klagenfurt Conf., June 10–13, 1982*, Hölder-Pichler-Tempsky, Wien, pp. 305–317.

Richter, G. and M. Stern [1984] Strongness in (semimodular) lattices of finite length, Wiss. Z. Univ. Halle 39, 73–77.

Rival, I. (ed.) [1982a] *Ordered Sets, Proc. NATO Advanced Study Inst., Conf. Held at Banff, Canada, Aug. 28–Sep. 12, 1981.*

Schmidt, E. T. [1990] Pasting and semimodular lattices, Algebra Universalis 27, 595–596.

Skorsky, M. [1992] Endliche Verbände. Diagramme und Eigenschaften, Ph.D. thesis, FB Mathematik, TH Darmstadt.

Stern, M. [1990b] The impact of Dilworth's work on the Kurosh–Ore theorem, in: Bogart et al. [1990], pp. 203–204.

Walendziak, A. [1994c] Strongness in lattices, Demonstratio Math. 27, 569–572.

Wille, R. [1982] Restructuring lattice theory: an approach based on hierarchies of concepts, in: Rival [1982a], pp. 445–470.

Wille, R. [1983] Subdirect decomposition of concept lattices, Algebra Universalis 17, 275–287.

Wille, R. [1985] Complete tolerance relations of concept analysis, in: *Contributions to General Algebra 3, Proc. Vienna Conf. June 21–24, 1984*, pp. 397–415.

5

The Covering Graph

5.1 Diagrams and Covering Graphs

Summary. We make some general remarks on the Hasse diagram and the covering graph of a finite poset. Then we turn to the question of orientations and reorientations of a covering graph. Birkhoff [1948] asked for necessary and sufficient conditions on a lattice L in order that every lattice M whose covering graph is isomorphic with the covering graph of L be lattice isomorphic to L. Solutions to this problem will be given in the subsequent sections.

The *Hasse diagram* (briefly: the diagram) of a finite poset $P = (P, \leq)$ is an oriented graph with the circles of P as its vertices and an edge $x \rightarrow y$ if y covers x in P. Usually the arrows on the edges are omitted and the graph is arranged so that all edges point upwards on the page. The diagram of a finite poset is the most common tool for representing the poset graphically. For example, the usual diagram of the lattice 2^3 of all subsets of a three-element set, ordered by set inclusion, is shown in Figure 1.3(c) (Section 1.2). The diagram of a finite poset P determines P up to isomorphism. The combinatorial interest in posets is largely due to two unoriented graphs associated with a given poset: the comparability graph (which we shall not consider here) and the covering graph. In this chapter we shall have a closer look at the covering graph of certain lattices.

For a finite poset P, its covering graph $G(P)$ was introduced in Section 1.9. As for diagrams, it is common to identify a pictorial representation of the covering graph with the covering graph itself. For example, Figure 5.1 illustrates the covering graph of the power set of $\{a, b, c\}$. More generally, the *n-dimensional cube* (*hypercube*) is the covering graph of the poset 2^n.

An *orientation* of a covering graph is a diagram with the same labeled covering graph. A *reorientation* of a finite poset P is an orientation of its covering graph. It is an important problem to enumerate all reorientations of the covering graph of a poset. It is clear that a reorientation differs from the original orientation only in reversing the direction of some of its edges. In other words, if a covers b in an edge of P, then a reversal makes b an upper cover of a in the reorientation.

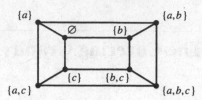

Figure 5.1

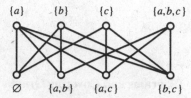

Figure 5.2

Hence the possible reorientations are precisely the subsets of edges of P that may be reversed.

Figure 5.2 shows the diagram of an orientation of the covering graph of 2^3 (of the power set of $\{a, b, c\}$).

In contrast to the diagram, the covering graph $G(P)$ does not determine P up to isomorphism: if $|P| \geq 3$ and $G(P)$ is connected, there is an orientation of $G(P)$ as the diagram of a poset that is neither isomorphic nor dually isomorphic to P. Thus the removal of the orientations in the Hasse diagram results in a loss of information: the orientation of P is hardly ever determined from its covering graph alone.

On the other hand, there are instances showing that the covering graph $G(P)$ still contains important information on the poset P, mainly in the case when P is a modular lattice. Some results concerning the modular case are mentioned below and in the subsequent sections.

Figure 5.3 indicates some interrelationships between the concepts mentioned above. All structures are assumed to be finite. The upper arrow indicates that every finite poset has a covering graph. The lower (double-headed) arrow indicates the one-to-one correspondence between finite posets and their Hasse diagrams.

Not every finite unoriented graph is the covering graph of a suitable finite poset. For example, it is obvious that a triangle (considered as a finite undirected graph) cannot be the covering graph of a poset.

Figure 5.4 shows *Grötzsch's graph* (see Grötzsch [1958] and Mycielski [1955]); it is the smallest triangle-free graph that is not a covering graph (for a proof of this see e.g. Mosesjan [1972 a, b, c]).

Ore [1962] posed the following problem: Characterize those finite unoriented graphs G for which there exists a poset P such that $G \cong G(P)$. This problem is

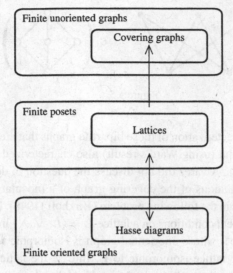

Figure 5.3

Figure 5.4   Grötzsch's graph

one of the major open problems of order theory (see Rival [1985b]). Not much is known for arbitrary posets. Much more is known about covering graphs of modular lattices. Two early results are:

(i) a modular lattice of finite length is distributive if and only if its covering graph contains no subgraph isomorphic to $K_{2,3}$ (the covering graph of M_3; cf. Figure 5.5) (Ward [1939]);

(ii) the covering graph of a modular lattice of finite length and of breadth n contains a subgraph isomorphic to the n-dimensional hypercube (Dilworth [1940]).

See also Section 5.6 for some more results of this kind.

Alvarez [1965] used path-length techniques in order to characterize those graphs that are orientable as the diagram of a modular lattice of finite length. He also showed that the covering graph of a modular lattice is bipartite. Moreover, he

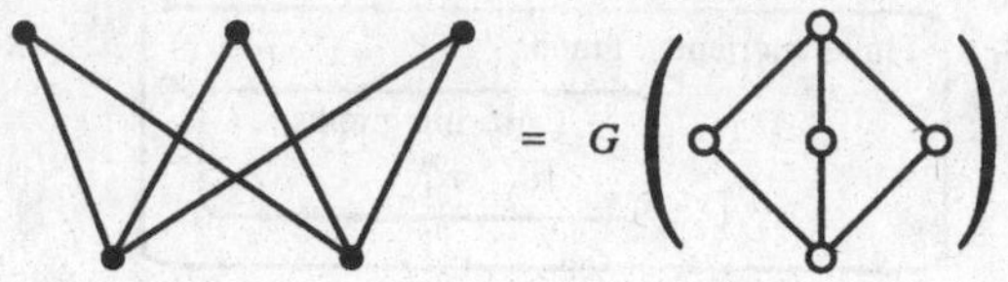

Figure 5.5

gave a complete characterization of those bipartite graphs that are covering graphs of modular lattices and (using Ward's result) also characterized covering graphs of distributive lattices. Alvarez did not discuss the question of deciding what the possible lattice orientations of the covering graph of a modular lattice are. This question has its roots in the following problem (Birkhoff [1948], Problem 8): Give necessary and sufficient conditions on a lattice $L = (L; \vee, \wedge)$ in order that every lattice $M = (L; \sqcup, \sqcap)$ whose unoriented graph is isomorphic with the covering graph $G(L)$ of L be lattice isomorphic to L. In case the lattices L and M are distributive or modular, the problem was solved by Jakubík & Kolibiar [1954] and Jakubík [1954a]. Jakubík [1975c] also showed that if one of L or M is modular (distributive) then so is the other. Duffus & Rival [1977] solved the problem for those graded lattices that are determined by the ordered subset of their atoms and coatoms. A common approach to the solutions of Jakubík [1954a] and Duffus & Rival [1977] will be given in Section 5.3.

Remark on the notation: If $P = (P, \leq)$ is a poset, we usually (but not always) write P for P and $G(P)$ instead of $G(P)$.

Notes

Some sharpenings of the results of Jakubík [1954a] were obtained by Jakubík [1975c].

Birkhoff [1982] proved that projectivity of prime intervals is in a certain sense invariant under graph isomorphisms of modular lattices and applied this result when investigating simple subdirect factors of modular lattices L and M of finite length having isomorphic covering graphs.

Isomorphisms of covering graphs of some types of posets (semilattices, multi-lattices) were studied by Jakubík [1956], Kolibiar [1982], and Tomková [1982]; in all these investigations certain covering conditions have been assumed. Jakubík's results have also been investigated in lattices with properties intermediate between modularity and distributivity. The most remarkable such class is that of the median lattices, which are characterized by a betweenness axiom. For this class Bandelt [1984] generalized the result of Duffus & Rival [1983].

For surveys of the problem of orientations and reorientations of graphs see Pretzel [1986] and Burosch [1989]. For surveys of the general role of diagrams see Rival [1985a], [1989]. For special properties of the comparability graph of semimodular lattices see the preprint J. D. Farley & S. E. Schmidt "A characterization of semimodular lattices whose comparability graphs are weakly connected" (manuscript, personal communication).

References

Alvarez, L. R. [1965] Undirected graphs as graphs of modular lattices, Canad. J. Math. 17, 923–932.

Bandelt, H.-J. [1984] Discrete ordered sets whose covering graphs are median, Proc. Amer. Math. Soc. 91, 6–8.

Birkhoff, G. [1948] *Lattice Theory* (2nd edition), Amer. Math. Soc. Colloquium Publications, Vol. 25, New York.

Birkhoff, G. [1982] Some applications of universal algebra, in: *Coll. Math. Soc. J. Bolyai, Vol. 29: Universal Algebra, Esztergom 1977*, North-Holland, Amsterdam, pp. 107–128.

Burosch, G. [1989] Hasse Graphen spezieller Ordnungen, in: *Graphentheorie. Band I: Anwendungen auf Topologie, Gruppentheorie und Verbandstheorie* (ed. K. Wagner and R. Bodendiek), BI Mannheim, pp. 157–245.

Dilworth, R. P. [1940] Lattices with unique irreducible decompositions, Ann. of Math. 41, 771–777.

Duffus, D. and I. Rival [1977] Path length in the covering graph of a lattice, Discrete Math. 19, 139–158.

Duffus, D. and I. Rival [1983] Graphs orientable as distributive lattices, Proc. Amer. Math. Soc. 88, 197–200.

Grötzsch, H. [1958] Ein Dreifarbensatz für dreikreisfreie Netze auf der Kugel, Wiss. Z. Martin-Luther-Univ. Halle, Math. Nat. Reihe 8, 109–119.

Jakubík, J. [1954a] On the graph isomorphism of lattices (Russian), Czech. Math. J. 4, 131–142.

Jakubík, J. [1956] Graph-isomorphism of multilattices (Slovak), Acta Fac. Rer. Nat. Univ. Comenianae Math. 1, 255–264.

Jakubík, J. [1975c] Unoriented graphs of modular lattices, Czech. Math. J. 25, 240–246.

Jakubík, J. and M. Kolibiar [1954] On some properties of a pair of lattices, Czech. Math. J. 4, 1–27.

Kolibiar, M. [1982] Semilattices with isomorphic graphs, in: *Colloq. Math. Soc. J. Bolyai, Vol. 29: Universal Algebra, Esztergom (Hungary) 1977*, North-Holland, Amsterdam, pp. 473–481.

Mosesjan, K. M. [1972a] Strongly basable graphs (Russian), Akad. Nauk Armjan. SSR Dokl. 54, 134–138.

Mosesjan, K. M. [1972b] Certain theorems on strongly basable graphs (Russian), Akad. Nauk Armjan. SSR Dokl. 54, 241–245.

Mosesjan, K. M. [1972c] Basable and strongly basable graphs (Russian), Akad. Nauk Armjan. SSR Dokl. 55, 83–86.

Mycielski, J. [1955] Sur le colorage des graphes, Colloq. Math. 3, 161–162.

Ore, O. [1962] *Theory of Graphs*, Amer. Math. Soc., Providence, R.I.

Pretzel, O. [1986] Orientations and reorientations of graphs, in: *Combinatorics and Ordered Sets, Proc. AMS–IMS SIAM Joint Summer Res. Conf., Arcata, Calif., 1985*, Contemporary Mathematics, Vol. 57, pp. 103–125.

Rival, I. [1985a] The diagram, in: *Graphs and Order* (ed. I. Rival) NATO ASI Seriés C: Mathematical and Physical Sciences, Vol. 147, Reidel, pp. 103–133.

Rival, I. [1985b] The diagram, "Unsolved Problems," Order 2, 101–104.

Rival, I. [1989] Graphical data structures for ordered sets, in: *Algorithms and Order* (ed.
    I. Rival) NATO ASI Series C: Mathematical and Physical Sciences, Vol. 255, Reidel,
    pp. 3–31.
Tomková, M. [1982] On multilattices with isomorphic graphs, Math. Slovaca 32, 63–74.
Ward, M. [1939] A characterization of Dedekind structures, Bull. Amer. Math. Soc. 45,
    448–451.

5.2 Path Length

Summary. We give some results concerning path length (distances) in graded posets and semimodular lattices. Some of these results will be needed as a further preparation for Section 5.3.

Let P be a poset, whose least and greatest elements (if they exist) will be denoted by 0_P and 1_P. Let $G = (V, H)$ be a finite connected graph without multiple edges, and let $a, b \in V$ be vertices of G. The notion of a shortest path from a to b was defined in Section 1.9. Recall also that if $a, b \in V$, then by $\delta(a, b)$ we denote the *distance* from a to b, that is, the length of a shortest path from a to b in G. By diam $G = \max\{\delta(a, b) : a$ and b vertices of $G\}$ we denote the diameter of G.

We first give four lemmas (for which see Duffus & Rival [1977]) concerning path length in graded posets (in fact, we shall consider only graded lattices). The results will be needed in the next section. The first statement (which is proved by induction) links the distance with the rank function by which the poset is graded.

Lemma 5.2.1 *Let P be a graded poset and $a, b \in P$. Then*

 (i) $\delta(a, b) \geq |r(a) - r(b)|$;
 (ii) $\delta(a, b) = r(a) - r(b)$ *if and only if $a \geq b$;*
 (iii) $\delta(a, b) \geq |r(a) - r(b)| + 2$ *if a and b are noncomparable.*

The next lemma says that an orientation of a graph as the diagram of a graded poset is determined by the choice for its least element.

Lemma 5.2.2 *Let P and P' be graded posets, and let ϕ be a graph isomorphism of $G(P)$ to $G(P')$. If $\phi(0_P) = 0_{P'}$, then $P \cong P'$.*

Proof. Let r' denote the rank function by which P' is graded. Let $x, y \in P$ and $x \leq y$. Repeated application of Lemma 5.2.1 (ii) yields

$$\delta(0_P, y) = \delta(0_P, x) + \delta(x, y) = \delta(\phi(0_P), \phi(x)) + \delta(\phi(x), \phi(y))$$
$$= \delta(0_{P'}, \phi(y)),$$

so that $r'(\phi(y)) = r'(\phi(x)) + \delta(\phi(x), \phi(y))$ or $\phi(x) \leq \phi(y)$. Similarly, as ϕ is a graph isomorphism, $\phi(x) \leq \phi(y)$ implies $x \leq y$. It follows that $P \cong P'$. ■

If only P, say, is graded in the hypothesis of Lemma 5.2.2, its conclusion in general fails: In Figure 5.6 (cf. Duffus & Rival [1977]) we have the diagrams of a

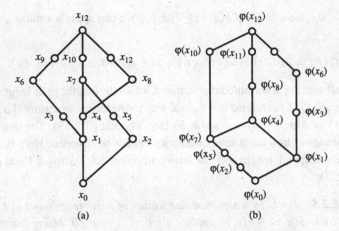

Figure 5.6

graded lattice L [Figure 5.6(a)] and a nongraded lattice L' [Figure 5.6(b)], which are nonisomorphic although there is a graph isomorphism φ of $G(L)$ to $G(L')$ and $\varphi(0_L) = 0_{L'}$ and even $\varphi(1_L) = 1_{L'}$.

The next result is contained in Alvarez [1965], Theorem 1.

Lemma 5.2.3 *Let P be a graded poset with a greatest element. Then* diam $G(P) = \delta(0_P, 1_P) = l(P)$.

Proof. For any $x, y \in P$, Lemma 5.2.1(ii) implies

$$2\delta(x, y) \le \delta(x, 1_P) + \delta(1_P, y) + \delta(x, 0_P) + \delta(0_P, y) = 2\delta(0_P, 1_P).$$

The second equality is an immediate consequence of Lemma 5.2.1(ii). ∎

Alvarez [1965] proved that any two elements that are a diameter apart in the covering graph of a modular lattice of finite length must be complementary. This is extended in

Lemma 5.2.4 *Let P be a graded poset with a greatest element, and let $a, b \in P$. If $\delta(a, b) = $ diam $G(P)$, then $a \vee b$ and $a \wedge b$ exist in P, and $a \vee b = 1_P$ and $a \wedge b = 0_P$.*

Proof. We may assume that a is noncomparable to b, since otherwise, Lemma 5.2.1(ii) gives $\{a, b\} = \{0_P, 1_P\}$. Let c and d be elements of P satisfying $a \le c$, $b \le c$, $d \le a$, and $d \le b$. It suffices to show that $c = 1_P$ and $d = 0_P$. Let us suppose that $\delta(a, c) \le \delta(d, b)$, say. In view of Lemma 5.2.3 we have diam $G(P) = \delta(0_P, 1_P) = \delta(0_P, d) + \delta(d, b) + \delta(b, c) + \delta(c, 1_P)$. It follows that

$$\text{diam } G(P) \ge \delta(0_P, d) + \delta(a, c) + \delta(b, c) + \delta(c, 1_P)$$
$$\ge \delta(0_P, d) + \text{diam } G(P) + \delta(c, 1_P),$$

whence $d = 0_P$ and $c = 1_P$. If $\delta(a, c) \geq \delta(d, b)$, we can apply a similar argument to

$$\text{diam } G(P) = \delta(0_P, 1_P) = \delta(0_P, d) + \delta(d, a) + \delta(a, c) + \delta(c, 1_P). \quad \blacksquare$$

We recall that for a semimodular lattice L of finite length, path length in the covering graph $G(L)$ is related in a natural way to the lattice structure: If $a, b \in L$, then $\delta(a, b) = \delta(a, a \vee b) + \delta(a \vee b, b)$ (see Theorem 1.9.19). The dual of the centered hexagon shows that semimodularity cannot be dropped. Here is another result connecting path length with the lattice structure (cf. Duffus & Rival [1977], Theorem 5.3):

Theorem 5.2.5 *Let L be a semimodular lattice of finite length and $a, b, c \in L$. Then $\delta(a, b) = \delta(a, c) + \delta(c, b)$ implies $a \wedge b \leq c \leq a \vee b$. Moreover, the converse holds if and only if L is distributive.*

Proof.   To establish the first implication it suffices to show that every path $a = c_0$, $c_1, \ldots, c_n = b$ of minimum length from a to b in $G(L)$ satisfies $a \wedge b \leq c_i \leq a \vee b$ for each $i = 0, 1, \ldots, n$. We proceed by induction on $\delta(a, b)$. Let us suppose that $c_1 \notin [a \wedge b, a \vee b]$. If $a \succ c_1$, then by Theorem 1.9.19 and Lemma 5.2.1(ii),

$$\delta(c_1, b) = \delta(c_1, c_1 \vee b) + \delta(c_1 \vee b, b) \geq \delta(a, a \vee b) + \delta(a \vee b, b) = \delta(a, b).$$

Hence $a \wedge b \leq c_i \leq a \vee b$. Finally, by the induction hypothesis,

$$a \wedge b \leq c_1 \wedge b \leq c_i \leq c_1 \vee b \leq a \vee b \qquad \text{for each} \quad i = 2, 3, \ldots, n.$$

A modular lattice L of finite length satisfies

$(+)\ r(a \vee b) + r(a \wedge b) = r(a) + r(b)$

for every $a, b \in L$ (see Section 1.9). Suppose now that L is distributive, and let $a, b, c \in L$ satisfy $a \wedge b \leq c \leq a \vee b$. Distributivity implies $(a \wedge c) \vee (b \wedge c) = c \vee (a \wedge b)$, whence $r((a \wedge c) \vee (b \wedge c)) = r(c \vee (a \wedge b))$. Several applications of $(+)$ yield

$$r(a \vee b) - r(a \wedge b) = r(a \vee c) - r(a \wedge c) + r(b \vee c) - r(b \wedge c),$$

which, by Lemma 5.2.1(ii), gives $\delta(a \wedge b, a \vee b) = \delta(a \wedge c, a \vee c) + \delta(b \wedge c, b \vee c)$, and (in view of Exercise 1) $\delta(a, b) = \delta(a, c) + \delta(c, b)$.

Finally, let us suppose that L is a semimodular lattice of finite length such that for each $a, b, c \in L$, $a \wedge b \leq c \leq a \vee b$ implies $\delta(a, b) = \delta(a, c) + \delta(c, b)$. If L is nonmodular, then by Theorem 3.1.10 it contains a cover-preserving sublattice $S \cong S_7$. However, if we choose $a, b, c \in S$ as indicated in Figure 5.7(a), we have $\delta(a, b) < \delta(a, c) + \delta(c, b)$. Hence L is modular. If L is nondistributive, then it

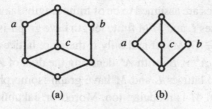

Figure 5.7

contains a cover-preserving sublattice $S \cong M_3$, and again, choosing $a, b, c \in S$ as indicated in Figure 5.7(b) gives a contradiction. Therefore L must be distributive. ∎

Notes

For further information on length and distances in lattices and posets we refer to Barbut & Monjardet [1970] and Haskins & Gudder [1972].

Exercises

1. Let L be a modular lattice of finite length and $a, b \in L$. Then $\delta(a, b) = \delta(a \wedge b, a \vee b)$ (Duffus & Rival [1977], Lemma 6). Give an example showing that this is not true in the semimodular case.
2. Show that the covering graph of a modular lattice is bipartite (Alvarez).

References

Alvarez, L. R. [1965] Undirected graphs as graphs of modular lattices, Canad. J. Math. 17, 923–932.

Barbut, M. and B. Monjardet [1970] *Ordre et classification,* Vols. 1, 2, Hachette, Paris.

Duffus, D. and I. Rival [1977] Path length in the covering graph of a lattice, Discrete Math. 19, 139–158.

Duffus, D. and I. Rival [1983] Graphs orientable as distributive lattices, Proc. Amer. Math. Soc. 88, 197–200.

Haskins, L. and S. Gudder [1972] Heights on posets and graphs, Discrete Math. 2, 357–382.

5.3 Graph Isomorphisms of Graded Balanced Lattices

Summary. Jakubík [1954a] proved that for discrete modular lattices all graph isomorphisms are given by certain direct-product decompositions. Duffus & Rival [1977] proved a similar theorem for graded lattices that are atomistic and coatomistic. Modifying some of the results of Duffus & Rival, we give a common generalization for lattices that are balanced and graded.

In this section all lattices are assumed to be of finite length. Jakubík [1954a] proved that two modular lattices L and M of finite length have graph-isomorphic covering graphs, that is, $G(L) \cong G(M)$, if and only if there are lattices A and B such that $L \cong A \times B$ and $M \cong A^* \times B$ (with A^* denoting the dual of A). Jakubík [1975c] also proved that if two lattices L and M have graph-isomorphic covering graphs and L is modular, then M is modular, too. Moreover, Jakubík [1954b] remarked that the above result is not true for semimodular lattices. Indeed, the centered hexagon S_7 (which is semimodular) and its dual (which is not semimodular) have the same covering graph.

On the other hand, there exist nonmodular semimodular lattices for which an analogue to Jakubík's theorem can be proved: Duffus & Rival [1977], Theorem 3.3, implies that such an analogue is true, in particular, for geometric lattices. Now geometric lattices and modular lattices of finite length are instances of semimodular lattices that are both strong and dually strong (see Section 4.6). The proof of Duffus & Rival [1977], Theorem 3.3, carried through for graded lattices that are both atomistic and coatomistic, indicates that it is not modularity that makes the proof work, but rather the properties of being graded as well as strong and dually strong (note that any atomistic lattice is trivially strong, and any coatomistic lattice is trivially dually strong). Lattices that are both strong and dually strong were called balanced by Reuter [1989] (see Section 4.6).

The ideas of Duffus & Rival [1977] can be modified in such a way as to prove a result for graded balanced lattices (see Theorem 5.3.3 below) yielding Duffus & Rival [1977], Theorem 3.3, and Jakubík's above-mentioned theorem as special cases. Aside from unifying both results, the present approach will also lead to a simpler proof of Jakubík's theorem.

Let us first recall that an element a of a lattice L is in the center of L if there is a direct-product decomposition $A \times B$ of L for which a corresponds to $(1_A, 0_B)$ or $(0_A, 1_B)$. Equivalently, a is in the center of L if the following three conditions are satisfied:

(i) there exists an element $b \in L$ such that $a \vee b = 1_L$ and $a \wedge b = 0_L$;
(ii) for each $x \in L$, $x = (x \wedge a) \vee (x \wedge b)$;
(iii) if $x, y \in L$ and $x \le a$, $y \le b$, then $x = (x \vee y) \wedge a$ and $y = (x \vee y) \wedge b$.

Duffus & Rival [1977], Lemma 3.1, is the crucial result for proving an analogue to Jakubík's result for graded lattices that are both atomistic and coatomistic. We modify this lemma in order to make it applicable to graded balanced lattices:

Lemma 5.3.1 *Let L and M be graded balanced lattices, let ϕ be a graph isomorphism of $G(L)$ to $G(M)$, and let $a, b \in L$ be elements such that $\phi(a) = 1_L$ and $\phi(b) = 0_L$. Then every join-irreducible j of L satisfies $j \le a$ or $j \le b$, and every meet-irreducible m of L satisfies $m \ge a$ or $m \ge b$.*

Proof. In view of Lemma 5.2.3, $\delta(\phi(a), \phi(b)) = \text{diam } G(M)$, whence by Lemma 5.2.4, $a \vee b = 1_L$ and $a \wedge b = 0_L$. In particular, it follows that $[0_L, a] \cap [0_L, b] = \{0_L\}$.

Suppose now $j \in J(L)$ such that $j \not\leq a$ and $j \not\leq b$. If we had $j > a$, we would get $j \succ j' \geq a$ and $j \leq b \vee j' = b \vee a = 1_L$ but $j \not\leq b$, contradicting strongness and therefore the property of being balanced. Thus j is incomparable to a and, similarly, j is incomparable to b. Lemma 5.2.1(iii) now implies that $\delta(j, a) = r(a) - r(j) + 2 = \delta(0_L, a) + 1$ and $\delta(j, b) = \delta(0_L, b) + 1$. On the other hand, $\phi(a) = 1_L$ and $\phi(b) = 0_L$ together with Lemma 5.2.1(ii) implies that

$$\delta(\phi(a), \phi(x)) + \delta(\phi(x), \phi(b)) = \delta(\phi(a), \phi(b)) = \delta(\phi(a), \phi(0_L)) + \delta(\phi(0_L), \phi(b)),$$

which, however, is a contradiction to $\delta(a, x) + \delta(x, b) = \delta(a, 0_L) + \delta(0_L, b) + 2$. The dual reasoning proves the corresponding assertion on meet-irreducibles. ∎

We can now proceed to prove Duffus & Rival [1977], Lemma 3.2, for graded balanced lattices:

Lemma 5.3.2 *Let L be a graded balanced lattice. Let M be a graded lattice, let ϕ be a graph isomorphism of $G(L)$ to $G(M)$, and let $a, b \in L$ such that $\phi(a) = 1_L$ and $\phi(b) = 0_L$. Then $L \cong [0_L, a] \times [0_L, b]$.*

Proof. We show that a is in the center of L. Then Lemma 5.2.3 implies that $\delta(\phi(a), \phi(b)) = \text{diam } G(M)$, which, in view of Lemma 5.2.4, yields $a \vee b = 1_L$ and $a \wedge b = 0_L$. To show that a is in the center of L, let $x \in L$, and let

$$A = \{j \in L : j \text{ join-irreducible}, j \leq a, \text{ and } j \leq x\}$$

and

$$B = \{j \in L : j \text{ join-irreducible}, j \leq b, \text{ and } j \leq x\}.$$

Since x is a join of join-irreducibles, we get by Lemma 5.3.1 that $\bigvee(A \cup B) = x$. Moreover, $a \wedge x = \bigvee A$. For, on the one hand, $a \geq \bigvee A$ and $x \geq \bigvee A$, that is, $a \wedge x \geq \bigvee A$. On the other hand, if $a \wedge x > \bigvee A$, then there exists a join-irreducible j such that $j \leq a \wedge x$ and $j \not\leq \bigvee A$. But $j \leq a \wedge x$ implies $j \leq a$ and $j \leq x$; hence $j \in A$, a contradiction. Similarly $b \wedge x = \bigvee B$. It follows that $x = \bigvee(A \cup B) = \bigvee A \vee \bigvee B = (a \wedge x) \vee (b \wedge x)$.

Let now $x, y \in L$ with $x \leq a$ and $y \leq b$. We have $(x \vee y) \wedge a \geq x$. Since every element of L is a meet of meet-irreducibles, $(x \vee y) \wedge a > x$ implies that there is a meet-irreducible m of L such that $m \geq x$ and $m \not\geq (x \wedge y) \vee a$. Hence $m \not\geq a$, which by Lemma 5.3.1 means that $m \geq b \geq y$. But then $m \geq x \vee y \geq (x \vee y) \wedge a$, a contradiction. Thus $(x \vee y) \wedge a = x$, and similarly $(x \vee y) \wedge b = y$. It follows that a is in the center of L and hence $L \cong [0_L, a] \times [0_L, b]$. ∎

Now we are able to prove the main result (cf. Stern [1996a]).

Theorem 5.3.3   *Let L and M be graded lattices with graph-isomorphic covering graphs. L is balanced if and only if M is balanced. Moreover, if this condition is satisfied, then there are sublattices A and B of L such that $L \cong A \times B$ and $M \cong A^* \times B$.*

Proof.   Let L be a graded lattice, and assume it is balanced. Let L' be a graded lattice, and ϕ a graph isomorphism of $G(L)$ to $G(M)$. Consider $a, b \in L$ such that $\phi(a) = 1_L$ and $\phi(b) = 0_L$. From Lemma 5.3.2 we get $L \cong [0_L, a] \times [0_L, b]$. The canonical isomorphism connected with the center property of $a \in L$ implies $L \cong [b, 1_L] \times [0_L, b]$. Let $K \cong [b, 1_L] \times [0_L, b]^*$. It is readily shown that K is balanced and graded. Moreover, there is a graph isomorphism ψ of $G(K)$ to $G(L)$ such that $\psi(0_K) = b$.

Finally, since $\phi\psi$ is a graph isomorphism of $G(K)$ to $G(M)$ such that $\phi(\psi(0_K)) = 0_L$, we get from Lemma 5.2.2 that $L' \cong K$.   ∎

As a consequence we get Duffus & Rival [1977], Theorem 3.3:

Corollary 5.3.4   *Let L and M be graded lattices with isomorphic covering graphs. Then L is atomistic and coatomistic if and only if M is atomistic and coatomistic. Moreover, if this condition is satisfied, then there are sublattices A and B of L such that $L \cong A \times B$ and $L' \cong A^* \times B$.*

Proof.   If L is atomistic and coatomistic, then it is balanced. From Theorem 5.3.3 it follows that M is balanced (in fact, we get that M is atomistic and coatomistic) and that the required direct-product decomposition exists.   ∎

Since a geometric lattice is graded, atomistic, and coatomistic, the conclusion of Corollary 5.3.4 holds, in particular, for geometric lattices. Reasoning as in the preceding corollary yields the main results of Jakubík [1954a], [1975c]:

Corollary 5.3.5   *Let L and M be graded lattices with graph-isomorphic covering graphs. Then L is modular if and only if M is modular. Moreover, if this condition holds, then there are sublattices A and B of L such that $L \cong A \times B$ and $M \cong A^* \times B$.*

According to Theorem 5.3.3, every strong (balanced) semimodular lattice L has the property that orientations of the covering graph $G(L)$ arise from direct-product decompositions of L. In particular, there are balanced semimodular lattices that are neither geometric nor modular and for which the conclusion of Theorem 5.3.3 holds, such as the lattice of Figure 4.23 (Section 4.6).

Notes

Related questions and extensions to semilattices, multilattices, and posets are investigated in a number of papers, for example, Birkhoff [1982], Gedeonová [1981], Jakubík [1954a], [1956], [1972], [1984], [1985c], [1986], Kolibiar [1982], [1985], Lee [1986].

Exercises

1. Give examples of nonisomorphic graded lattices L and L' with graph-isomorphic covering graphs, such that L^* is not isomorphic with L' and neither L nor L' has a nontrivial direct product decomposition (Duffus & Rival [1977], Fig. 3).

2. Let L and L' be lattices with isomorphic covering graphs. Show that if L is semimodular and has finite length, then L' also has finite length. Give an example showing that semimodularity cannot be dropped. (Lee [1986].)

3. If P is a finite poset with $G(P)$ connected and if a is any element of P, then there is an orientation of $G(P)$ as the diagram of a poset with a as its least element (Sands, cf. Duffus & Rival [1977]).

References

Birkhoff, G. [1948] *Lattice Theory* (2nd edition), Amer. Math. Soc. Colloq. Publ. 25, Providence, R. I.

Birkhoff, G. [1982] Some applications of universal algebra, in: *Colloq. Math. Soc. J. Bolyai, Vol. 29: Universal Algebra, Esztergom 1977*, North-Holland, Amsterdam, pp. 107–128.

Duffus, D. and I. Rival [1977] Path length in the covering graph of a lattice, Discrete Math. 19, 139–158.

Gedeonová, E. [1981] The orientability of the direct product of graphs, Math. Slovaca 31, 71–78.

Jakubík, J. [1954a] On the graph isomorphism of lattices (Russian), Czech. Math J. 4, 131–142.

Jakubík, J. [1954b] On the graph isomorphism of semimodular lattices (Slovak), Mat. Fyz. Časopis 4, 162–177.

Jakubík, J. [1956] Graph-isomorphism of multilattices (Slovak), Acta Fac. Rer. Nat. Univ. Comenianae Math. 1, 255–264.

Jakubík, J. [1972] Weak product decompositions of partially ordered sets, Colloq. Math. 25, 13–26.

Jakubík, J. [1975c] Unoriented graphs of modular lattices, Czech. Math. J. 25, 240–246.

Jakubík, J. [1984] On lattices determined up to isomorphisms by their graphs, Czech. Math. J. 34, 305–314.

Jakubík, J. [1985c] On weak direct product decompositions of lattices and graphs, Czech. Math. J. 35, 269–277.

Jakubík, J. [1986] Covering graphs and subdirect decompositions of partially ordered sets, Math. Slovaca 36, 151–162.

Kolibiar, M. [1982] Semilattices with isomorphic graphs, in: *Colloq. Math. Soc. J. Bolyai, Vol. 29: Universal Algebra, Esztergom (Hungary) 1977*, North-Holland, Amsterdam, pp. 473–481.

Kolibiar, M. [1985] Graph isomorphisms of semilattices, in: *Contributions to General Algebra 3, Proc. Vienna Conf., June 21–24, 1984*, pp. 225–235.

Lee, Jeh Gwon [1986] Covering graphs of lattices, Bull. Korean Math. Soc. 23, 39–46.

Reuter, K. [1989] The Kurosh–Ore exchange property, Acta Math. Hungar. 53, 119–127.

Stern, M. [1996a] On the covering graph of balanced lattices, Discrete Math. 156, 311–316.

5.4 Semimodular Lattices with Isomorphic Covering Graphs

Summary. We briefly indicate another generalization of Jakubík's result (Corollary 5.3.5) and an application to semimodular lattices. In connection with this, Jakubík [1985b] posed a problem, which was solved in the affirmative by Ratanaprasert & Davey [1987].

Let us first reformulate Jakubík's result using the more precise notation mentioned in the beginning of Section 5.1.

Let $\mathbf{L} = (L; \leq)$ and $\mathbf{M} = (M; \leq^\circ)$ be lattices, whose covering graphs are denoted by $G(\mathbf{L})$ and $G(\mathbf{M})$, respectively. We say that $G(\mathbf{L})$ is isomorphic with $G(\mathbf{M})$ if there is a bijection $f : L \rightarrow M$ such that for all $a, b \in L$, $\{a, b\}$ is an edge of $G(\mathbf{L})$ if and only if $\{f(a), f(b)\}$ is an edge of $G(\mathbf{M})$. Throughout this section we assume, without loss of generality, that $\mathbf{L} = \mathbf{M}$ and f is the identity map whenever $G(\mathbf{L})$ is isomorphic to $G(\mathbf{M})$, whence $G(\mathbf{L}) = G(\mathbf{M})$.

Jakubík proved (cf. Corollary 5.3.5) that for discrete modular lattices $\mathbf{L}$ and $\mathbf{M}$ on the same underlying set L, the graphs $G(\mathbf{L})$ and $G(\mathbf{M})$ are isomorphic if and only if the following condition holds:

(a) There exist lattices $\mathbf{A} = (A; \leq)$, $\mathbf{B} = (B; \leq)$ and a direct-product representation $\psi : L \rightarrow A \times B$ via which $\mathbf{L}$ is isomorphic with $\mathbf{A} \times \mathbf{B}$ and $\mathbf{M}$ is isomorphic with $\mathbf{A}^* \times \mathbf{B}$.

By means of the concept of a proper cell (for which see below) Jakubík generalized his result from the discrete modular case to arbitrary discrete lattices.

Let $u, v, x_1, x_2, \ldots, x_m, y_1, y_2, \ldots, y_n$ be distinct elements of L such that

(i) $u \prec x_1 \prec x_2 \prec \cdots \prec x_m \prec v, u \prec y_1 \prec y_2 \prec \cdots y_n \prec v$ and
(ii) either $x_1 \vee y_1 = v$ or $x_m \wedge y_n = u$.

The set $C = \{u, v, x_1, x_2, \ldots, x_m, y_1, y_2, \ldots, y_n\}$ is said to be a *cell* in $\mathbf{L}$. If $x_1 \vee y_1 = v$, we call C a cell of type $V(m, n)$ (see Figure 5.8). A cell C is called *proper* if either $m > 1$ or $n > 1$.

Let $h : L \rightarrow L_1$ be any bijection, and let $T \subseteq L$. The set T is said to be *preserved* [*reversed*] *under* h if, whenever $x_1, x_2 \in T$ and $x_1 < x_2$, one has $h(x_1) < h(x_2)$ [or $h(x_1) > h(x_2)$, respectively]. If $G(\mathbf{L}) = G(\mathbf{M})$, then a set $C \subseteq L$ is said to be *preserved* if, whenever $a, b \in C$ and $a \prec b$, one has $a \prec^\circ b$.

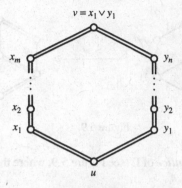

Figure 5.8

Jakubík [1985a] proved that for discrete lattices condition (a) is equivalent to

(b) **L** and **M** have isomorphic graphs, and all proper cells of **L** and all proper cells of **M** are either preserved or reversed.

Since a modular lattice has no proper cells, it follows that Jakubík's solution (see Corollary 5.3.5) to Birkhoff's problem No. 8 is a special case of the equivalence (a) $\Leftrightarrow$ (b).

Consider now the following condition for lattices **L** and **M** (and for the bijection $h : L \to M$):

($\alpha 1$) All sublattices of type S_7 of **L** are preserved under h, and all sublattices of type S_7 of **M** are preserved under h^{-1}.

Jakubík [1954b], Theorem 2, proved the following result: If **L** and **M** are semi-modular lattices and h is a graph isomorphism of **L** onto **M**, then ($\alpha 1$) $\Rightarrow$ (a). He assumed that **L** and **M** are finite, but the proof remains valid if both lattices are locally finite (discrete). Jakubík [1985b], Lemma 3, also showed that if **L** and **M** are discrete semimodular lattices, then (a) $\Rightarrow$ ($\alpha 1$). Both results imply Jakubík [1985b], Theorem 4:

Theorem 5.4.1 *Let* **L** *and* **M** *be discrete semimodular lattices, and let h be a graph isomorphism of* **L** *onto* **M**. *Then condition* ($\alpha 1$) *holds if and only if condition* (a) *holds.*

In a semimodular lattice, all proper cells are sublattices of type S_7. In fact, all proper cells of a semimodular lattice L are c_1-sublattices in the following sense: Let $T = (T; \leq)$ be a sublattice of **L**. If there is an isomorphism ψ of S_7 onto **T** such that

$$\psi(u) \prec \psi(x_1) \prec \psi(w), \qquad \psi(u) \prec (y_1) \prec \psi(w),$$
$$\psi(x_2) \prec \psi(v), \quad \text{and} \quad \psi(y_2) \prec \psi(v),$$

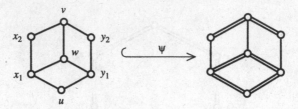

Figure 5.9

then **T** is called a c_1-*sublattice* of **L** (see Figure 5.9, where the double lines indicate the covering relation).

Jakubík [1985b] proved that for discrete semimodular lattice **L** and **M** condition (b) is equivalent to the following condition:

(c) **L** and **M** have isomorphic graphs, and all c_1-sublattices of **L** and all c_1-sublattices of **M** are preserved (this is condition (α_{11}) in Jakubík [1985b]).

The question arises whether this equivalence remains valid if "c_1-sublattice" in condition (c) is replaced by "cover-preserving sublattice." This question was answered by Ratanaprasert & Davey [1987], who showed that for discrete semi-modular lattice **L** and **M** condition (b) is equivalent to the condition

(d) **L** and **M** have isomorphic graphs, and all cover-preserving sublattices of **L** of type S_7 are preserved.

Hence, for discrete semimodular lattice **L** and **M** , conditions (a) and (d) are also equivalent. This solves in the affirmative the problem raised by Jakubík [1985b].

Notes

Several results obtained by Jakubík have been extended to semilattices; see, for example, Kolibiar [1982], [1985].

References

Birkhoff, G. [1948] *Lattice Theory* (2nd ed.), Amer. Math. Soc. Colloquium Publications, Vol. 25, New York.

Jakubík, J. [1954a] On the graph isomorphism of lattices (Russian), Czech. Math. J. 4, 131–142.

Jakubík, J. [1954b] On the graph isomorphism of semimodular lattices (Slovak), Mat. Fyz. Časopis 4, 162–177.

Jakubík, J. [1975a] Modular lattices of locally finite length, Acta Sci. Math. (Szeged) 37, 79–82.

Jakubík, J. [1975b] Sublattices with saturated chains, Czech. Math. J. 25, 442–444.

Jakubík, J. [1975c] Unoriented graphs of modular lattices, Czech. Math. J. 25, 240–246.

Jakubík, J. [1984] On lattices determined up to isomorphisms by their graphs, Czech. Math. J. 34, 305–314.

Jakubík, J. [1985a] On isomorphisms of graphs of lattices, Czech. Math. J. 35, 188–200.
Jakubík, J. [1985b] Graph isomorphisms of semimodular lattices, Math. Slovaca 35, 229–232.
Kolibiar, M. [1982] Semilattices with isomorphic graphs, in: *Colloq. Math. Soc. J. Bolyai, Vol. 29: Universal Algebra, Esztergom (Hungary) 1977*, North-Holland, Amsterdam, pp. 473–481.
Kolibiar, M. [1985] Graph isomorphisms of semilattices, in: *Contributions to General Algebra 3, Proc. Vienna Conf.*, June 21–24, 1984, pp. 225–235.
Ratanaprasert, C. and B. Davey [1987] Semimodular lattices with isomorphic graphs, Order 4, 1–13.

5.5 Centrally Symmetric Graphs and Lattices

Summary. The results of this section are related to a conjecture of Kotzig to the effect that a graph is centrally symmetric if and only if it is, for some integer n, the covering graph of the direct product of lattices denoted by K_{2n}. That conjecture turned out to be false. On the other hand, Zelinka showed that if L is a finite modular lattice having n atoms and a centrally symmetric covering graph, then $L \cong 2^n$. This result was sharpened by Duffus & Rival, who replaced modularity by semimodularity. Here we give a further sharpening using only that the underlying lattice is graded and that its dual is strong.

We recall that for a finite connected and unoriented graph $G = (V, H)$ and a vertex $v \in V$ there is at most one orientation of G as the diagram of a graded poset with v as its least element (cf. Lemma 5.2.2). The notion of a symmetric graph is due to Kotzig [1968a]. We shall use the following equivalent formulation (cf. Gedeonová [1980, 1990]).

A finite connected graph $G = (V, H)$ is a *symmetric graph* (*S graph*, for short) if for every vertex $v \in V$ there exists a graded poset $(V, \leq_v)$ with least element v such that the covering graph of $(V, \leq_v)$ is isomorphic to the graph G.

An S graph is called *centrally symmetric* (CS, for short) if $(V, \leq_v)$ is a graded lattice for each $v \in V$. The notion of a centrally symmetric graph was introduced by Duffus & Rival [1977].

If the covering graph of a lattice is an S graph or a CS graph, then this lattice is called an *S lattice* or a *CS lattice*, respectively. It is clear that every CS lattice is an S lattice. The converse is not true, as the lattice L of Figure 5.10 shows

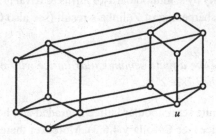

Figure 5.10

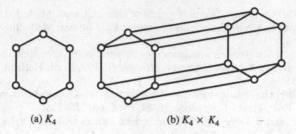

(a) K_4 (b) $K_4 \times K_4$

Figure 5.11

(Gedeonová [1983]): L is an S lattice, but the graded poset with the least element u and the same covering graph is not a lattice, that is, L is not a CS lattice.

The two-element path is a centrally symmetric graph. This property is preserved under direct products, and hence every hypercube is centrally symmetric, too. Equivalently, the Boolean lattice 2^n with n atoms is a CS lattice. Nonmodular examples of CS lattices are the lattice K_4 [Figure 5.11(a)] and the direct product $K_4 \times K_4$ [Figure. 5.11(b)].

More generally we have the following result: For an integer n, let K_{2n} denote the lattice on $\{a_0, a_1, \ldots, a_n, b_0, b_1, \ldots, b_n\}$ determined by $0 \prec a_0 \prec a_1 \prec \cdots \prec a_n = 1$ and $0 \prec b_0 \prec b_1 \prec \cdots \prec b_n = 1$. Then K_{2n} is a CS lattice, and so is the direct product of finitely many copies of K_{2n}.

Kotzig [1968a, b] conjectured that a graph is symmetric if and only if it is the covering graph of the direct product of lattices, each isomorphic to K_{2n}, for some integer n. Gedeonová [1983] pointed out that this conjecture is false: the lattice of Figure 5.10 shows that not every S graph can be obtained as a direct product of simple even-length cycles (there is no orientation of the covering graph of this lattice which has a nontrivial direct product decomposition). On the other hand, Zelinka [1970] proved

Theorem 5.5.1 *Let L be a finite modular lattice with n atoms. If $G(L)$ is centrally symmetric, then $L \cong 2^n$.*

Replacing modularity by semimodularity, Duffus & Rival [1977], Theorem 4.2, proved the following sharpening of Zelinka's result (see also Gedeonová [1983], Theorem 11).

Theorem 5.5.2 *Let L be a finite semimodular lattice with n atoms. If $G(L)$ is centrally symmetric, then $L \cong 2^n$.*

Now recall that a finite semimodular lattice is graded and has the property that its dual is a strong lattice (see Corollary 4.6.3). Moreover there are graded lattices that are not semimodular and have the property that their duals are strong lattices (e.g. the dual of any nonmodular geometric lattice).

Using the results of Zelinka [1970], Duffus & Rival [1977], and Gedeonová [1980], we now give a further sharpening, replacing "semimodular" by "graded + dually strong." To do this, we shall use the following facts:

(a) Every CS lattice is self-dual (Gedeonová [1980], Theorem 8 (iii)).

(b) The greatest element of a CS lattice L is the irredundant join of all atoms of L (Duffus & Rival [1977], Section 4).

(c) Let L be a lattice whose greatest element is a join of atoms. Then L is coatomistic if and only if its dual lattice L^* is strong (Stern [1991c], Theorem 4).

(d) Let $G = (V, H)$ be a CS graph. Then for every vertex $v \in V$ there exists a unique vertex $\bar{v}$ satisfying $\delta(v, \bar{v}) = \operatorname{diam}(G)$ (see Duffus & Rival [1977], Section 4).

(e) If L is a CS lattice and p is an atom of L, then for every $z \in L$ either $p \leq z$ or $z \leq \bar{p}$ (Gedeonová [1980], Theorem 8(iv)).

(f) Let L be a CS lattice. Then m is a dual atom if and only if $\bar{m}$ is an atom (Duffus & Rival [1977], Section 4).

(g) Let L be a CS lattice, b be a join of atoms, and $0 \prec p \leq b$. Then $b \wedge \bar{p} \prec b$ (Gedeonová [1983], Lemma 2).

Now we can prove the following result (cf. Stern [1996b]):

Theorem 5.5.3 *Let L be a finite (graded) lattice with n atoms such that its dual lattice L^* is strong. If $G(L)$ is centrally symmetric, then $L \cong 2^n$.*

Proof. Let L be a finite lattice with n atoms such that the dual lattice L^* is strong, and assume moreover that the covering graph $G(L)$ is centrally symmetric. Then L is graded (this follows from the definition of a CS graph). It suffices to show that L is modular, whence the assertion follows from Theorem 5.5.1. Since L is self-dual by (a), we first observe that the strongness of L^* implies the strongness of L. By (b) the greatest element of L is a join of atoms. This implies by (c) that L is dually atomistic. From the self-duality of L it follows that it is also atomistic. Next we show that L is lower semimodular. Since L is dually atomistic, it suffices to prove that, for every $z \in L$ and for every dual atom m that is incomparable with z, the relation $z \wedge m \prec z$ holds. Now if $z \in L$ and if m is a dual atom that is incomparable with z, then $m \not\geq z$. Hence it follows by the dual of property (e) that $z \geq \bar{m}$. Condition (f) implies that $\bar{m}$ is an atom. Moreover z is a join of atoms, since L is atomistic. Hence the assumptions of condition (g) are satisfied with $p = \bar{m}$ and $b = z$. Since $\bar{p} = m$, we conclude by (g) that $z \wedge m \prec z$, that is, L is lower semimodular. Upper semimodularity follows, since L is self-dual. Thus L is modular. ∎

A finite atomistic lattice L is strong (see Section 4.6); if, in addition, $G(L)$ is centrally symmetric, then L is graded and its dual L^* is also strong. Hence Theorem 5.5.3 implies the following result of Duffus & Rival [1977], Theorem 4.1.

Corollary 5.5.4   *Let L be a finite atomistic lattice with n atoms. If $G(L)$ is centrally symmetric, then $L \cong 2^n$.*

Notes

Gedeonová [1985] derived a result (see Exercise 1) that may be used to prove the following: If a CS lattice L is p-modular, then L is a direct product of lattices each isomorphic to K_{2n} for some nonnegative integer n (Gedeonová [1985], Theorem 14).

Gedeonová [1990] presented a method of constructing S lattices of arbitrary finite length. In the construction described in this paper, the covering graph of the S lattice is a special type of S graph, called an ST graph. [An ST graph is an S graph such that for every a and b where the distance between a and b is 1, the set $(a, \bar{b})$ is a convex subgraph of G.] Some open questions posed in this paper are the following. Is the covering graph of every S lattice an ST graph? Is every ST graph a covering graph of some S lattice?

Exercises

1. Let L be a CS lattice, and let for every three distinct atoms $p, q, s \in L$ either $r(p \vee q) = 2$ or $r(q \vee s) = 2$ (r denoting the rank). Then L is a direct product of lattices each isomorphic to the lattice K_{2n} for some nonnegative integer n (Gedeonová [1985], Theorem 13).
2. If a, b, c are distinct atoms of an S lattice, then the sublattice generated by a, b, c is a Boolean lattice (Gedeonová [1980], Theorem 13).

References

Duffus, D. and I. Rival [1977] Path length in the covering graph of a lattice, Discrete Math. 19, 139–158.

Gedeonová, E. [1980] Lattices whose covering graphs are S-graphs, Colloq. Math. Soc. J. Bolyai 33, 407–435.

Gedeonová, E. [1983] Lattices with centrally symmetric covering graphs, in: *Contributions to General Algebra, Proc. Klagenfurt Conf., June 10–13, 1982*, pp. 107–113.

Gedeonová, E. [1985] Central elements in CS-lattices, in: *Contributions to General Algebra, Proc. Vienna Conf., June 21–24, 1984*, pp. 143–155.

Gedeonová, E. [1990] Constructions of S-lattices, Order 7, 249–266.

Kotzig, A. [1968a] Centrally symmetric graphs (Russian), Czech. Math. J. 18, 606–615.

Kotzig, A. [1968b] Problem, in: *Beiträge zur Graphentheorie*, (ed. H. Sachs) Teubner, Leipzig, p. 394.

Stern, M. [1991c] Dually atomistic lattices, Discrete Math. 93, 97–100.

Stern, M. [1996b] On centrally symmetric graphs, Math. Bohemica 121, 25–28.

Zelinka, B. [1970] Centrally symmetric Hasse diagrams of finite modular lattices, Czech. Math. J. 20, 81–83.

5.6 Subgraphs of the Covering Graph

Summary. We have indicated that sometimes important properties of a lattice L can be derived from an analysis of the subgraphs of the covering graph $G(L)$. For example, Ward proved that a modular lattice L of finite length is distributive if and only if $G(L)$ contains no subgraph graph-isomorphic to $G(M_3) = K_{23}$. Some properties of a lattice L can be obtained from the subgraphs of $G(L)$ even when $G(L)$ has lattice orientations that are not determined by direct-product decompositions of L. This is the case, for instance, if L is a finite, semimodular, and dismantlable lattice.

Let L be a lattice of finite length, and let S be a subgraph of $G(L)$ graph-isomorphic to $G(M_3)$. It is easy to verify that the vertices of S considered as elements of L determine a cover-preserving sublattice of L isomorphic to M_3. In particular, this implies that a modular lattice L of finite length is distributive if and only if $G(L)$ contains no subgraph graph-isomorphic to $G(M_3)$ (Ward [1939]).

The analogue for arbitrary finite semimodular lattices is not true. That is, for a finite semimodular lattice L the covering graph $G(L)$ may contain a subgraph that is graph-isomorphic to $G(S_7)$ and yet L may be modular. Indeed, 2^3 is modular, while $G(2^3)$ contains a subgraph graph-isomorphic to $G(S_7)$. Nevertheless, for an important class of finite semimodular lattices such an analogue can be proved.

According to Rival [1974], we call a finite lattice L of order n *dismantlable* if there is a chain $L = L_0 \supset L_1 \supset \cdots \supset L_n = \emptyset$ of sublattices of L such that $|L_i - L_{i+1}| = 1$ for each $i = 0, 1, \ldots, n-1$. Note that for each i, $L_i - L_{i+1}$ consists of an element *doubly irreducible* in L_i, that is, an element that is both join-irreducible and meet-irreducible in L_i. It can be shown that a finite lattice is dismantlable in this sense if and only if it is dismantlable in the sense defined in Section 4.2.

Kelly & Rival [1974] proved that a finite lattice is dismantlable if and only if it contains no crown: For an integer $n \geq 6$, a subset $C = \{c_1, c_2, \ldots, c_n\}$ is a *crown* provided that $c_1 < c_2, c_2 > c_3, c_3 < c_4, \ldots, c_{n-1} < c_n, c_n > c_1$ are the only comparability relations that hold in C (see Figure 5.12).

Duffus & Rival [1977], Theorem 6.1, proved the following analogue of Ward's theorem.

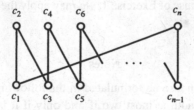

Figure 5.12

Theorem 5.6.1   *Let L be a finite, dismantlable, semimodular lattice. Then L is modular if and only if its covering graph $G(L)$ contains no subgraph graph-isomorphic to $G(S_7)$.*

Proof.   In view of Theorem 3.1.10, L must be modular if $G(L)$ contains no subgraph graph-isomorphic to $G(S_7)$. Let now S be a subgraph of $G(L)$ graph-isomorphic to $G(S_7)$. Since L contains no crowns, it is easy to verify that the vertices of S considered as elements of L constitute a cover-preserving sublattice of L isomorphic to S_7 or S_7^*. It follows that L is nonmodular.   ∎

Duffus & Rival [1977], Theorem 6.4, also proved the following subgraph characterization of dismantlability for finite semimodular lattices.

Theorem 5.6.2   *Let L be a finite semimodular lattice. Then L is dismantlable if and only if its covering graph $G(L)$ contains no subgraph isomorphic to $G(2^3)$.*

Sketch of proof.   If L is a finite lattice and S is a subgraph of $G(L)$ graph-isomorphic to $G(2^3)$, then it is easily shown that the vertices of S, considered as elements of L, determine a cover-preserving sublattice of L isomorphic to 2^3. If L is dismantlable, then it contains no crown (Kelly & Rival [1974]). In particular, L contains no sublattice isomorphic to 2^3. It follows that $G(L)$ contains no subgraph graph-isomorphic to $G(2^3)$.

Assume now that L is a finite semimodular lattice and $G(L)$ contains no subgraph graph-isomorphic to $G(2^3)$. We show by induction on $|L|$ that L is dismantlable. Because of Exercise 2 we may assume without loss of generality that the set $D(L)$ of doubly irreducible elements of L is not empty. By the induction hypothesis we may also suppose that L is *linearly indecomposable* (that is, there is no pair A, B of nonempty subsets of L such that $L = A \cup B$ and $a > b$ for every $a \in A$ and every $b \in B$). Let a be doubly irreducible in L, and let a^* (a') denote the unique upper cover (lower cover) of a in L. Since L is linearly indecomposable, there exists an element $x \in L$ noncomparable to a. If a' is meet irreducible then $x \vee a' = x \vee a$ and $x \wedge a = x \wedge a'$ implies, by Mac Lane's condition (Mac) (see Section 3.1), that there exists an element $d \in L$ satisfying $x \wedge a < d \leq x$ and $(a' \vee d) \wedge a = a'$. Hence a' must be meet-reducible. Let b be an element of L distinct from a such that $b \succ a'$. Then $b \vee a' = a^*$, and it follows that $G(L - \{a\})$ is a subgraph of $G(L)$. Because of Exercise 1, we may apply the induction hypothesis to complete the proof.   ∎

Notes

The preceding result can also be formulated in the following way: A finite semimodular lattice has breadth at most two if and only if it is dismantlable (Rival [1976 b], Corollary 6). (For the notion of breadth see Section 8.4.)

Exercises

1. Let L be a finite semimodular lattice, and let $a \in L$ be doubly irreducible in L. Then $L - \{a\}$ is a semimodular sublattice of L (Rival [1976b]; cf. also Duffus & Rival [1977], Lemma 6.2).

2. Let L be a semimodular lattice of finite length. Then either $D(L)$, the set of all doubly irreducible elements of L, is empty, or L contains a cover-preserving sublattice isomorphic to 2^3. (This result is implicit in Rival [1976b]; cf. also Duffus & Rival [1977], Lemma 6.3).

3. Let L and L' be graded lattices with graph-isomorphic covering graphs. If L is semimodular and dismantlable, then L' is dismantlable (Duffus & Rival [1977], Theorem 6.5).

References

Duffus, D. and I. Rival [1977] Path length in the covering graph of a lattice, Discrete Math. 19, 139–158.

Kelly, D. and I. Rival [1974] Crowns, fences and dismantlable lattices, Canad. J. Math. 26, 1257–1271.

Rival, I. [1974] Lattices with doubly irreducible elements, Canad. Math. Bull. 17, 91–95.

Rival, I. [1976b] Combinatorial inequalities for semimodular lattices of breadth two, Algebra Universalis 6, 303–311.

Ward, M. [1939] A characterization of Dedekind structures, Bull. Amer. Math. Soc. 45, 448–451.

6

Semimodular Lattices of Finite Length

6.1 Rank and Covering Inequalities

Summary. We first recall the matching problem for finite lattices and then proceed to outline Kung's general approach to this problem. This approach leads to a number of new rank and covering inequalities for certain finite lattices. For example, one obtains an extension of the Dowling–Wilson inequalities from finite geometric lattices to arbitrary finite semimodular lattices. Moreover, the approach developed by Kung enabled him to prove a strengthening of Dilworth's covering theorem and to give an affirmative answer to Rival's matching conjecture for modular lattices.

Let J and M be subsets of a finite lattice L. We ask: Does there exist an injection $\sigma : J \to M$ such that $j \leq \sigma(j)$ holds for all $j \in J$? If such an injection exists, then it is called a *matching* of J into M. Note that J and M need not coincide with the sets of join-irreducibles and meet-irreducibles, respectively, of the lattice L. However, in our applications we shall mostly encounter the situation when J is (a subset of) the set of join-irreducibles and M (a subset of) the set of meet-irreducibles of the lattice L.

Let 0_L and 1_L denote the least and the greatest element, respectively, of a finite lattice L. If there is a matching from

$$J(1, L) = J(L) \cup \{0_L\} \quad \text{into} \quad M(1, L) = M(L) \cup \{1_L\},$$

we briefly say that the lattice L has a matching (of the set of join-irreducibles into the set of meet-irreducibles).

Let J and M be subsets of a finite lattice L. The *incidence matrix* $I(M \mid J)$ *of J versus M* is the matrix with rows indexed by M and columns indexed by J whose m, j entry is 1 if $j \leq m$ and 0 otherwise. For example, consider the centered hexagon S_7 in Figure 6.1, and let $J = J(1, S_7) = \{j_0, j_1, j_2, j_3, j_4\}$ and $M = M(1, S_7) = \{m_0, m_1, m_2, m_3\}$ be the subsets of join-irreducibles and meet-irreducibles, respectively.

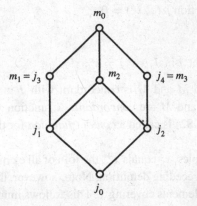

Figure 6.1

The incidence matrix $I(M \mid J)$ of J versus M is

	j_0	j_1	j_2	j_3	j_4
m_0	1	1	1	1	1
m_1	1	1	0	1	0
m_2	1	1	1	0	0
m_3	1	0	1	0	1

In the lattice S_7 there is no matching of $J(1, S_7)$ into $M(1, S_7)$. However, if we restrict ourselves to the subset of consistent join-irreducibles $\{j_0, j_1, j_2\}$, then the map $\sigma : j_0 \rightarrow m_0, \ j_1 \rightarrow m_1, \ j_2 \rightarrow m_2$ is a matching of this subset into the set of meet-irreducibles (compare with Theorem 6.1.12, according to which such a matching always exists). The following result is due to Kung [1985], Theorem 3.

Theorem 6.1.1 *If the incidence matrix $I(M \mid J)$ has rank $|J|$, then there exists a matching from J into M, and hence $|J| \leq |M|$.*

Proof. If the incidence matrix $I(M \mid J)$ has rank $|J|$, there exists a square submatrix H of size $|J|$ in $I(M \mid J)$ that is nonsingular. Since the determinant of H is not zero, there exists a nonzero term in the expansion of the determinant of H. This nonzero term yields an injection $\sigma : J \rightarrow M$ such that for all j, the $\sigma(j), j$ entry is 1, which means that $\sigma(j) \geq j$. $\blacksquare$

Kung provided a general sufficient condition for two subsets J and M of a finite lattice L to have an incidence matrix $I(M \mid J)$ of rank $|J|$ (see below Theorem 6.1.2). This condition involves the notion of a *concordant pair of subsets*, which we are going to describe first. Let J and M be subsets of a finite lattice L. The subset J is said to be *concordant* with M if for every element x in L, either x is in M, or there exists an element x° such that:

(CS1) the Möbius function $\mu(x, x^\circ) \neq 0$

and

(CS2) for every element j in J, $x \vee j \neq x^\circ$.

If J is concordant with M and M is concordant with J in the order dual L^* of L, then we say that J and M are *concordant*. A function $x \to x^\circ$ $(x \in L - M)$ satisfying (CS1) and (CS2) is called a *contact function* for the concordance from J to M.

In many of our examples, x° equals x^+, the join of all elements covering x. This is not required by the preceding definition. Note, however, that for (CS1) to hold, x° must be the join of elements covering x. This follows immediately from Rota's cross-cut theorem (Theorem 4.1.6).

Any finite semimodular lattice satisfies (CS1) (cf. Lemma 4.1.9), and any finite consistent lattice satisfies (CS2) (by the definition of consistency).

The importance of the concept of concordance is seen from the following result on the existence of matchings due to Kung [1987], Theorem 2.2.

Theorem 6.1.2   *Let J and M be subsets of a finite lattice, and let J be concordant with M. Then the incidence matrix $I(M \mid J)$ has rank $|J|$.*

Hence by Theorem 6.1.1 there exists an injection $\sigma : J \to M$ such that $j \leq \sigma(j)$ for all $j \in J$, that is, a matching from J into M. In particular, it follows that $|J| \leq |M|$.

Sketch of proof.   This proof of the theorem uses the finite Radon transform (for details see Kung [1987]). Let L be a finite lattice, and $f : L \to k$ a function from L to a field k. The *Radon transform Tf* of the function f is the function $Tf : L \to k$ defined by

$$Tf(x) = \sum_{y : y \leq x} f(y).$$

In what follows let L^v denote the vector space of functions from L to k, let M^v denote the space of functions from M to k, and let J^v denote the subspace of L^v consisting of functions $f : L \to k$ such that $f(x) \neq 0$ only if $x \in J$. The Radon transform induces a linear transformation $T : J^v \to L^v$, $f \to Tf$. By restricting the domain of Tf to M, we obtain a linear transformation $T^\# : J^v \to M^v$ given by $T^\# f = Tf|_M$. We call $T^\# f$ the upper Radon transform of f. The situation is visualized in Figure 6.2, where R denotes a linear transformation from M^v to L^v. The next step in the proof of Theorem 6.1.2 is to show that if J is concordant with M, then there exists a linear transformation from M^v to L^v, such that the triangle commutes. This is equivalent to showing that the Radon transform Tf can be reconstructed from the upper Radon transform $T^\# f$.

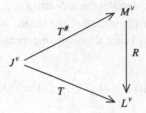

Figure 6.2

If R is given, one may proceed as follows. Using Möbius inversion, the Radon transform $T^{\#}: J^v \to M^v$ can be inverted. It follows that T has rank $|J|$, the dimension of the vector space J^v. Both R and $T^{\#}$ have rank at least $|J|$, because $T = R \circ T^{\#}$. This implies the assertion of Theorem 6.1.2, since the matrix of $T^{\#}$ relative to the standard basis is $I(M \mid J)$. ∎

We refer to Kung [1986b] and Bolker [1987] for an introduction to the finite Radon transform.

Let us now have a look at rank inequalities. For a finite graded lattice, the *Whitney numbers (of the second kind)* are defined by $W_k =$ number of elements of rank k. (For a survey on Whitney numbers see Aigner [1987].) Concerning Whitney numbers of geometric lattices we recall two hitherto unsolved conjectures, the unimodality conjecture and the top-heaviness conjecture:

Unimodality Conjecture (cf. Harper & Rota [1971], p. 209).   Let L be a finite geometric lattice of rank n. Then there exists a positive integer n such that

$$W_0 \le W_1 \le W_2 \le \cdots \le W_{k-1} \le W_k \quad \text{and} \quad W_k \ge W_{k+1} \ge \cdots \ge W_{n-1} \ge W_n.$$

Top-heaviness Conjecture.   Let L be a finite geometric lattice of rank n. Then $W_k \le W_{n-k}$ holds for $0 \le k \le n/2$.

Concerning the latter conjecture, there are some partial affirmative results. For example, the special case $W_1 \le W_{n-1}$ (hyperplane theorem) has been proved independently and in different contexts by several authors (Motzkin [1951], Basterfield & Kelly [1968], Greene [1970], [1975], Heron [1973], Woodall [1976]). This special case has been extended by the *Dowling–Wilson inequalities* (Dowling & Wilson [1975]), for which see Corollary 6.1.4 below.

Here we outline Kung's generalization of the Dowling–Wilson inequalities from finite geometric lattices to finite semimodular lattices. Let us first recall some notation and facts.

For an element x of a finite lattice L, x^+ denotes the join of all upper covers of x, and x_+ the meet of all lower covers of x. L^x denotes the *sublattice of L generated by the elements covering x. L^x is a geometric lattice with least element*

x and greatest element x^+. L_x denotes the *sublattice of L generated by the elements covered by x*. L_x is a coatomistic semimodular lattice with greatest element x and least element x_+. Note that L_x is not atomistic in general. The *reach of* the element x is defined by

$$\text{reach}(x) = r(x^+) - r(x) = r(L^x)$$

where r stands for the rank. In a geometric lattice, $\text{reach}(x) = \text{corank}(x)$, that is, the rank of x in the dual lattice. Dually, the *coreach* of x is defined by

$$\text{coreach}(x) = r(x) - r(x_+) = r(L_x).$$

In a geometric lattice, $\text{coreach}(x) = r(x)$. We shall be concerned with the following subsets of a finite semimodular lattice L:

$$L_k = \{x : r(x) \le k\} \qquad \text{(the } \textit{subset of lower elements} \text{ of } L\text{)};$$
$$U_k = \{x : r(x) \ge n - k\}$$
$$\quad = \{x : \text{corank}(x) \le k\} \qquad \text{(the } \textit{subset} \text{ of } \textit{upper elements} \text{ of } L\text{)};$$
$$R_k = \{x : \text{reach}(x) \le k\};$$
$$S_k = \{x : \text{coreach}(x) \le k\}.$$

Note that $R_1 = M(L) \cup \{1\}$ and in a finite geometric lattice we have $L_k = S_k$ and $U_k = R_k$.

The following extension of the Dowling–Wilson inequalities was proved by Kung [1987], Theorem 6.1. [Recall that $\text{Mod}(L)$ denotes the set of (right) modular elements of L.]

Theorem 6.1.3   Let L be a finite semimodular lattice. Then $L_k \cup (S_k \cap \text{Mod}(L))$ is concordant with R_k. In particular, $|L_k \cup (S_k \cap \text{Mod}(L))| \le |R_k|$.

Proof.   We have to show that the sets in question satisfy conditions (CS1) and (CS2) in the definition of concordant subsets. Suppose $x \notin R_k$. We define x° to be x^+. By Lemma 4.1.9, $\mu(x, x^+) \ne 0$ and hence (CS1) is satisfied. To prove (CS2), let j be an element in $L_k \cup (S_k \cap \text{Mod}(L))$. If $j \in L_k$, then we get (by submodularity) $r(x \vee j) - r(x) \le r(j) - r(x \wedge j) \le k < r(x^+) - r(x)$ and hence $x \vee j \ne x^+$.

Next suppose that $j \in S_k \cap \text{Mod}(L)$ and $x \vee j = x^+$. Since j is a right modular element, the intervals $[x \wedge j, x]$ and $[x, x \vee j] = [x, x^+]$ are isomorphic (by Corollary 2.1.5). Since $[x, x^+]$ contains the geometric lattice L^x and geometric lattices are coatomistic, x is the meet of all the elements covered by x^+. Hence $x \wedge j$ is the meet of all elements covered by j. Thus $x \wedge j \ge j_+$ and we obtain $\text{coreach}(j) = r(j) - r(j_+) \ge r(j) - r(x \wedge j) = r(x^+) - r(x) = \text{reach}(x) > k$, contradicting the assumption that j is in S_k. Hence $x \vee j \ne x^+$. ∎

If L is a geometric lattice, then $L_k = S_k$ and $U_k = R_k$, and Theorem 6.1.3 yields the Dowling–Wilson inequalities (Dowling & Wilson [1975]):

Corollary 6.1.4 *Let L be a finite geometric lattice of rank n. Then L_k is concordant with U_k. In particular, we have $W_0 + W_1 + \cdots + W_k \leq W_{n-k} + W_{n-k+1} + \cdots + W_{n-1} + W_n$ (for $0 \leq k \leq n/2$).*

For what follows we need some additional concepts. The *geometric coreach* (*g-coreach* for short) of an element x of a finite semimodular lattice L is the maximum rank of a geometric lattice that can be embedded into an interval of the form $[b, x]$ in L. We have g-coreach $(x) \leq \text{coreach}(x)$, since geometric lattices are coatomistic. In addition to the above-introduced subsets L_k, U_k, R_k, S_k we also consider the subset $G_k = \{x : g\text{-coreach}(x) \leq k\}$ and the *subset of the modular Whitney numbers M_k* [= number of (right) modular elements in L of rank k]. We have (cf. Kung [1987], Theorem 6.1):

Theorem 6.1.5 *Let L be a finite semimodular lattice. Then, in the order dual of L, $R_k \cap \text{Mod}(L)$ is concordant with G_k. In particular, $|R_k \cap \text{Mod}(L)| \leq |G_k|$.*

Proof. Let x be an element not in G_k. Then there exists an element $b \leq x$ in L and a geometric lattice G such that G embeds into $[b, x]$ and $r(G) = r(x) - r(b) > k$. Choose one such element b, and define x° to be b. By Lemma 4.1.9 we have $\mu(x^\circ, x) \neq 0$, which means that (CS1) is satisfied in the order dual of L. To prove (CS2), let j be an element in $R_k \cap \text{Mod}(L)$, and suppose $j \wedge x = x^\circ$. Since j is a modular element, the interval $[j, j \vee x]$ is isomorphic to $[x^\circ, x]$ (by Corollary 2.1.5). Since $[x^\circ, x]$ contains a geometric lattice of rank $r(x) - r(x^\circ)$ as a subinterval and geometric lattices are atomistic, x is the join of elements covering x°. Hence $j \vee x$ is the join of elements covering j, and $j \vee x \leq j^+$. This implies that reach$(j) = r(j^+) - r(j) \geq r(j \vee x) - r(j) > k$, contradicting the assumption that x is in R_k. It follows that (CS2) holds in the order dual of L. ∎

For the special case of geometric lattices we obtain (cf. Kung [1987], Corollary 6.4)

Corollary 6.1.6 *Let L be a finite geometric lattice of rank n. Then, in the order dual of L, $U_k \cap \text{Mod}(L)$ is concordant with L_k. In particular,*

$$W_0 + W_1 + \cdots + W_k \geq M_{n-k} + M_{n-k+1} + \cdots + M_n.$$

Observing that $G_1 = S_1$, $R_1 = M(L) \cup \{1_L\} = M(1, L)$, and $S_1 = J(L) \cup \{0_L\} = J(1, L)$, we obtain

Corollary 6.1.7 *In a finite semimodular lattice, the set of modular meet-irreducibles is concordant with the set of join-irreducibles in the order dual. In particular,* *the number of modular meet-irreducibles is less than or equal to the number of join-irreducibles.*

Corollary 6.1.8   *In a finite modular lattice,* $|S_k| = |R_k|$.

For $k = 1$ this means that in a finite modular lattice the number of join-irreducibles equals the number of meet-irreducibles (which is also a consequence of Dilworth's covering theorem – see Theorem 6.1.9 below).

Let us now turn to covering inequalities. We shall again follow Kung's approach, who generalized Dilworth's covering theorem and simultaneously solved Rival's matching conjecture. Surveys can be found in Kung [1986b] and Rival [1990]. Dilworth's covering theorem for modular lattices (Dilworth [1954]), at which we hinted in Section 1.6, is the fundamental result in the theory of covering relations:

Theorem 6.1.9   *In a finite modular lattice, the number of elements covering precisely k elements equals the number of elements covered by precisely k elements.*

We introduce some more notation. Let $J_k(L)$ denote the set of elements having exactly k lower covers, and $M_k(L)$ the set of elements having exactly k upper covers. Thus we have $J_1(L) = J(L)$ and $M_1(L) = M(L)$ [the subsets $J(L)$ and $M(L)$ were introduced in Section 1.3]. With this notation the statement of Dilworth's covering theorem is: In a finite modular lattice L, $|J_k(L)| = |M_k(L)|$ for any integer $k \geq 0$.

In particular, this theorem settled the conjecture that, in a finite modular lattice L, the number of join-irreducibles ($\neq 0$) equals the number of meet-irreducibles ($\neq 1$), that is, $|J(L)| = |M(L)|$. Elementary proofs of Dilworth's covering theorem were given by Ganter & Rival [1973], Kurinnoi [1973], and Reuter [1985].

In 1972 Rival posed the following problem (see Rival [1976b]), which was mentioned in Section 1.6: Let L be a finite modular lattice. Is there a bijection $\sigma : J(1, L) \to M(1, L)$ such that $x \leq \sigma(x)$ holds for all $x \in J(1, L)$? Such a mapping would be a matching in the sense defined above. More precisely, it would be a simultaneous matching of the set of join-irreducibles into the set of meet-irreducibles and of the set of meet-irreducibles into set of join-irreducibles.

Rival conjectured that the above problem has an affirmative answer (this is *Rival's matching conjecture*). He had already observed that such a bijection always exists in finite distributive lattices. An affirmative answer was also known for the lattice of subspaces of a finite projective incidence geometry and thus for all finite complemented modular lattices. Duffus [1982] made an attack on the general modular case (see also Duffus [1985]).

There were also affirmative answers concerning the existence of matchings of $J(1, L)$ into $M(1, L)$ for certain classes of finite semimodular lattices L. For example, the hyperplane theorem $W_1 \le W_{n-1}$ mentioned above means the existence of a matching of the set of join-irreducibles (atoms) into the set of meet-irreducibles (coatoms) for finite geometric lattices. Moreover, Rival [1976b] proved that a finite semimodular lattice of breadth at most 2 possesses a matching of the set of meet-irreducibles into the set of join-irreducibles (see Theorem 8.4.3). On the other hand, we had already seen that there are nonmodular semimodular lattices L that do not have matchings of $J(1, L)$ into $M(1, L)$, such as S_7 (see Figure 6.1).

We are now going to describe Kung's approach to the matching conjecture. Generalizing the notion of join-irreducibles and meet-irreducibles we first introduce some more notation. Let L be a finite lattice and let k be a positive integer. An element j is said to be k-*covering* if it covers *at most* k elements. Dually, an element m is said to be k-*covered* if it is covered by *at most* k elements. Let $J(k, L)$ be the set of k-covering elements and $M(k, L)$ be the set of k-covered elements.

Kung developed an inductive technique for reconstructing Radon transforms and proved thereby the following strengthening of both Dilworth's covering theorem and Rival's matching conjecture (see Kung [1986b], Theorem 3.1.3, and Kung [1987], Corollary 7.3).

Theorem 6.1.10 Let L be a finite modular lattice. Then $J(k, L)$ and $M(k, L)$ are concordant. It follows that there exists a matching from $J(k, L)$ in $M(k, L)$ and vice versa. In particular, $|J(k, L)| = |M(k, L)|$.

This result can be derived from the following generalization for semimodular lattices (see Kung [1986b], Proposition 3.1.4).

Theorem 6.1.11 In a finite semimodular lattice L the incidence matrix $I(M(k, L)$ $|J(k, L) \cap \mathrm{Mod}(L))$ has rank $|J(k, L) \cap \mathrm{Mod}(L)|$.

Sketch of proof. We want to apply Theorem 6.1.2 and show therefore that the subset $J(k, L) \cap \mathrm{Mod}(L)$ is concordant with the subset $M(k, L)$. To see this, assume that $x \notin M(k, L)$, that is, x is covered by at least $k + 1$ elements. We show that (CS1) and (CS2) are satisfied for $x^\circ = x^+$, the join of all elements in L covering x. The interval $[x, x^+]$ contains the geometric sublattice L^x generated by the elements covering x. By Lemma 4.1.9 we have $\mu(x, x^+) \ne 0$, and hence (CS1) holds. Since the geometric lattice L^x contains at least $k + 1$ atoms, we get from Corollary 6.1.4 (case $k = 1$) that L^x contains at least $k + 1$ dual atoms. By Kung [1986b], Proposition 2.4.4, these dual atoms are covered by x^+. Hence x^+ covers at least $k + 1$ elements.

Consider now an element $j \in J(k, L) \cap \mathrm{Mod}(L)$, and assume $j \vee x = x^+$, that is, (CS2) is violated. Now $j \in \mathrm{Mod}(L)$ implies $x M j$ for all $x \in L$, and this yields $j M x$ (by M-symmetry) and thus also $j M^* x$ (by Maeda & Maeda [1970], Lemma 1.2).

By Corollary 2.1.5 the intervals $[x \wedge j, j]$ and $[x, x^+]$ are isomorphic. Because x^+ covers at least $k + 1$ elements, this isomorphism implies that j, too, covers at least $k + 1$ elements, contradicting the assumption that $j \in J(k, L)$. Thus for all $j \in J(k, L) \cap \mathrm{Mod}(L)$ we have $j \vee x \neq x^+$, that is, (CS2) holds. ∎

Theorem 6.1.10 follows immediately from the preceding result, since every element of a modular lattice is a modular element and the dual of a modular lattice is also modular.

Originally Kung [1985] used a different approach in his solution of Rival's matching conjecture, working with consistent join-irreducibles. This approach leads to covering inequalities for arbitrary finite lattices (i.e. without the additional assumption of upper semimodularity). For the following result cf. Kung [1986b], Theorem 3.1.5, and Kung [1987], Theorem 9.1.

Theorem 6.1.12 *Let $J_c(1, L)$ be the set of consistent join irreducibles in a finite lattice L. Then the incidence matrix $I(M(1, L)|J_c(1, L))$ has rank $|J_c(1, L)|$, and hence $|J_c(1, L)| \leq |M(1, L)|$.*

Proof. In order to be able to apply Theorem 6.1.2 one has to show that the subset $J_c(1, L)$ is concordant with the subset $M(1, L)$. In other words, we have to show that, for $x \notin M(1, L)$, the conditions (CS1) and (CS2) hold with a suitably chosen element x°. Now if x is not a meet-irreducible, then there exists an element u such that u does not cover x and $\mu(x, u) \neq 0$. To see this recall the following properties of the Möbius function: $\sum_{y \geq x} \mu(x, y) = 0$, $\mu(x, x) = 1$, and $\mu(x, y) = -1$ if y covers x (cf. Section 4.1). Since there are at least two elements covering x, there must be an element u not covering x such that $\mu(x, u) \neq 0$. Choose such a u, and define x° to be u. Then (CS1) holds by definition. To show (CS2), note that if u is a join-irreducible in $[x, 1]$, then $\mu(x, u) = 0$. It follows that u is not a join-irreducible in $[x, 1]$. Consistency therefore implies $j \vee x \neq u$. ∎

For an elementary proof of this result see Kung [1985].

Corollary 6.1.13 *Let L be a finite consistent lattice. Then the incidence matrix $I(M(1, L)|J(1, L))$ has rank $|J(1, L)|$, and hence there exists a matching from $J(1, L)$ into $M(1, L)$.*

This result (cf. Kung [1986b], Corollary 3.1.6, and Kung [1987], Corollary 9.2) implies Rival's conjecture, since finite modular lattices are consistent. Recall, however, that finite semimodular lattices are not consistent in general. Therefore Corollary 6.1.13 and the case $k = 1$ of Theorem 6.1.11 generalize the case $k = 1$ of Dilworth's covering theorem in different directions.

Another application of Theorem 6.1.12 is the following generalization (due to Kung [1986b], Corollary 3.1.7) of the hyperplane theorem $W_1 \leq W_{n-1}$ from finite geometric lattices to finite coatomistic semimodular lattices:

Corollary 6.1.14 *Let L be a finite coatomistic semimodular lattice of rank n. Then $W_1 \leq W_{n-1}$, that is, the number of atoms is at most the number of coatoms.*

Proof. Semimodularity implies that the atoms are consistent join-irreducibles. Since the lattice is coatomistic, the assertion follows from the preceding theorem. ∎

A further application of Theorem 6.1.12 concerns semimodular lattices of breadth 2 (cf. Rival [1976b]). For this we refer to Section 8.4.

In the modular case many of the rank or covering inequalities considered above become equalities. Conversely, the following result has been proved (cf. Greene [1970]): If L is a finite geometric lattice of rank n such that $W_1 = W_{n-1}$ then L is modular. This is generalized by the Dowling–Wilson equality: Let L be a finite geometric lattice of rank n in which $W_0 + W_1 + \cdots + W_k = W_{n-k} + W_{n-k+1} + \cdots + W_{n-1} + W_n$ for some k, $0 \leq k \leq n/2$. Then L is modular (Dowling & Wilson [1975]).

It follows that for finite geometric lattices this equality is a necessary and sufficient condition for modularity. On the other hand, for a finite consistent lattice L we have $|J(1, L)| \leq |M(1, L)|$ (see Theorem 6.1.12), but the equality $|J(1, L)| = |M(1, L)|$ does not imply modularity, as the lattice N_5 shows. These and other facts raise the following general problem: For which inequalities is equality a necessary and sufficient condition for a finite lattice to be modular? The following positive result is due to Race [1986]: Let L be a finite semimodular lattice in which the number of consistent join-irreducibles equals the number of meet-irreducibles. Then L is modular.

We also refer to Section 8.5, where we consider modularity in consistent lower semimodular lattices.

Notes

Kung has also found another variation on the Dowling–Wilson inequalities by looking at the modular Whitney numbers M_k ($=$ number of modular elements of rank k). If L is a finite geometric lattice of rank n, then for $0 \leq k \leq n/2$, one has $W_0 + W_1 + \cdots + W_k \geq M_{n-k+1} + \cdots + M_{n-1} + M_n$ (Kung [1986b], Theorem 2.5.9). The special case $k = 1$ was proved by Brylawski [1975].

A finite lattice L is said to be *linearly indecomposable* if there do not exist $x, y \in L$ with $x < y$ such that $z \leq x$ or $z \geq y$ for all $z \in L - \{x, y\}$. This condition means that the diagram of L cannot be disconnected by removing one edge. Reuter

[1987] proved that if L is a linear indecomposable modular lattice, then there is a matching from $J_1(L)$ to $M_1(L)$.

Exercises

1. Show that in a finite semimodular lattice the number of modular join-irreducibles is less than or equal to the number of meet-irreducibles (Kung [1986b], Corollary 4.2.3).
2. A lattice is called k-consistent if coreach$(x) \leq k$ implies coreach$(x \vee y)$ in $[y, 1]$ is less than or equal to coreach(x) for all $y \in L$. Show that this condition is only apparently stronger than consistency: a consistent semimodular lattice of finite length is k-consistent for each k (Race [1986], Lemma IV.4.1).
3. Let L be a consistent semimodular lattice of finite length. L is modular if and only if its dual is consistent (Race [1986], Lemma IV.5.1).
4. Prove that direct products preserve matchings (Duffus [1982]).

References

Aigner, M. [1987] Whitney numbers, in: White [1987], pp. 139–160.

Basterfield, J. G. and L. M. Kelly [1968] A characterization of sets of n points which determine n hyperplanes, Proc. Cambridge Phil. Soc. 64, 585–588.

Bogart, K. P., R. Freese, and J. P. S. Kung (eds.) [1990] *The Dilworth Theorems. Selected Papers of Robert P. Dilworth*, Birkhäuser, Boston.

Bolker, E. [1987] The finite Radon transform, Contemp. Math. 63, 27–50.

Brylawski, T. [1975] Modular constructions for combinatorial geometries, Trans. Amer. Math. Soc. 203, 1–44.

Dilworth, R. P. [1954] Proof of a conjecture on finite modular lattices, Ann. Math. 60, 359–364.

Dowling, T. A. and R. M. Wilson [1975] Whitney number inequalities for geometric lattices, Proc. Amer. Math. Soc. 47, 504–512.

Duffus, D. [1982] Matching in modular lattices, J. Combin. Theory Ser. A 32, 303–314.

Duffus, D. [1985] Matching in modular lattices, Order 1, 411–413.

Ganter, B. and I. Rival [1973] Dilworth's covering theorem for modular lattices: a simple proof, Algebra Universalis 3, 348–350.

Greene, C. [1970] A rank inequality for finite geometric lattices, J. Combin. Theory 9, 357–364.

Greene, C. [1975] An inequality for the Möbius function of a geometric lattice, Stud. Appl. Math. 54, 71–74.

Harper, L. H. and G.-C. Rota [1971] Matching theory, an introduction, in: *Advances in Probability 1* (ed. P. Ney), Marcel Dekker, New York, pp. 169–215.

Heron, A. P. [1973] A property of the hyperplanes of a matroid and an extension of Dilworth's theorem, J. Math. Anal. Appl. 42, 119–131.

Kung, J. P. S. [1985] Matchings and Radon transforms in lattices I. Consistent lattices, Order 2, 105–112.

Kung, J. P. S. [1986b] Radon transforms in combinatorics and lattice theory, in: *Combinatorics and Ordered Sets*, Contemporary Mathematics, Vol. 57 (ed. I. Rival), pp. 33–74.

Kung, J. P. S. [1987] Matchings and Radon transforms in lattices II. Concordant sets, Math. Proc. Camb. Phil. Soc. 101, 221–231.

Kurinnoi, G. C. [1973] A new proof of Dilworth's theorem (Russian), Vestnik Khar'kov Univ. Mat.-Mech. 38, 11–15.

Maeda, F. and S. Maeda [1970] *Theory of Symmetric Lattices*, Springer-Verlag, Berlin.

Motzkin, T. S. [1951] The lines and planes connecting the points of a finite set, Trans. Amer. Math. Soc. 70, 451–464.

Race, D. M. [1986] Consistency in lattices, Ph.D. thesis, North Texas State Univ., Denton, Tex.

Reuter, K. [1985] Counting formulas for glued lattices, Order 1, 265–276.

Reuter, K. [1987] Matchings for linearly indecomposable modular lattices, Discrete Math. 63, 245–247.

Rival, I. [1976b] Combinatorial inequalities for semimodular lattices of breadth two, Algebra Universalis 6, 303–311.

Rival, I. [1990] Dilworth's covering theorem for modular lattices, in: Bogart et al. [1990], pp. 261–264.

White, N. L. (ed.) [1987] *Combinatorial Geometries*, Encyclopedia of Mathematics and Its Applications, Vol. 29, Cambridge Univ. Press, Cambridge.

Woodall, D. R. [1976] The inequality $b \geq v$, in: *Proc. Fifth British Combinatorial Conf. (University of Aberdeen, 1975)*, Congressus Numerantium 15, Utilitas Math., Winnipeg, Manitoba, pp. 661–664.

6.2 Embeddings

Summary. Following the approach of Grätzer & Kiss [1986], we consider the question of embedding a finite lattice endowed with a so-called pseudorank function isometrically into a finite semimodular lattice. By their approach Grätzer and Kiss also prove earlier results of Dilworth and Finkbeiner.

A map $f : K \to L$ between two posets K and L is an *order embedding* if $a \leq b$ is equivalent to $f(a) \leq f(b)$ for all $a, b \in K$ (in particular, f is injective). An order embedding of K into L is a one-to-one, order-preserving map of K to L whose inverse is also order-preserving. Recall that an order embedding is cover-preserving if $a \prec b$ implies $f(a) \prec f(b)$. An order embedding f between two lattices K and L is a *join embedding* [*meet embedding*] if $f(a \vee b) = f(a) \vee f(b)$ [$f(a \wedge b) = f(a) \wedge f(b)$] holds for all $a, b \in K$. It is a *lattice embedding* if it is both a join and a meet embedding. A lattice embedding will also briefly be called an *embedding*. Thus a lattice embedding of a lattice K into a lattice L is a one-to-one map f of K to L such that, for each $a, b \in K$, $f(a \wedge b) = f(a) \wedge f(b)$ and $f(a \vee b) = f(a) \vee f(b)$. We say that K is lattice-embeddable in L if there is a lattice embedding of K into L. Usually we identify a lattice embedding f of K into L with the sublattice $f(K)$ of L. For a survey on lattice embeddings see Rival & Stanford [1992].

Dilworth observed the following construction. Consider the Boolean lattice of all subsets of a four-element set (Figure 6.3), and remove the atoms (Figure 6.4). The lattice in Figure 6.4 is atomistic but not semimodular; it is graded by its rank

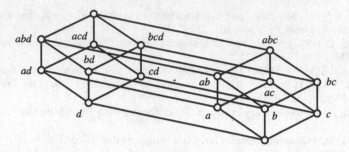

Figure 6.3

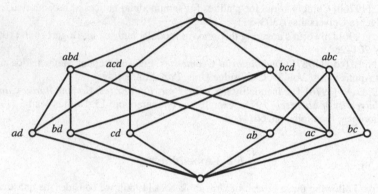

Figure 6.4

function r, and it is *semimodular above the atoms*, that is,

$$a \wedge b \neq 0 \quad \text{implies} \quad r(a \vee b) + r(a \wedge b) \leq r(a) + r(b).$$

Atomistic lattices of finite length that are semimodular above the atoms have been called *quasimodular* by Dilworth [1944]. The quasimodular lattice of Figure 6.4 arises from a "truncation" of the Boolean lattice 2^4 shown in Figure 6.3.

More generally one gets natural examples of quasimodular lattices by truncating geometric lattices instead of Boolean lattices: take a geometric lattice L of rank n, and identify all the elements of rank less than a fixed positive integer k. Using Dilworth's construction, one obtains a geometric lattice $D_k(L)$ of rank $n - k + 1$, which contains a copy of the upper $n - k$ levels of L. The lattice $D_k(L)$ is now called the kth *Dilworth truncation* of L. For example, the second Dilworth truncation of the Boolean lattice of all subsets of an n-element set S is isomorphic to the lattice of partitions on S.

Dilworth [1944] proved that any finite quasimodular lattice can be embedded in a finite geometric lattice. Dilworth's construction represents the elements of a quasimodular lattice as closed sets (flats) of a matroid. This representation yields an injection of the quasimodular lattice into the geometric lattice of closed sets of the

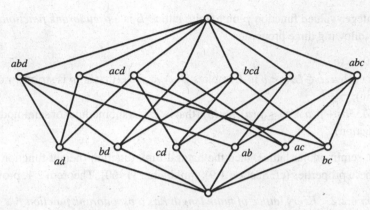

Figure 6.5

matroid that preserves the rank and meets, but not necessarily joins. In other words, this injection is a rank-preserving meet embedding, but it is not a lattice embedding.

The lattice of Figure 6.4 is embedded by this procedure in the lattice shown in Figure 6.5, which is isomorphic to the partition lattice Π_4 on a set of four elements.

Dilworth also used his construction to show that every modular lattice of rank 3 can be embedded as a sublattice into a complemented modular lattice of rank 3. This result is a counterpart to the examples of modular lattices (of rank at least 4) that cannot be embedded into complemented modular lattices (cf. Hall & Dilworth [1944]).

For more details on Dilworth truncations of geometric lattices see Mason [1977] and Kung [1990a]. The role of Dilworth's construction (the Dilworth completion) in combinatorial optimization is analyzed by Nguyen [1986], Faigle [1990], and Fujishige [1991].

The general problem behind Dilworth [1944] is the question whether a finite lattice can be (lattice-)embedded into a finite geometric lattice. Around 1950 Dilworth succeeded in proving this result, which was published much later in Crawley & Dilworth [1973], Theorem 5.3.6:

Theorem 6.2.1 *Every finite lattice can be lattice-embedded in a finite geometric lattice.*

The proof is based on the ideas used by Dilworth [1944] to show that any finite quasimodular lattice can be embedded in a finite geometric lattice. For a sketch of the proof see also Kung [1990b].

Another approach to embedding finite lattices into finite semimodular lattices is due to Finkbeiner [1960]. Finkbeiner introduced what he called a pseudorank function. Such a function is a special submodular (semimodular) function.

An integer-valued function p on a finite lattice L is a *pseudorank function* if it has the following three properties:

(i) $p(0) = 0$;
(ii) for each $a, b \in L$, $a < b$ in L implies $p(a) < p(b)$ (that is, p is strictly monotone);
(iii) $p(a \vee b) + p(a \wedge b) \leq p(a) + p(b)$ (that is, p is a submodular or semimodular function).

If L is a semimodular lattice, then the natural rank function (height function) on L has these properties (cf. Section 1.9). Finkbeiner [1960], Theorem 2.1, proved

Theorem 6.2.2 *Every lattice of finite length has a pseudorank function.*

In fact, if L is a lattice of finite length with rank function r, then by setting $p(1) = 2^{r(1)-1}$ and $p(a) = 2^{r(1)-1} - 2^{r(1)-r(a)-1}$ for $a \neq 1$ we obtain a pseudorank function on L.

The following example of Grätzer & Kiss [1986] shows that the preceding pseudorank function is the smallest possible one: For each $n \geq 1$ one can construct a lattice L_n of length n such that, for each pseudorank function p on L_n one has $p(1) \geq 2^{n-1}$. The lattice L_n is constructed inductively (Figure 6.6 shows L_4). Let L_1 be the two-element chain. To construct L_{n+1}, consider two disjoint copies of L_n, add a new 0, and glue the two unit elements together. Applying the submodularity of p to the two zeros of the two copies of L_n in L_{n+1}, we see by induction that L_n satisfies the requirement.

Let p be a pseudorank function on the lattice L, and let L be a sublattice of a semimodular lattice S. We say that L is embedded into S *isometrically* with respect to p if p is the restriction of the rank function of S to L. Finkbeiner's main result is (cf. Finkbeiner [1960])

Theorem 6.2.3 *Every finite lattice L with pseudorank function p can be lattice-embedded in a finite, semimodular lattice K such that $p = r \mid L$, where $r \mid L$ is the*

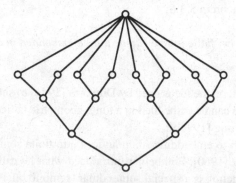

Figure 6.6

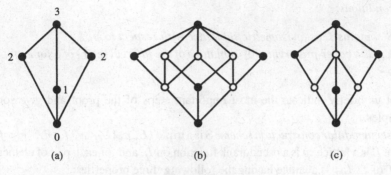

Figure 6.7

restriction of the rank function r of K to L. In other words: Every lattice L has an isometric embedding with respect to any of its pseudorank functions.

Finkbeiner calls his construction *normal embedding* (cf. Notes). Let's illustrate this by an example. Consider the lattice L of Figure 6.7(a): it is the lattice M_3 endowed with the pseudorank function indicated by the numbers written beside the elements. This pseudorank function is *not* the rank function (height function) on the modular lattice M_3.

Figure 6.7(b) shows a semimodular lattice S into which L is isometrically embedded with respect to the pseudorank function given in Figure 6.7(a). This is Finkbeiner's normal embedding (it is an isometric embedding of the given lattice into a semimodular lattice).

One might conjecture that S is isomorphic to a sublattice of any semimodular lattice that contains L as a sublattice and preserves the pseudorank function originally defined on L. However, the lattice in Figure 6.7(c) reveals that no general embedding exists that is minimal in this sense: This lattice is clearly the smallest embedding lattice posssible, and it is also a lattice into which M_3 is isometrically embedded with respect to the pseudorank function given in Figure 6.7(a). However, neither of the lattices shown in Figure 6.7(b) and Figure 6.7(c) is a sublattice of the other.

Grätzer & Kiss succeeded in blending the results of Dilworth and Finkbeiner:

Theorem 6.2.4 *Every finite lattice L with pseudorank function p can be lattice-embedded in a finite geometric lattice S such that $p = r \mid L$, where r is the rank function of S.*

More precisely, the main result of Grätzer & Kiss [1986] is

Theorem 6.2.5 *Let p be a pseudorank function on a finite lattice L. For each interval $[e, f]$ of L, let a finite semimodular lattice L_{ef} of length at most $p(f) - p(e)$ be given. Then there is a finite geometric lattice S satisfying the following* *two conditions:*

(i) *S contains L as an isometric sublattice with respect to p;*
(ii) L_{ef} *is a cover-preserving {0}-sublattice of the interval* $[e, f]$ *of S for e, f ∈ S,
     e < f.*

Let us briefly indicate the most important steps of the proof and give some examples.

A *semimodular construction scheme* S is a triple $\langle L, p, \{L_{ef} : e, f \in L, \ e < f\}\rangle$ where L is a lattice, p is a pseudorank function on L, and for each pair of elements $e < f$ of L, L_{ef} is a lattice having the following three properties:

(I) L_{ef} is semimodular;
(II) L_{ef} has a unique dual atom;
(III) for each join-irreducible f ($\neq 0$) of L with lower cover f', the condition

$$(r(L_{ef}) - 1 : e < f) \geq p(f) - p(f') - 1$$

holds.

The lattice L is called the *frame* of S, and the lattices L_{ef} are called the *blocks* of S. In Finkbeiner's approach all L_{ef} are chains; this leads to his isometric embedding (normal embedding) of L into a semimodular lattice. Condition (II) is a technical one, while condition (III) secures that the L_{ef} are large enough.

Let $S = \langle L, p, \{L_{ef} : e, f \in L, \ e < f\}\rangle$ be a semimodular construction scheme. With this scheme the embedding is performed in three steps:

Step 1. Make a set P by inserting for all $e, \ f \in L, e < f$, the lattice L_{ef} into the interval $[e, f]$ of L. Form the disjoint union of L and of all L_{ef}; identify the zero of L_{ef} with e and the unit of L_{ef} with f. The resulting set P can be partially ordered in the following way: For all $x, y \in P, x \leq y$ if and only if either x and y are in the same L_{ef} and $x \leq y$ holds in this lattice, or $\bar{x} \leq y$ holds in L. (Notation: for $x \in L_{ef}$ and $x \neq e, f$ we put $\underline{x} = e$ and $\bar{x} = f$; for $a \in L$ we put $\underline{a} = \bar{a} = a$.) P is called the poset corresponding to S, and it is also denoted by $P(S)$. The poset P is even a lattice, but it is not semimodular, in general.

Step 2. Let now $I (\neq \emptyset)$ be a special order ideal (hereditary subset) of P with the property that I intersects L as well as each L_{ef} in a principal ideal or in the empty set. These order ideals form a lattice $\mathfrak{I}(P)$ with respect to set inclusion. Next a suitable extension $\hat{p}$ of p to $\mathfrak{I}(P)$ is defined.

Step 3. In the third step, we select a meet subsemilattice S of $\mathfrak{I}(P)$, which consists of those ideals $I \in \mathfrak{I}(P)$ that are maximal in the sense that $J > 1$ implies that $\hat{p}(J) > \hat{p}(I)$. Ideals with this special property are called *trimmed ideals*. The meet subsemilattice of the trimmed ideals is denoted by $T_P(\mathfrak{I}(P))$. Finally it is shown

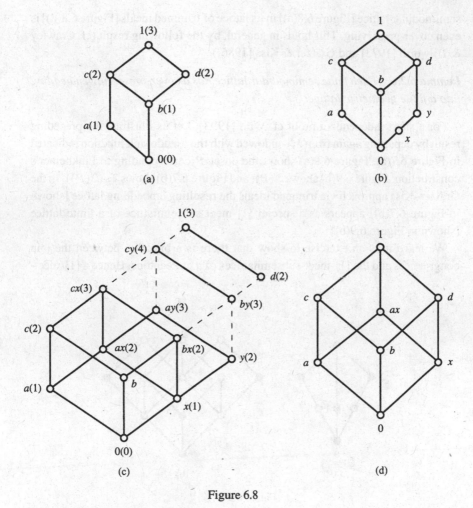

Figure 6.8

that $T_P(\Im(P))$ is a geometric lattice into which the original lattice L is isometrically embedded (with respect to the pseudorank function given on L).

For example, consider the nongeometric semimodular (in fact, distributive) lattice L shown in Figure 6.8(a) together with the pseudorank function p given by the numbers in parentheses. Figure 6.8(b) shows the poset P, and Figure 6.8(c) shows the lattice $\Im(P)$ together with the extension $\hat{p}$ of the pseudorank function p given by the numbers in parentheses. Finally, Figure 6.8(d) shows the geometric lattice $T_P(\Im(P))$ of trimmed ideals, which is a 1-meet-subsemilattice of the lattice in Figure 6.8(c). With respect to the given pseudorank function, the original lattice of Figure 6.8(a) is isometrically embedded into the geometric (in fact, Boolean) lattice of Figure 6.8(d). In fact, we see that the embedding of the original semimodular lattice [Figure 6.8(a)] in its lattice of trimmed ideals [Figure 6.8(d)] is even cover-preserving. This holds in general, by the following result (cf. Crawley & Dilworth [1973] and Grätzer & Kiss [1986]).

Lemma 6.2.6 *Each finite semimodular lattice has a cover-preserving embedding into a finite geometric lattice.*

For a short independent proof cf. Wild [1993]. Let us illustrate the preceding result by departing again from M_3 endowed with the pseudorank function indicated in Figure 6.7(a). Figure 6.9(a) shows the poset P corresponding to Finkbeiner's construction. Figure 6.9(b) shows $\mathfrak{S}(P)$, and Figure 6.7(b) shows $T_P(\mathfrak{S}(P))$. In the Grätzer–Kiss approach via trimmed ideals the resulting embedding lattice [shown in Figure 6.7(b)] appears as a special {1}-meet-subsemilattice of a finite lattice [shown in Figure 6.9(b)].

We leave it as an exercise to show that there is a bijection between the join congruences and the {1}-meet-subsemilattices of a finite lattice. Hence a {1}-meet-

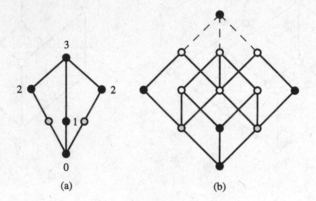

(a)   (b)

Figure 6.9

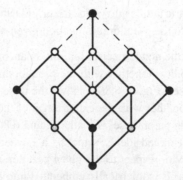

Figure 6.10

subsemilattice of a finite lattice can be identified with a join congruence of this lattice.

Instead of considering the join congruence induced by trimmed ideals, one may take another join congruence that leads to a different isometric embedding of M_3 with the given pseudorank function. For example, the join congruence shown in Figure 6.10 leads to the embedding lattice shown in Figure 6.7(c).

Grätzer & Kiss [1986], Problem 1, ask whether it is possible to obtain their results by considering another "natural" join congruence on $\Im(P)$.

Notes

Let L be a finite lattice, and let p be a pseudorank function on L. For $e,\ f \in L$, $e < f$, define the lattice L_{ef} to be a chain of length $p(f) - p(e)$ if f is join-irreducible ($\neq 0$) and e is the lower cover of f, and let L_{ef} be the two-element lattice otherwise. Let P be the poset associated with this construction scheme. A subset N of the set $J(P)$ of all join-irreducibles of the lattice P is called *normal* if $|N| = p(\bigvee N)$ and if, for each $M \subseteq N$, the inequality $|M| \leq (\bigvee M)$ holds. A subset I of P is said to be closed if it is hereditary and $N \in I$ for every normal subset N of I. The closed subsets of P form a lattice, and the map assigning to each $a \in L$ the closed subset (a] of P is Finkbeiner's normal embedding of L with respect to p. Grätzer & Kiss [1986], Proposition 19, prove that a subset I of P is a trimmed ideal if and only if it is a closed subset. This shows that Finkbeiner's embedding is a special case of the Grätzer–Kiss approach.

Whitman [1946] proved that every lattice is lattice-embeddable in a partition lattice and conjectured that every finite lattice is lattice-embeddable in a finite partition lattice. This is the lattice embedding problem for partition lattices. Pudlák & Tůma [1977], [1980] settled this longstanding conjecture of Whitman affirmatively. However, the earlier results of Dilworth remain important because of the embedding techniques.

It is obvious that many rank functions can be defined on a lattice. In fact, many submodular (semimodular) rank functions can be given [cf. e.g. Figure 6.7(a): M_3 is endowed with a submodular rank function distinct from the natural rank function (height function)]. Grätzer & Kiss [1986], Problem 2, ask: Is there a way to survey all possible pseudorank functions on a finite lattice?

It seems that the distance of a lattice L from being semimodular is measured by the existence of a "small" pseudorank function p on L. Let n be the length of the *shortest* maximal chain between 0 and 1 of L, and define the *deviance* dev(L) of L to be min$\{p(1) - n : p$ is a pseudorank function on $L\}$. We have dev(L) = 0 if and only if L is semimodular (cf. Section 1.9). Grätzer & Kiss [1986], Problem 3, ask for a characterization of finite lattices with small deviances.

Exercises

1. Prove that among all finite semimodular lattices that contain a given lattice as an isometric sublattice, Finkbeiner's normal embedding lattice has the fewest atoms and also the smallest possible number of join-irreducible elements (Finkbeiner [1960]).
2. Show that if a finite lattice L is semimodular, then the normal embedding lattice based on the natural rank function for L is isomorphic to L (Finkbeiner [1960]).
3. Show that there is a bijection between the join congruences and the $\{1\}$-meet-subsemilattices of a finite lattice.

References

Bogart, K. P., R. Freese, and J. P. S. Kung (eds.) [1990] *The Dilworth Theorems. Selected Papers of Robert P. Dilworth*, Birkhäuser, Boston.

Crawley, P. and R. P. Dilworth [1973] *Algebraic Theory of Lattices*, Prentice-Hall, Englewood Cliffs, N.J.

Dilworth, R. P. [1944] Dependence relations in a semimodular lattice, Duke Math. J. 11, 575–587.

Dilworth, R. P. [1990c] Background to Chapter 5, in: Bogart et al. [1990], pp. 265–267.

Faigle, U. [1990] Dilworth's completion, submodular functions, and combinatorial optimization, in: Bogart et al. [1990], pp. 287–294.

Finkbeiner, D. T. [1960] A semimodular imbedding of lattices, Canad. J. Math. 12, 582–591.

Fujishige, S. [1991] *Submodular Functions and Optimization*, Annals of Discrete Mathematics, Vol. 47, Elsevier Science.

Grätzer, G. and E. Kiss [1986] A construction of semimodular lattices, Order 2, 351–365.

Hall, M. and R. P. Dilworth [1944] The imbedding problem for modular lattices, Ann. Math. 45, 450–456.

Kung, J. P. S. [1990a] Dilworth truncations of geometric lattices, in: Bogart et al. [1990], pp. 295–297.

Kung, J. P. S. [1990b] Dilworth's proof of the embedding theorem, in: Bogart et al. [1990], pp. 458–459.

Mason, J. H. [1977] Matroids as the study of geometrical configurations, in: *Higher Combinatorics* (ed. M. Aigner), Reidel, Dordrecht, pp. 133–176.

Nguyen, H. Q. [1986] Semimodular functions, in: White [1986], pp. 272–279.

Pudlák, P. and J. Tůma [1977] Every finite lattice can be embedded in the lattice of all equivalences over a finite set, Comment. Math. Carolinae 18, 409–414.

Pudlák, P. and J. Tůma [1980] Every finite lattice can be embedded in a finite partition lattice, Algebra Universalis 10, 74–95.

Rival, I. and M. Stanford [1992] Algebraic aspects of partition lattices, in: White [1992], pp. 106–122.

White, N. L. (ed.) [1992] *Matroid Applications*, Encyclopedia of Mathematics and its Applications, Vol. 40, Cambridge Univ. Press, Cambridge.

Whitman, P. M. [1946] Lattices, equivalence relations, and subgroups, Bull. Amer. Mat. Soc. 52, 507–522.

Wild, M. [1993] Cover preserving embedding of modular lattices into partition lattices, Discrete Math. 11, 207–244.

6.3 Geometric Closure Operators

Summary. In this section we have a look at Faigle's geometries, a concept extending the notion of a matroid. The idea is to consider certain closure operators (possessing the so-called geometric exchange property) on partially ordered sets rather than on totally unordered sets. These closure operators lead to finite semimodular lattices and vice versa. We also indicate Faigle's theory of morphisms for semimodular lattices, which is equivalent to the theory of closure operators having the geometric exchange property.

Faigle [1980a] extended the concept of matroids on finite sets to finite posets. Faigle's geometries are defined in terms of closure operators on finite posets. He was guided by the following example. Denote by P the collection of directly indecomposable subgroups of a finite abelian group G. On the partially ordered set $(P, \subseteq)$ one can define a closure operator cl in a natural way as follows: For any $S \subseteq P$, $\text{cl}(S) = \{u \in P : u \subseteq \langle S \rangle\}$, where $\langle S \rangle$ is the subgroup of G generated by S.

Consider, for example, the group $G = Z_2 \times Z_4$, that is, the direct product of the cyclic groups of order 2 and order 4. Its lattice $L(G)$ of subgroups is shown in Figure 6.11(a), while Figure 6.11(b) exhibits the set of nontrivial directly irreducible subgroups of G with the induced partial ordering. The subsets $\{b\}$, $\{c\}$, $\{a, d\}$, $\{a, e\}$, and $\{a, b, c\}$ of P can be interpreted as the hyperplanes of a certain "modular geometry" on P. The flats (closed sets) of this geometry are $\emptyset$, $\{a\}$, $\{b\}$, $\{c\}$, $\{a, d\}$, $\{a, e\}$, $\{a, b, c\}$, and $\{a, b, c, d, e\}$. The lattice of flats is isomorphic to $L(G)$.

The closure operator indicated in the preceding example satisfies some specific properties (as compatibility and geometric exchange) that were investigated by Faigle [1980a] in a general setting. Before describing these properties, we introduce some additional notation.

If $P = (P, \leq)$ is a finite partially ordered set and $S \subseteq P$, we denote by $\hat{S}$ the smallest order ideal containing the set S, that is, $\hat{S} = \{u \in P : u \leq s$ for some $s \in S\}$. Thus a subset $S \subseteq P$ is an order ideal if and only if $S = \hat{S}$. For any $u \in P$, $[u] = \hat{u} - u = \{v \in P : v < u\}$.

A closure operator cl on P is *compatible* if, for all $u \in P$ and $S \subseteq P$, $u \in S$ implies $[u] \in S$. Compatibility ensures that a closed set is always an order ideal.

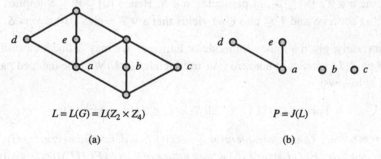

$$L = L(G) = L(Z_2 \times Z_4)$$

$$P = J(L)$$

(a)   (b)

Figure 6.11

A compatible closure operator on P is *pregeometric* if it satisfies the following *geometric exchange property*:

(GEP) For all $u, v \in P$ and $S \subseteq P$ such that $[u] \in \mathrm{cl}(S)$, the relations $v \in \mathrm{cl}(S \cup u)$
  and $v \notin \mathrm{cl}(S)$ imply $u \in \mathrm{cl}(S \cup v)$.

For a trivially ordered (totally unordered) set P, this exchange property reduces to the Steinitz–Mac Lane exchange property (cf. Section 1.7). A *pregeometry* or *Faigle geometry* $F(P)$ is any finite partial order P together with a pregeometric closure operator. A pregeometry is a *geometry* if it has the following two additional properties:

for all $u \in P$, $\emptyset \neq S \subseteq P$, $\mathrm{cl}(u) = \mathrm{cl}(S)$ implies $\mathrm{cl}(u) = \mathrm{cl}(S)$ for some $s \in S$, and
for all $u \in P$, $u \in \mathrm{cl}([u])$ implies $u \in \mathrm{cl}(\emptyset)$.

Finally, a geometry is said to be *proper* if $\emptyset = \mathrm{cl}(\emptyset)$ and, for all $u \in P$, $\mathrm{cl}(u) = \hat{u}$.

  The name *Faigle geometry* was coined by Björner & Ziegler [1992]. Faigle [1980a] used the name *pregeometry*, Crapo [1984] *quasigeometry*, and Korte et al. [1991] *ordered geometry*. If $F(P)$ is a Faigle geometry, we denote by $L(F(P))$ its lattice of flats. In what follows we describe the connection between Faigle geometries and semimodular lattices.

Lemma 6.3.1 *For any* $T, S \in L(F(P))$, T *covers* S *in* $L(F(P))$ *if and only if* $T = \mathrm{cl}(S \cup u)$ *for some* $u \in P$, $u \notin S$ *so that* $v \in S$ *for all* $v < u$.

Proof. If T covers S, choose any $u \in T - S$. Then $S \subset \mathrm{cl}(S \cup u) \subseteq T$, that is, $T = \mathrm{cl}(S \cup u)$. Conversely, let $u \notin S$ such that $[u] \subseteq S$. Then by (GEP) the relations $v \in \mathrm{cl}(S \cup u)$ and $v \notin \mathrm{cl}(S)$ imply $u \in \mathrm{cl}(S \cup v)$. Thus $T = \mathrm{cl}(S \cup u)$ covers S in $L(F(P))$. ∎

Lemma 6.3.2 *If* $F(P)$ *is a Faigle geometry, then its lattice of flats* $L(F(P))$ *is semimodular.*

Proof. Let $S, T \in L(F(P))$, such that $T \succ S \wedge T$ in $L(F(P))$. We have to show that $S \vee T \succ S$. Let $W = S \cap T = S \wedge T$. By Lemma 6.3.1 we have $T = \mathrm{cl}(W \cup u)$ for some $u \notin W$, $[u] \subseteq W$. In particular, $u \notin S$. Hence $[u] \subseteq W \subseteq S$ implies that $\mathrm{cl}(S \cup u)$ covers S and $T \subseteq \mathrm{cl}(S \cup u)$ yields that $S \vee T = \mathrm{cl}(S \cup u)$ covers S. ∎

  Conversely, given a finite semimodular lattice L, we may define a closure operator on $J(L)$ (the set of nonzero join-irreducibles of L) with the induced partial order as follows:

$$\text{For} \quad S \subseteq J(L), \qquad \mathrm{cl}(S) = \{u \in J(L) : u \leq \bigvee S\}.$$

Theorem 6.3.3 *The closure operator* $S \to \mathrm{cl}(S)$, *as defined above, gives rise to a proper geometry* $F(J(L))$ *on* $J(L)$ *whose lattice of flats* $L(F(J(L)))$ *is isomorphic with the given semimodular lattice* L.

Sketch of proof. We show first that (GEP) holds. Let $u \in J(L)$ and $S \subseteq J(L)$ such that $[u] \subseteq \mathrm{cl}(S)$, $u \notin \mathrm{cl}(S)$. Note that u covers $\sup[u]$ in L. Hence $\sup[u] =$ $(\sup S) \wedge u$. Since L is semimodular, $(\sup S) \vee u$ covers $\sup S$ in L. Assume now that $v \in J(L)$ satisfies $v \leq \sup S \vee u$ and $v \nleq \sup S$. Then $\sup S \vee v$ and $\sup S \vee u$ must coincide, that is, $u \in \mathrm{cl}(S \cup v)$. The isomorphism between L and $L(F(J(L)))$ follows from the fact that each element x of L may be identified with the flat $J(x)$ (i.e. the set of all nonzero join-irreducibles of L that are less than or equal to x). This isomorphism also implies that $F(J(L))$ is a proper geometry (see Faigle [1980a], Theorem 1 (c)). ∎

In the lattice of flats, the *geometric exchange property* (GEP) takes the following form: For all $u, v \in J(L)$ and $x \in L$ such that $u' \leq x$, the relations $v \leq x \vee u$ and $v \nleq x$ imply $u \leq x \vee v$.

This is equivalent to the formulation given in Theorem 3.3.5(ii). In the lattice of flats of a matroid, the condition $u' \leq x$ is trivially satisfied, since $u' = 0$. In this case the geometric exchange property coincides with the Steinitz–Mac Lane exchange property in its lattice-theoretic form (see Section 1.7). The above example (cf. Figure 6.11) shows that the condition $u' \leq x$ is an essential one: for $u = e$, $v = c$, and $x = b$ we have $v \leq x \vee u$ and $v \nleq x$ but $u \nleq x \vee v$ (in this case the condition $u' \leq x$ is violated).

Faigle [1980a] considers a number of other topics, such as deletion and contraction (as constructions on Faigle geometries) and strong geometries (characterized as those geometries for which every contraction results in a geometry). He shows that among the geometries precisely the strong geometries possess the Kurosh–Ore exchange property. He also gives cryptomorphic descriptions of Faigle geometries in terms of flats, hyperplanes, rank function, independent, and B-independent sets.

Faigle [1980b] also developed a theory of morphisms for semimodular lattices and proved that it is equivalent to the theory of closure operators possessing the above geometric exchange property (GEP). Faigle's morphisms and weak morphisms extend the concept of strong maps and maps of geometric lattices to the class of semimodular lattices of finite length. The foundations of the theory of strong maps were laid down by Crapo [1965], [1967] and Higgs [1968] (see also Kung [1986d]). Weak maps (or maps, as they were first called) were discovered by Higgs [1966] (see also Kung & Nguyen [1986]).

The concept of morphism can be motivated by the following example. A (vector-space) homomorphism f of a finite-dimensional vector space V_n into a finite-dimensional vector space V_m induces a mapping $f^\circ : L(V_n) \to L(V_m)$ between the associated lattices of subspaces of V_n and V_m. For two arbitrary subspaces x, y of V_n we have $f^\circ(x \vee y) = f^\circ(x) \vee f^\circ(y)$. Moreover, if x is in addition a subspace of y, then the dimension (dim, for short) satisfies the inequality

$$\dim(y) - \dim(x) \geq \dim(f^\circ(y)) - \dim(f^\circ(x)).$$

In what follows we mean by $y \downarrow x$ that $y \succ x$ or $x = y$. A map $f : L_1 \to L_2$ between two semimodular lattices L_1 and L_2 of finite length is said to be a *morphism* if the following two conditions are satisfied:

(M1) $f(x \vee y) = f(x) \vee f(y)$ holds for all $x, y \in L_1$;

(M2) $y \downarrow x$ implies $f(y) \downarrow f(x)$ for all $x, y \in L_1$.

Thus morphisms do not necessarily satisfy $f(x \wedge y) = f(x) \wedge f(y)$, that is, they are not lattice morphisms in the usual sense. Condition (M2) means in particular that a morphism reduces the length of chains. Since L_1 does not contain chains of infinite length, it is immediate that (M1) is equivalent to

(M1') $f(\sup A) = \sup\{f(a) : a \in A\}$ for all subsets $A \subseteq L_1$.

A map $f : L_1 \to L_2$ between two semimodular lattices L_1 and L_2 of finite length is said to be a *weak morphism* if the following two conditions are satisfied:

(WM1) $f(x) = \sup\{f(u) : u \in J(x)\}$ for all $x \in L_1$;

(WM2) $r(f(x)) \leq r(x)$ (with r denoting the rank function).

Every morphism is obviously a weak morphism, but not conversely (example?).

The usual composition of weak morphisms does not yield a weak morphism in general. For this reason we define the composition of two weak morphisms $f_1 : L_1 \to L_2$ and $f_2 : L_2 \to L_3$ as the map $f : L_1 \to L_3$ defined by $f(x) = \sup\{f_2(f_1(p)) : p \in J(x)\}$.

Lemma 6.3.4

(a) *The composition of two weak morphisms is a weak morphism.*
(b) *The composition of two morphisms is a morphism and coincides with the usual composition.*

Proof. We have to show that f has property (WM2). Consider an arbitrary element $x \in L_1$. For every $p \in J(x)$ we have $f_2(f_1(p)) \leq f_2(f_1(x))$, that is,

$$f(x) = \sup\{f_2(f_1(p)) : p \in J(x)\} \leq f_2(f_1(x)).$$

By assumption we have $r(x) \geq r(f_1(x)) \geq r(f_2(f_1(x)))$. It follows that $r(x) \geq r(f(x))$. If f_1 and f_2 are morphisms, then (M1') implies $f(x) = \sup\{f_2(f_1(p)) : p \in J(x)\} = f_2(\sup\{f_1(p) : p \in J(x)\}) = f_2(f_1(\sup\{p : p \in J(x)\})) = f_2(f_1(x))$. If $y \downarrow x$ holds in L_1, then $f_1(y) \downarrow f_1(x)$ holds in L_2 and $f(y) = f_2(f_1(y)) \downarrow f_2(f_1(x))$ in L_3.

Let now L be a semimodular lattice of finite length, and consider an element $x \in L$. The *reduction* of L onto the interval $[0, x]$ is the map $f_{\text{red}} : L \to [0, x]$

defined by $f_{red}(y) = y \wedge x$. The *contraction* of L onto the interval $[x, 1]$ is the map $f_{con} : L \to [x, 1]$ defined by $f_{con}(y) = y \vee x$. ∎

Lemma 6.3.5

(a) *The reduction is a weak morphism.*
(b) *The contraction is a morphism.*

Proof.   (a): For every element $z \in L$ the equality $f_{red}(z) = z$ holds if and only if $z \leq x$. Thus we have $f_{red}(y) = \sup\{J(x) \cap J(y)\} = \sup\{f_{red}(p) : p \in J(y)\}$. It is obvious that f_{red} cannot increase the rank.

(b): Property (M1) holds because of the associativity of the supremum operation. Applying Exercise 2, we see that the contraction preserves the $\downarrow$-relation. ∎

Mappings satisfying conditions (M1) and (M2) can also be defined between lattices that are not semimodular. Lemma 6.3.5(b) and the definition of semimodularity then show that a lattice of finite length is semimodular if and only if every contraction satisfies (M1) and (M2) (if $x \succ x \wedge y$, consider the contraction onto $[y, 1]$). The categorical character of morphisms in the class of semimodular lattices is even more emphasized in

Lemma 6.3.6 *Let $f : L_1 \to L_2$ be a surjective map of a semimodular lattice L_1 of finite length onto a lattice L_2. If f satisfies conditions (M1) and (M2), then L_2 is semimodular.*

Proof.   We have to show that $a \succ a \wedge b$ implies $a \vee b \succ b$ for arbitrary $a, b \in L_2$. It is sufficient to show $(a \vee b) \downarrow b$. Since f is onto, we can choose an element z having maximal rank in L_1 such that $f(z) = a \wedge b$. Let t be an element of L_1 such that $f(t) = a$, and denote by x an upper cover of z in the interval $[z, z \vee t]$. Then $f(z) < f(x) \leq f(z \vee t) = a$. Similarly we get the existence of an element $y \in L_1$ such that $z \leq y$ and $f(y) = b$. Thus we have $x \succ x \wedge y = z$, and the semimodularity of L_1 implies $x \vee y \succ y$. Now (M2) yields $a \vee b = f(x \vee y) \downarrow f(y) = b$. ∎

The assertion of Lemma 6.3.6 is not true for weak morphisms, as the example of Figure 6.12 shows.

An important property of semimodular lattices is given in

Lemma 6.3.7 *Let $f : L_1 \to L_2$ be a surjective morphism of a semimodular lattice L_1 of finite length onto a semimodular lattice L_2 such that $r(L_1) = r(L_2)$. Then L_1 and L_2 are isomorphic.*

Proof.   We have to show that f is injective and that f^{-1} is order-preserving. Suppose that $x \not\geq y$ are elements of L_1 such that $f(x) = f(y)$. Then $f(x \vee y) = f(x) \vee f(y) = f(x)$. If K is a maximal chain in L_1 containing x and $x \vee y$, then (M2)

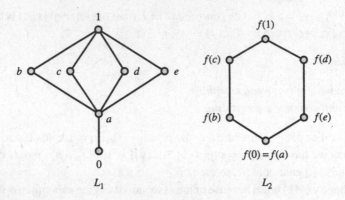

Figure 6.12

implies that $f(K)$ is a maximal chain of shorter length in L_2, contradicting the assumption $r(L_1) = r(L_2)$. Hence f is injective. If $a \leq b$ are elements in L_2 such that $f^{-1}(a) \nleq f^{-1}(b)$, then $f^{-1}(a) \vee f^{-1}(b) \neq f^{-1}(b)$ and thus $f(f^{-1}(a) \vee f^{-1}(b)) \neq b$, contradicting (M1).   ■

By means of weak morphisms Faigle [1980b] characterized modularity and distributivity within the class of semimodular lattices of finite length. To do so, some more properties of these morphisms will be needed.

A weak morphism $f : L_1 \to L_2$ between two semimodular lattices L_1 and L_2 of finite length is said to be *regular* if it has the following property: If $y \downarrow x$ are two elements of L_1 such that $f(y) \neq f(x)$, then there exists a $p \in J(y)$ satisfying $y = x \vee p$ and $f(x) \vee f(p) \succ f(x)$. If, in addition, for all $p, q \in J(L_1)$ and $x \in L_1$ the relation $(p \vee q) \downarrow x$ implies $f(p \vee q) \downarrow f(x)$, then the weak morphism f is called *nonsingular*. (Note that here "nonsingular" has a somewhat more general meaning than the corresponding notion as defined by Higgs [1968].)

If L_1 is a geometric lattice, then each atom of L_1 is mapped under a weak morphism onto an atom or 0 [because of (WM2)]. It follows that every weak morphism defined on a geometric lattice is regular. Further examples of regular weak morphisms are provided by reductions:

Lemma 6.3.8

(a) *Each reduction of a semimodular lattice of finite length is semimodular.*
(b) *Each reduction defined on a geometric lattice is nonsingular.*
(c) *Each regular weak morphism defined on a finite distributive lattice is nonsingular.*

Proof.   (a): Let $y \downarrow x$ and z be elements of a semimodular lattice L of finite length, and consider the reduction of L onto $[0, z]$. If $p \in J(y \wedge z)$ implies $p \in J(x \wedge z)$, then $x \wedge z = y \wedge z$. Otherwise we can choose an element $p \in J(y \wedge z) - J(x \wedge z)$

such that $r \leq x \wedge z$ holds for all $r \in J(p) - p$. In particular we have $p \notin J(x)$ and thus (by Exercise 2) we get $y = x \vee p$ and $p \vee (x \wedge z) \succ x \wedge z$.

(b): Let L be a geometric lattice and p, q two arbitrary atoms of L. The case $p = q$ is trivial. Suppose therefore $p \neq q$. If $x \leq p \vee q$ is another atom and z an arbitrary element of L, then either $x \wedge z = x$ and hence $(z \wedge (p \vee q)) \downarrow x$ [because of $2 = r(p \vee q) \geq r(z \wedge (p \vee q))$] or $x \wedge z = 0$. If we had $r(z \wedge (p \vee q)) = 2$ in the latter case, then $x \leq p \vee q = z \wedge (p \vee q)$, contradicting the assumption.

(c): Let now L be a finite distributive lattice. Then $J(p \vee q) = J(p) \cup J(q)$, and we may assume without loss of generality that $J(x) = J(p \vee q) - p$. Thus p is the uniquely determined element in $J(p \vee q)$ such that $p \vee q = x \vee p$. Now regularity implies the assertion. ∎

Lemma 6.3.9 *A semimodular lattice L of finite length is modular if every weak morphism $f : L \to L'$ into an arbitrary semimodular lattice L' has property* (M2), *that is, preserves the relation* $\downarrow$.

Proof. If L is not modular, then there exist elements $x, y \in L$ such that $x \vee y \succ y$ but x is not an upper cover of $x \wedge y$. The reduction of L onto $[0, z]$ is then a regular weak morphism that does not satisfy (M2) for the elements x and y. ∎

The converse to Lemma 6.3.9 is not true, as the example of Figure 6.13 shows.

Lemma 6.3.10 *Let L be a modular lattice of finite length. Then every nonsingular weak morphism $f : L \to L'$ of L into an arbitrary semimodular lattice L' has property* (M2).

Proof. Let $y \downarrow x$ be two distinct elements of L. We apply induction on the rank $r(x)$. If $r(x) = 1$, then $r(y) = 2$ and there exist elements $u, v \in J(L)$ such that $u \vee v = y$. Since f is nonsingular, it follows that $f(y) \downarrow f(x)$. Suppose now that (M2) holds for all elements of rank $< n$, and let $r(x) = n$. If $f(u) \leq f(x)$ for all $u \in J(y)$, then $f(x) = f(y)$ and there remains nothing to show. If $f(x) \neq f(y)$, then the regularity of f implies the existence of an element $u \in J(y) - J(x)$ such that $f(x) \vee f(u) \succ f(x)$. Let now c be an arbitrary element in $J(y)$. Putting

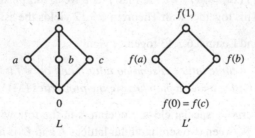

Figure 6.13

$k = c \vee u$, we get by modularity $r(k \wedge x) = r(k) + r(x) - r(k \vee x) = r(k) + r(x) - r(y) = r(k) - 1$, that is, $k \downarrow (k \wedge x)$.

If $k \wedge x < x$, then we have by induction hypothesis $f(k) \downarrow f(k \wedge x)$. Hence $f(c) \leq f(k) \leq f(p) \vee f(k \wedge x) \leq f(p) \vee f(x)$.

If $k \wedge x = x$ then $k = (c \vee u) \downarrow x$. Since f is nonsingular, we get $f(c \vee u) \downarrow f(x)$. Because $f(c \vee u) \geq f(u) \vee f(x) \succ f(x)$, it follows that $f(c) \leq f(c \vee u) = f(p) \vee f(x)$.

In both cases we have $f(y) = \sup\{f(c) : c \in J(y)\} \leq f(p) \vee f(x)$, that is, $f(y) \downarrow f(x)$. ∎

Corollary 6.3.11 *A geometric lattice L is modular if and only if every nonsingular weak morphism $f : L \to L'$ of L into an arbitrary semimodular lattice L' has property* (M2).

Proof. Observing Lemma 6.3.8(b) the assertion follows from Lemma 6.3.9 and Lemma 6.3.10. ∎

Theorem 6.3.12 *A finite semimodular lattice L is distributive if and only if every regular weak morphism $f : L \to L'$ of L into an arbitrary semimodular lattice L' has property* (M2).

Proof. If L is not distributive, then there are elements $x, y, z \in L$ such that $x \wedge (y \vee z) \neq (x \wedge y) \vee (x \wedge z)$. This means that the reduction of L onto $[0, z]$ is a regular weak morphism that does not satisfy (M1). If L is distributive, then it is modular and by Lemma 6.3.8(c) every regular weak morphism is nonsingular. By Lemma 6.3.10 we have only to show that (M1) holds. For arbitrary elements $x, y \in L$ we have $J(x \vee y) = J(x) \cup J(y)$ and thus $f(x \vee y) = \sup\{f(p) : p \in J(x) \cup J(y)\} = \sup\{f(p) : p \in J(x)\} \vee \sup\{f(p) : p \in J(y)\} = f(x) \vee f(y)$. ∎

Lemma 6.3.13 *Every finite semimodular lattice L is the image of a distributive lattice F under a morphism f.*

Proof. Let F be the distributive lattice associated with the poset $J(L)$ (see Section 1.3). Then $J(F) = J(L)$. The identity map $i : J(F) \to J(L)$ induces a map $f : F \to L$ in the following way: to an element $x \in F$ we assign $f(x) = \sup_L J(x)$. From $r(f(x)) \leq |J(x)| = r(x)$ we see that f is a weak morphism that is regular (by Exercise 2). This together with Theorem 6.3.12 yields the assertion. ∎

Lemma 6.3.6 and Lemma 6.3.13 together yield

Theorem 6.3.14 *A finite lattice is semimodular if and only if it is the image of a finite distributive lattice under a map having the properties* (M1) *and* (M2).

Morphisms determine special closure operators in the following way. A morphism $f : L \to L'$ between two semimodular lattices L and L' is associated with a map $f^\nabla : L' \to L$ given by $f^\nabla(y) = \sup\{x \in L : f(x) \leq y\}$. The map $x \to \mathrm{cl}(x) =$

$f^\nabla(f(x))$ is a closure operator on L (the kernel of f). The set Q of the closed elements of L (under this closure operator) is order-isomorphic with $f(L)$. Lemma 6.3.6 implies that $f(L)$ (and hence also Q) is a semimodular lattice.

Lemma 6.3.15 *Let L be a semimodular lattice of finite length. A closure operator* cl $: L \to L'$ *is the kernel of a morphism f if and only if it has the geometric exchange property* (GEP).

The proof is left as an exercise. It follows that the theory of morphisms on a semimodular lattice L of finite length is equivalent to the theory of closure operators possessing the geometric exchange property. A morphism thus defines a geometric structure on L. For example, let PG be a finite-dimensional projective geometry over a finite field, and denote by B(PG) the Boolean lattice of the subsets of points of PG. By Lemma 6.3.13 there exists a natural morphism of B(PG) into the modular lattice of the projective subspaces of PG. The kernel of this morphism is a geometric closure operator on B(PG) that determines the structure of PG.

Faigle [1980b] also constructed a special quotient, the Higgs lift, and used it to show that each morphism decomposes into so-called elementary morphisms and that each morphism factors into an injection and a contraction.

Notes

The first example of this section (derived from finite abelian groups) is an instance of a "projective geometry on a partially ordered set" in the sense of Faigle & Herrmann [1981].

In the classical approach of Veblen & Young [1916], a projective space (on a totally unordered set) may be viewed as an incidence structure involving "points" and "lines" that satisfy certain requirements. If a projective space is finite-dimensional, its lattice of linear subsets is modular, complemented, and of finite length. In fact, every modular complemented lattice of finite length may be identified (up to isomorphism) with the lattice of linear subsets of a unique finite-dimensional projective space (see Birkhoff [1967], p. 93).

Baer [1942] extended the scope of projective geometry and developed a theory that includes the theory of classical projective geometry and the theory of finite abelian groups as special cases (see also Inaba [1948] and Jónsson & Monk [1969]). Baer's theory gives rise to the observation that not only the atoms but also other join-irreducible elements of certain modular lattices (primary lattices) behave like points in a projective geometry.

On the other hand, Kurinnoi [1975] observed that two finite modular lattices are isomorphic if and only if their partially ordered sets of all elements that are joins of at most two join-irreducible elements are isomorphic. For complemented modular lattices, Kurinnoi's result follows immediately from the fact that a projective space is determined by the incidence structure of its points and lines.

The facts listed above gave rise to the following question: Does there exist a theory of projective incidence geometry that, in particular, allows us to understand the set of all join-irreducible elements of a modular lattice of finite length as the set of points of some finite-dimensional projective space? Faigle & Herrmann [1981] present a system of axioms for such projective geometries on partially ordered sets of points and lines. They show that there is a correspondence between projective geometries (defined on posets) and modular lattices of finite length in the same way as between classical projective geometries (defined on unordered sets) and complemented modular lattices of finite length. Thus, in the same way as a matroid may be thought of as a generalization of a projective geometry, Faigle's geometries are generalizations of projective geometries (on posets). There are also other attempts to generalize matroids to posets (cf. Dunstan et al. [1972]).

The geometric exchange property was discovered by Finkbeiner [1951], who studied the properties a closure operator on a poset P must have in order that the closed subsets of P form a semimodular lattice whose set of join irreducible elements with the induced order is isomorphic to P. Note, however, that Finkbeiner [1951] investigates posets that are not necessarily finite, while Faigle [1980a] studies slightly more general closure operators (on finite posets).

Faigle [1980c] investigates the possibility of extending a geometric closure operator in the following sense: for a given geometric closure operator on a poset P' and a poset P such that $P' = P - e$ for some maximal element e in P, he determines the geometric closure operators on P whose restriction to P' is equal to the original operator. The principal tools are "E-filters," which are the analogues of modular cuts for matroids (see Crapo [1965]). For matroids, Faigle's construction of extensions reduces to Crapo's construction.

Faigle [1986b] investigates exchange properties in the general context of combinatorial closure spaces: he considers closure operators whose lattices of closed sets are not necessarily semimodular. We recall that matroids may be axiomatized via systems of bases with a certain exchange property or, equivalently, via closure operators with a certain exchange property. For bases, this property generalizes the Steinitz exchange property for bases of vector spaces, whereas for closure operators the relevant exchange property is a special case of Mac Lane's [1938] property of semimodular lattices (see Section 3.1). Thus it has become customary to view the matroid exchange properties above just as manifestations of the Steinitz–Mac Lane exchange property (as we used it in Section 1.7). The purpose of Faigle's paper consists in investigating properties of closure spaces which relate to the Steinitz or to the Mac Lane exchange property. Within the setting of closure spaces, these properties are not equivalent (as in the matroid case). Faigle et al. [1984] study the Steinitz exchange property with respect to the class of semimodular lattices.

Morphisms can be defined between arbitrary lattices. In particular, a mapping f of a lattice L into itself is called a join endomorphism if it satisfies (M1) for all

$x, y \in L$. Let E be the set of all join endomorphisms of a given lattice L. If $f, g \in E$, we put $f \leq g$ if $f(x) \leq g(x)$ holds for all $x \in L$. Grätzer & Schmidt [1958b] and Jakubík [1958] independently proved the following results:

(i) If L is a complete lattice, then E is also a complete lattice.
(ii) There exists a finite lattice L such that the lattice E is not semimodular.
(iii) There exists a lattice L such that E is not a lattice.

The second result answers Birkhoff [1948], Problem 93, p. 209 (for finite lattices). In particular, Grätzer & Schmidt [1958b], Theorem 2, proved that the lattice of all join endomorphisms of a nondistributive lattice with finite bounded chains is not semimodular. The third result was already stated by Dilworth [1950], implying that the answer to Birkhoff's problem is negative for general lattices.

Exercises

1. Give an example of a weak morphism that is not a morphism.
2. Let x, y be elements of a semimodular lattice L of finite length. Then $x \prec y$ holds if and only if there exists an element $u \in J(L) - J(x)$ such that $y = x \vee u$ and $v \leq x$ for all $v \in J(u) - u$ (Faigle [1980b], Lemma 2.1).
3. Give a proof of Lemma 6.3.15 (Faigle [1980b]).

References

Baer, R. [1942] A unified theory of projective spaces and finite abelian groups, Trans. Amer. Math. Soc. 52, 283–343.

Birkhoff, G. [1948] *Lattice Theory* (2nd edition), Amer. Math. Soc. Colloquium Publications, Vol. 25, New York.

Birkhoff, G. [1967] *Lattice Theory*, Amer. Math. Soc. Colloquium Publications, Vol. 25, Providence, R.I. (reprinted 1984).

Björner, A. and G. M. Ziegler [1992] Introduction to greedoids, in: White [1992], pp. 284–357.

Crapo, H. H. [1965] Single-element extensions of matroids, J. Res. Nat. Bur. Standards Sect. B 69B, 55–65.

Crapo, H. H. [1967] Structure theory for geometric lattices, Rend. Sem. Mat. Univ. Padova. 38, 14–22.

Crapo, H. H. [1984] Selectors: a theory of formal languages, semimodular lattices, and branching and shelling processes, Adv. in Math. 54, 233–277.

Dilworth, R. P. [1950] Review of Birkhoff [1948], Bull. Amer. Math. Soc. 56, 204–206.

Dunstan, F. D. J., A. W. Ingleton, and D. J. A. Welsh [1972] Supermatroids, in: *Proc. Conf. Combin. Math.*, Math. Inst., Oxford, pp. 72–122.

Faigle, U. [1980a] Geometries on partially ordered sets, J. Combin. Theory Ser. B 28, 26–51.

Faigle, U. [1980b] Über Morphismen halbmodularer Verbände, Aequationes Math. 21, 53–67.

Faigle, U. [1980c] Extensions and duality of finite geometric closure operators, J. Geometry 14, 23–34.

Faigle, U. [1986b] Exchange properties of combinatorial closure spaces, Discrete Applied Math.15, 249–260.

Faigle, U. and C. Herrmann [1981] Projective geometry on partially ordered sets, Trans. Amer. Math. Soc. 266, 319–332.

Faigle, U., G. Richter, and M. Stern [1984] Geometric exchange properties in lattices of finite length, Algebra Universalis 19, 355–365.

Finkbeiner, D. T. [1951] A general dependence relation for lattices, Proc. Amer. Math. Soc. 2, 756–759.

Grätzer, G. and E. T. Schmidt [1958b] On the lattice of all join-endomorphisms of a lattice, Proc. Amer. Math. Soc. 9, 722–726.

Higgs, D. [1966] Maps of geometries, J. London Math. Soc. 41, 612–618.

Higgs, D. [1968] Strong maps of geometries, J. Combin. Theory. Ser. A 5, 185–191.

Inaba, E. [1948] On primary lattices, J. Fac. Sci. Hokkaido Univ. 11, 39–107.

Jakubík, J. [1958] Note on the endomorphisms of a lattice, Čas. Pešt. Mat. 83, 226–229.

Jónsson, B. and G. Monk [1969] Representation of primary arguesian lattices, Pacific J. Math. 30, 95–139.

Korte, B., L. Lovász, and R. Schrader [1991] *Greedoids*, Springer-Verlag, Berlin.

Kung, J. P. S. [1986d] Strong maps, in: White [1986], Chapter 8.

Kung, J. P. S. and H. Q. Nguyen [1986] Weak maps, in: White [1986], Chapter 9.

Kurinnoi, G. C. [1975] Condition for isomorphism of finite modular lattices (Russian), Vestnik Har'kov. Univ. Mat. Meh. 40, 45–47.

Mac Lane, S. [1938] A lattice formulation for transcendence degrees and p-bases, Duke Math. J. 4, 455–468.

Veblen, O. and J. W. Young [1916] *Projective Geometry*, Vol. 1, Ginn, New York.

White, N. L. (ed.) [1986] *Theory of Matroids*, Encyclopedia of Mathematics and its Applications, Vol. 26, Cambridge Univ. Press, Cambridge.

White, N. L. (ed.) [1992] *Matroid Applications*, Encyclopedia of Mathematics and its Applications, Vol. 40, Cambridge Univ. Press, Cambridge.

6.4 Semimodular Lattices and Selectors

Summary. Crapo [1984] developed a theory of certain hereditary languages, the so-called selectors, which turned out to be part of the theory of greedoids in the sense of Korte and Lovász. Here we have a look at the representation of finite semimodular lattices by selectors. For the special case of upper locally distributive (join-distributive) lattices we refer to Section 7.2. All results and examples of this section are due to Crapo.

We consider set systems over a finite ground set $E (\neq \emptyset)$, that is, pairs (E, F) where $F \subseteq 2^E$ is a collection of subsets. The ground set E is also called an *alphabet*, its elements are the *letters*. Let E^* denote the set of all sequences $\alpha = x_1 x_2 \ldots x_k$ of elements $x_i \in E$, $1 \le i \le k$; the elements of E^* are called *words* or *strings*. The number of letters in a word α is its length $\lambda(\alpha)$. The symbol $\emptyset$ is also used to denote the empty word. A collection of words $\Lambda \subseteq E^*$ is called a *language* over the alphabet E. The *concatenation* $\alpha\beta$ *of two words* α and β is the string α followed by the string β; α is then also called a *prefix* of $\alpha\beta$. We assume that the language Λ is *simple*, that is, no letter is repeated in any word. A language Λ is *hereditary* if it has the following two properties:

(S0) $\emptyset \in \Lambda$,

(S1) $\alpha\beta \in \Lambda$ implies $\alpha \in \Lambda$.

The binary relation $\in$ between letters and words establishes a Galois connection between the Boolean lattice $B(E)$ of subsets of E and the Boolean lattice $B(\Lambda)$ of Λ. We denote this Galois connection by the mappings

$$\mathbf{v} : B(E) \to B(\Lambda) \quad \text{and} \quad \mathbf{a} : B(\Lambda) \to B(E).$$

For any set $S \subseteq E$ of letters, $\mathbf{v}(S)$, the S-vocabulary, is the set of words that can be spelled using only letters of S. For any set $\Gamma \subseteq \Lambda$ of words, $\mathbf{a}(\Gamma)$, the Γ-alphabet, is the set of letters used in any of the words in Γ. The map $\mathbf{a}$ preserves joins, and the map $\mathbf{v}$ preserves meets. The image $\mathbf{A}$ of the map $\mathbf{a}$ is exactly the set of subsets $S \subseteq E$ *coclosed* with respect to the *coclosure operator* $S \to \mathbf{a}(\mathbf{v}(S)) = \text{int}(S)$. The coclosed subsets are called *partial alphabets* for the hereditary language Λ.

The notion of length for words is extended to arbitrary partial alphabets A by defining $\lambda(A)$ as the greatest length of any word in the A-vocabulary, that is, in the set of words which can be spelt using only letters of A. The proof of the following result (Crapo [1984], Theorem 1) is left as an exercise.

Theorem 6.4.1 *The set A_p of partial alphabets of a hereditary language Λ, ordered by set inclusion, is an upper locally distributive (join-distributive) lattice. The rank (height) of any partial alphabet $A \in A_\mathrm{p}$ in this lattice is equal to its cardinality $|A|$.*

Consider for example the hereditary language $\Lambda = \{\emptyset, a, c, ab, abc\}$. Figure 6.14 (cf. Crapo [1984], Figure 1) shows the lattice A_p of partial alphabets together with the values λ of the extended word length in parentheses. The function λ is integer-valued and monotone, but it is neither cover-preserving [since $ac \prec abc$ but $\lambda(ac) = 1$ and $\lambda(abc) = 3$] nor submodular [since $\lambda(ab \wedge ac) + \lambda(ab \vee ac) = \lambda(a) + \lambda(abc) = 1 + 3 > 2 + 1 = \lambda(ab) + \lambda(ac)$].

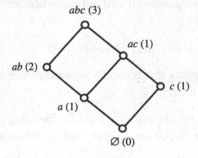

Figure 6.14

For any partial alphabet A in a hereditary language Λ on a set E, a word $\alpha \in \Lambda$ is an *A-basis* if and only if every letter in α is in the set A and $\lambda(\alpha) = \lambda(A)$. A *selector* is a hereditary language Λ on a set E satisfying the following properties:

(S2) Every letter in E appears in some word in Λ.

(S3) Let A be a partial alphabet and $x \notin A$ a letter such that $\alpha x \in \Lambda$ for some A-word α. If $\lambda(A) < \lambda(A \cup x)$, then for any A-basis β, βx is an $(A \cup x)$-basis.

Axiom (S2) says that the entire ground set E is a partial alphabet ($E = \operatorname{int} E$). In a selector, an E-basis is simply called a *basis*. Björner [1983] has shown that axiom (S3) is equivalent to the following property: If α and β are words in Λ, with $\lambda(\alpha) < \lambda(\beta)$, then there exists a subsequence γ of $\lambda(\alpha) - \lambda(\beta)$ letters (not necessarily consecutive), which can be chosen from β in the induced order, such that $\alpha\beta$ is also a word in Λ. This shows that selectors are special greedoids in the sense of Korte et al. [1991].

The following results deal with some properties of the function λ.

Theorem 6.4.2 *If C and D are partial alphabets in a selector Λ, and D covers C in the lattice A_p of partial alphabets of Λ, then $\lambda(D) - \lambda(C) = 0$ or 1, that is, λ is a unit increase function on the lattice A_p.*

Proof. $\lambda(C)$ and $\lambda(D)$ are integers with $\lambda(C) \leq \lambda(D)$. Assume that $\lambda(C) < \lambda(D)$. Since C and D are partial alphabets and D covers C in A_p, D also covers C in $B(E)$ (by Theorem 6.4.1) and $D = C \cup x$ for some element $x \in E$. Since C is a partial alphabet, there is a C-word using the letter x. Dropping all letters after x in this word, one obtains a C-word α such that $\alpha x \in \Lambda$. If β is a C-basis, then [by Axiom (S3)] βx is a $(C \cup x)$-basis. But then β is a C-word of length $\lambda(\beta x) - 1 = \lambda(D) - 1$. $\lambda(C) < \lambda(D)$ implies that β is a C-basis and $\lambda(C) = \lambda(D) - 1$. ∎

An integer-valued function λ on a lattice L of finite length is called *locally submodular* (or *locally semimodular*) if and only if, whenever y and z are elements covering an element x in L,

$$\lambda(x) = \lambda(y) = \lambda(z) \quad \text{implies} \quad \lambda(x) = \lambda(y \vee z).$$

Any unit increase function λ that is locally submodular on a semimodular lattice is also submodular (cf. Exercise 2).

Theorem 6.4.3 *Let Λ be a selector on a set E. Then the rank function λ of Λ is (locally) submodular on the lattice A_p of partial alphabets of Λ.*

Proof. Assume that A, $A \cup x$, and $A \cup y$ are partial alphabets for Λ, all having the same rank $\lambda(A) = k$. Any union of partial alphabets is itself a partial alphabet and thus $A \cup x \cup y \in A_p$. Let α be an A-basis. Since $\lambda(A \cup x) = \lambda(A)$, α is also an $(A \cup x)$-basis. If $\lambda(A \cup x) < \lambda(A \cup x \cup y)$, then [by axiom (S3)] αy is an $(A \cup x \cup y)$-basis of length $k + 1$ contained in the set $A \cup y$. This contradicts the assumption that $\lambda(A) = \lambda(A \cup x \cup y)$. It follows that λ is locally submodular on A_p and hence submodular. ∎

We have the following characterization of words in a selector.

Theorem 6.4.4 *Let Λ be a selector on a set E. A string $\alpha = x_1 \ldots x_n$ of letters in E is a word in Λ if and only if the letters in every initial segment $x_1 \ldots x_i (1 \le i \le n)$ of α form a partial alphabet of rank i in Λ.*

Proof. By axiom (S1), any initial segment $x_1 \ldots x_i$ of a word $\alpha = x_1 \ldots x_n$ in Λ is also in Λ and its letters form a partial alphabet of rank $\lambda(x_1 \ldots x_i) = i$. The converse will be proved by induction. The empty string is a word in Λ. Assume now that the initial segment $\alpha_{i-1} = x_1 \ldots x_{i-1}$ is a word in Λ and a basis for the partial alphabet $\{x_1 \ldots x_{i-1}\}$. Since $\{x_1 \ldots x_i\}$ is a partial alphabet of rank i, we get by axiom (S3) that the $\{x_1 \ldots x_{i-1}\}$-basis α_{i-1} can be extended to an $\{x_1 \ldots x_i\}$-basis of the form $\alpha_{i-1}x = x_1 \ldots x_i$. Induction implies that $\alpha = x_1 \ldots x_n$ is a word. ∎

In terms of the rank function λ of a selector, we define a closure operator on the lattice A_p of partial alphabets:

$$A \to \mathrm{cl}(A) = \bigcup \{B \in A_p : A \subseteq B \text{ and } \lambda(A) = \lambda(B)\}.$$

A partial alphabet A is said to be *closed* if $\mathrm{cl}(A) = \mathrm{cl}(\mathrm{cl}(A))$. Closed partial alphabets are also called the *flats* of the selector Λ. By $L(\Lambda)$ we denote the lattice of flats of Λ.

Theorem 6.4.5 *The lattice $L(\Lambda)$ of flats of a selector Λ, ordered by inclusion, is semimodular and λ coincides with the rank function (height function) on $L(\Lambda)$.*

Sketch of proof. We leave it as an exercise to show that the operator $A \to \mathrm{cl}(A)$ is indeed a closure operator on the lattice A_p of partial alphabets. The quotient set $L(\Lambda)$ of closed partial alphabets is therefore a complete lattice (with respect to set inclusion). Next we show that the lattice $L(\Lambda)$ is semimodular. Let A, B, C be flats such that B covers A in the lattice $L(\Lambda)$. Let A' be a partial alphabet covering A in the lattice A_p and contained in B. Since A and B are closed, $\mathrm{cl}(A')$ strictly contains A and is contained in B. But B covers A in $L(\Lambda)$, and thus $\mathrm{cl}(A') = B$. In the semimodular lattice A_p, $A' \cup C$ covers or is equal to $A \cup C$, and thus $\lambda(A' \cup C) - \lambda(A \cup C) = 0$ or 1. The closure $\mathrm{cl}(A' \cup C)$ is equal to the closure $\mathrm{cl}(B \cup C)$, and thus it is equal to the join $B \cup C$ in $L(\Lambda)$. Since the rank of any partial alphabet is equal to the rank of its closure, we get that $\lambda(B \vee C) - \lambda(A \vee C) = 0$ or 1. Since

λ is strictly increasing on any chain in $L(\Lambda)$, the flat $B \vee C$ covers or is equal to the flat $A \vee C$ and hence the lattice $L(\Lambda)$ is semimodular. Since λ is strictly increasing on chains in the quotient lattice, and is unit-increase and normalized on A_p, λ coincides with the height function in $L(\Lambda)$. ■

Let us now give examples of selectors.

The Faigle geometry $F(P)$ on a poset P was defined in Section 6.3. Faigle [1980a] defines the rank $r(S)$ of any subset $S \subseteq P$ as the rank of the flat $cl(S)$ in the lattice $L(F(P))$ of flats of $F(P)$. He then defines a *B-independent string* $x_1 \ldots x_n$ as a sequence of distinct elements of P such that $r(\{x_1, \ldots, x_i\}) = i$ for $i = 1, \ldots, n$. If $\alpha = x_1 \ldots x_n$ is a B-independent string, the closures $S_i = cl(\{x_1 \ldots x_i\})$, $i = 0, \ldots, n$, of its initial segments form a maximal chain in the lattice $L(F(P))$ of flats of $F(P)$ from 0 to the flat S_n. Conversely, if $0 = S_0 \le S_1 \le \cdots \le S_n = S$ is a maximal chain from 0 to a flat S in L, then any choice $x_i \in P$ from each difference set $S_i - S_{i-1}$, $i = 1, \ldots, n$ will produce a B-independent string.

Theorem 6.4.6 *The B-independent strings of elements in a Faigle geometry form a selector.*

The proof is left as an exercise (Exercise 3). Consider, for example, the Faigle geometry shown in Figure 6.15(a). Its lattice of order ideals is shown in Figure 6.15(b).

The selector Λ of B-independent strings of this Faigle geometry is given by

λ	Λ
0	$\emptyset$
1	a, b
2	ac, ab, ba, bd
3	$acb, acd, abc, abd, bac, bad, bda, abd$

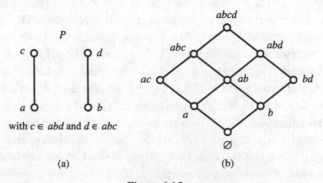

with $c \in abd$ and $d \in abc$

(a)　　　　　　(b)

Figure 6.15

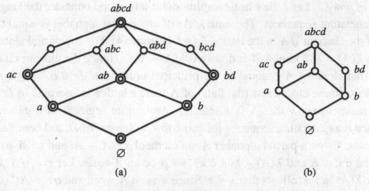

Figure 6.16

Figure 6.16(a) shows the lattice A_p of partial alphabets of the selector Λ of B-independent strings, while Figure 6.16(b) shows the lattice of flats of this selector.

In particular, matroids are selectors:

Theorem 6.4.7 *Given a matroid G, let Λ be the set of (arbitrary) linear orderings of independent sets of elements of G. Then Λ is a selector.*

Proof. This is a special case of a selector determined by a Faigle geometry. The ordered set P is simply the antichain of unordered points of G, and the words in Λ defined above are the B-independent strings in that Faigle geometry. Thus Λ is a selector. ∎

For other examples (arising from convexity) see Chapter 7.4.

Let now L be a finite semimodular lattice. A sequence $\alpha = x_1, \ldots, x_n$ of distinct elements of L is called a *generating sequence* if the joins $y_i = x_1 \vee \cdots \vee x_i$, $i = 0, \ldots, n$, form a maximal chain from 0 to y_n in L.

Lemma 6.4.8 *Let Λ be the language of generating sequences in a finite semimodular lattice. Then for every partial alphabet $A \in A_p$ and every maximal word $x_1 \ldots x_n$ in Λ formed by elements of A one has $\bigvee x_i = \bigvee \{a : a \in A\}$.*

Proof. Let $x_1 \ldots x_n$ be a maximal word formed by elements of A, and let y be any letter in A. If $y \not\leq x = x_1 \vee \cdots \vee x_n$, then let $y_1 \ldots y_k$ be a word in A with $y_k = y$ and y_i the first letter for which $y_i \not\leq x$. Then $y_1 \vee \cdots \vee y_i$ covers $y_1 \vee \cdots \vee y_{i-1} \leq x$. Semimodularity implies that $y_i \vee x$ covers x and thus $x_1 \ldots x_n y_i$ is a word in Λ, contradicting the maximality of $x_1 \ldots x_n$. Hence $x \in A$ is an upper bound for A and thus $x = \bigvee \{a : a \in A\}$. ∎

We have the following converse to Theorem 6.4.5.

Theorem 6.4.9 *Any finite semimodular lattice L is isomorphic to the lattice of flats of a selector Λ, where Λ consists of all generating sequences of flats in L.*

Sketch of proof.   Let L be a finite semimodular lattice and consider the language Λ of generating sequences. The rank $\lambda(A)$ of any partial alphabet is equal to the rank of the element $\bigvee A$ in the lattice L (see Exercise 4). For any partial alphabets $A \subseteq B$, $\lambda(A) = \lambda(B)$ holds if and only if $\bigvee A = \bigvee B$. Hence the closure $\mathrm{cl}(A)$ of any partial alphabet A is equal to the principal order ideal $\{x \in L : x \leq \bigvee A\}$. It follows that these closed sets (the flats) of Λ form a lattice isomorphic to L.

It remains to show that Λ is a selector. Any initial segment of a generating sequence is a generating sequence for some lower element of L, and hence axiom (S1) holds. Given a partial alphabet A, an element $x \in X - A$, and an A-word α such that $\alpha x \in \Lambda$ and $\lambda(A) < \lambda(A \cup x)$, let β be an A-basis. Let $z = \bigvee A$. From $\lambda(A) < \lambda(A \cup x)$ it follows that $x \not\leq z$. Since α is an A-word and $\alpha x \in \Lambda$, $\bigvee(\alpha x)$ covers $\bigvee \alpha$ in L and $\bigvee \alpha \leq z$. Thus $x \vee z$ covers z, and βx is a generating sequence (a word) and an $(A \cup x)$-basis. ∎

It is not necessary to use the entire lattice L as the set of letters in the language Λ. It is sufficient to consider the nonzero join-irreducible elements, and the set of words could moreover be restricted to those generating sequences $x_1, \ldots, x_n$ where each x_i is a minimal join-irreducible generating the next covering. This type of reduced representation is illustrated in the following example of a selector representation of a finite semimodular lattice. Consider the centered hexagon S_7 labeled as in Figure 6.17(a). Then $J(S_7) = \{b, c, d, f\}$, and the selector representation of S_7 is given by

λ	Λ
0	$\emptyset$
1	b, c
2	bd, bc, cb, cf
3	$bdc, bcd, bcf, cbd, cbf, cfb$

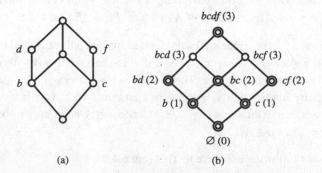

(a)   (b)

Figure 6.17

Thus, the language Λ of generating sequences with $E = J(S_7) = \{b, c, d, f\}$ is

$$\{\emptyset, b, c, bd, bc, cb, cf, bdc, bcd, cbd, cbf, cfb\},$$

and the lattice of partial alphabets of this language is shown in Figure 6.17(b) with λ written in parentheses. The lattice $L(\Lambda)$ of flats of Λ is indicated by concentric circles in Figure 6.17(b).

Notes

For the theory of greedoids see Korte et al. [1991] and Björner & Ziegler [1992]. In particular, the interrelationships between classes of selectors and classes of semimodular lattices is visualized in Korte et al. [1991], Chapter 7.

Exercises

1. Prove Theorem 6.4.1: The set A_p of partial alphabets of a hereditary language Λ, ordered by inclusion, is an upper locally distributive (join-distributive) lattice. The rank (height) of any partial alphabet $A \in A_p$ in this lattice is equal to its cardinality $|A|$. (Crapo [1984], Theorem 1.)
2. Any unit increase function λ that is locally submodular on a semimodular lattice is also submodular.
3. Show that the B-independent strings of elements in a Faigle geometry form a selector. (Crapo [1984], Theorem 18.)
4. If Λ is the set of generating sequences in a finite semimodular lattice L, then for any partial alphabet $A \subseteq L$, the join of the letters in any A-basis α is equal to the join of the entire subset $A \subseteq L$. (Crapo [1984], Lemma after Theorem 17.)

References

Björner, A. [1983] On matroids, groups, and exchange languages, in: *Matroid Theory and Its Applications, Proc. Conf. Szeged 1982* (ed. L. Lovász and A. Recski), North-Holland, Amsterdam, pp. 25–60.

Björner, A. and G. M. Ziegler [1992] Introduction to greedoids, in: White [1992], pp. 284–357.

Crapo, H. H. [1984] Selectors: a theory of formal languages, semimodular lattices, and branching and shelling processes, Adv. in Math. 54, 233–277.

Korte, B., L. Lovász, and R. Schrader [1991] *Greedoids*, Springer-Verlag, Berlin.

White, N. L. (ed.) [1992] *Matroid Applications*, Encyclopedia of Mathematics and Its Applications, Vol. 40, Cambridge Univ. Press, Cambridge.

6.5 Consistent Semimodular Lattices

Summary. Consistent lattices were introduced in Section 4.5. For semimodular lattices of finite length it was proved in Section 4.6 that consistency and strongness are equivalent (cf. Theorem 4.6.8). In the present section we consider some more properties of consistent semimodular lattices.

We characterized strongness in lattices of finite length by means of certain forbidden pentagons (see Lemma 4.6.1). If the lattices in question are in addition semimodular, then we have the following result (cf. Faigle et al. [1984], Theorem 2), whose proof is left as an exercise (Exercise 1).

Theorem 6.5.1   *A semimodular lattice L of finite length is consistent if and only if it does not contain a sublattice of the form shown in Figure 6.18 with $j \in J(L)$ and with five covers indicated by double lines.*

Let L be a lattice of finite length, and let $a \in L$, $a \neq 0$. Recall that $a_+ = \bigwedge (c \in L : c \prec a)$ and $a' = \bigvee (j' : j \in J(L), \ j \leq a)$. Figure 4.18 (Section 4.6) shows that the elements a' and a_+ are not comparable in general. If a lattice L of finite length is strong, then $a' \leq a_+$ holds for all $a \in L$, $a \neq 0$ (cf. Lemma 4.6.2). We shall show now that, in a semimodular lattice of finite length, the two elements are always connected by the reverse inequality $a_+ \leq a'$. This is a consequence of the following "butterfly lemma" (Richter & Stern, unpublished; see also Stern [1991a], Theorem 10.1):

Lemma 6.5.2   *Let L be a semimodular lattice of finite length, and for $a \in L$, $a \neq 0$ let $a = j_1 \vee \cdots \vee j_n [j_i \in J(L)$, $i = 1, \ldots, n]$ be an irredundant join representation of $a (\in L)$ as a join of join-irreducible elements. Then $a_+ \leq j'_1 \vee \cdots \vee j'_n$.*

Proof.   (For an illustration see Fig. 6.19).

For the sake of brevity we introduce the notation $\overline{j_{i_1,\ldots,i_r}} = j'_{i_1} \vee \cdots \vee j'_{i_r} \vee j_{i_{r+1}} \vee \cdots \vee j_{i_n}$ where $1 \leq i_j \leq n$. After suitable renumbering of the subscripts we may suppose that

$$\overline{j_{1,\ldots,s}} = a \tag{1}$$

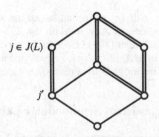

Figure 6.18

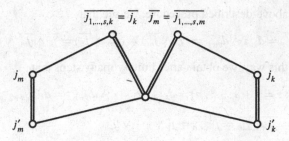

Figure 6.19

and

$$\overline{j_{1,\ldots,s,k}} < a \qquad \text{for } s < k \le n. \tag{2}$$

It may happen that there exist no s such that (1) holds. From (2) it follows by semimodularity that

$$\overline{j_k} = \overline{j_{1,\ldots,s,k}} \prec a \qquad (s < k \le n), \tag{3}$$

and this implies

$$a_+ = \bigwedge(\overline{j_i} : i = s+1,\ldots,n). \tag{4}$$

Now consider the element $\overline{j_m} = \overline{j_{1,\ldots,s,m}}$, where $s < m \le n$ and $m \ne k$. As in (3), we have $\overline{j_m} \prec a$. We also have

$$j_m \not\le \overline{j_m}, \tag{5}$$

since otherwise $a = j_m \vee \overline{j_m} \le \overline{j_m} \prec a$, a contradiction. From (5) it follows that

$$j_m \wedge \overline{j_m} = j'_m \prec j_m. \tag{6}$$

Next consider the element $d_{k,m} = \overline{j_{1,\ldots,s,k,m}}$, where $s < k$, $m \le n$, and $k \ne m$. Since $d_{k,m} \le \overline{j_m}$, we get from (6) the relation

$$j_m \wedge d_{k,m} = j'_m \prec j_m, \tag{7}$$

which implies by semimodularity the relation

$$d_{k,m} \prec d_{k,m} \vee j_m = \overline{j_{1,\ldots,s,k}} = j_k \tag{8}$$

(see the "butterfly" in Figure 6.19). Similarly we obtain

$$d_{k,m} \prec j_m. \tag{9}$$

Noting that $k \ne m$, the relations (8) and (9) together yield $d_{k,m} = \overline{j_k} \wedge \overline{j_m}$. Now consider the elements $d_{k,m,p} = \overline{j_{1,\ldots,s,k,m,p}}$ where $s < k$, m, $p \le n$ and $p \ne k$, $p \ne m$.

Repeating the above-described procedure, we see that

$$d_{k,m,p} = d_{k,m} \wedge d_{m,p} = (\overline{j_k} \wedge \overline{j_m}) \wedge (\overline{j_m} \wedge \overline{j_p}) = \overline{j_k} \wedge \overline{j_m} \wedge \overline{j_p}.$$

Continuing in this way, we obtain after finitely many steps that

$$\bigwedge (\overline{j_i} : i = s + 1, \ldots, n) = d_{s,s+1} \wedge \cdots \wedge d_{n-1,n} = d_{s+1,s+2,\ldots,n-1,n}$$

$$= \overline{j_{n+1,\ldots,n}} = \overline{j_{1,\ldots,n}} = j'_1 \vee \cdots \vee j'_n. \tag{10}$$

Now (10) together with (4) implies $a_+ \le j'_1 \vee \cdots \vee j'_n$, which was to be shown.

∎

Corollary 6.5.3   *In a semimodular lattice L of finite length, $a' \ge a_+$ holds for all $a \in L$, $a \ne 0$.*

Proof.   In a lattice L of finite length every element $a \in L$, $a \ne 0$, has an irredundant join representation $a = u_1 \vee \cdots \vee u_n [u_i \in J(L), i = 1, \ldots, n]$. By definition of a' we have $a' \ge u'_1 \vee \cdots \vee u'_n$. Applying the preceding lemma, we get $u'_1 \vee \cdots \vee u'_n \ge a_+$.   ∎

Corollary 6.5.4   *In a consistent semimodular lattice L of finite length, $a' = a_+$ holds for all $a \in L$, $a \ne 0$.*

Proof.   Follows from Corollary 6.5.3 and Lemma 4.6.2, since "consistent" means the same as "strong" for semimodular lattices of finite length (cf. Theorem 4.6.8).

∎

For modular lattices the butterfly lemma (Lemma 6.5.2) was proved by Soltan [1975]; this result has its roots in investigations on the lattice-theoretic background of the Jordan normal form for matrices (cf. Soltan [1973]). Richter [1991] extended the butterfly lemma to a class of semimodular lattices that are not of finite length in general. For the following result see Stern [1990a].

Theorem 6.5.5   *A semimodular lattice L of finite length is consistent if and only if $a' = a_+$ holds for all $a \in L$, $a \ne 0$.*

Proof.   Let L be a consistent semimodular lattice of finite length. Then $a' = a_+$ holds by Corollary 6.5.4. Conversely, let L be a semimodular lattice finite length that is not consistent. By Theorem 6.5.1, L contains a hexagon sublattice of the form shown in Figure 6.18. In what follows we refer to the notation of this figure. For the element $a = b \vee j = b \vee j'$ we have $a_+ \le (d \vee j') \wedge b = d$. On the other hand we have $a' > j'$. Now equality would imply $j' < a' = a_+ \le d$, contradicting the fact that j' and d are incomparable. Thus $a' \ne a_+$.   ∎

Next we relate strongness (and thus consistency) in semimodular lattices of finite length to an exchange property that reduces in the case of geometric lattices to the Steinitz–Mac Lane exchange property (EP) (see Section 1.7). This *strong*

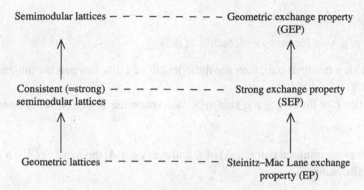

Figure 6.20

exchange property (SEP) (cf. Stern [1982]) is a special case of the geometric exchange property (GEP) considered in Section 6.3: Let L be a lattice of finite length, $u, v \in J(L)$ and $b \in L$. We say that L has the *strong exchange property* if

(SEP) $v \leq b \vee u$ and $v \nleq b \vee u'$ imply $u \leq b \vee v$.

Whereas modular lattices of finite length and geometric lattices have (SEP), this is not so for arbitrary semimodular lattices of finite length: the centered hexagon S_7 is a semimodular lattice that does not possess (SEP). This indicates that (SEP) might characterize strong semimodular lattices. In fact, we have

Theorem 6.5.6   *A lattice of finite length is a strong (and hence consistent) semimodular lattice if and only if it satisfies the strong exchange property* (SEP).

The proof is left as an exercise (Exercise 2). Figure 6.20 visualizes the interrelationships between classes of lattices and the corresponding exchange properties.

An exchange property seemingly different from (SEP) was investigated by Dlab [1962], [1966]. Gaskill & Rival [1978] used it to characterize modularity in lattices of finite length by means of what they call "join symmetry" (to be defined below). We shall identify the join-symmetric lattices as exactly the consistent semimodular lattices.

Let L denote a lattice of finite length, and $J(L)$ its set of join-irreducibles ($\neq 0$). Let A_1 and A_2 denote antichains of $J(L)$. We write $A_1 \leq A_2$ if and only if for each $a_1 \in A_1$ there exists $a_2 \in A_2$ such that $a_1 \leq a_2$. This clearly induces a partial order on the set of antichains in $J(L)$.

The following notion of a minimal pair is essentially the same as in Gaskill & Rival [1978].

A *minimal pair* (p, A) of a lattice L of finite length is an antichain A of $J(L)$ together with an element $p \in J(L)$ such that the following three conditions hold:

(Mp1) $p \in A$;

(Mp2) $p \leq \bigvee A$ (in the lattice L);

(Mp3) $p \nleq \bigvee \tilde{A}$ for every antichain $\tilde{A} < A$.

If (p, A) is a minimal pair, then the finite length of L implies that the antichain A is finite.

A lattice L of finite length is said to be *join-symmetric* if the following condition holds:

(Js) For every minimal pair (p, A) of L and every $q \in A$, the pair $(q, (A - q) \cup p)$ is also minimal.

Geometric lattices are clearly join-symmetric. In fact, if L is a geometric lattice, then (p, A) is a minimal pair of L if and only if $A \cup p$ is a circuit (that is, a minimal dependent subset) of the combinatorial geometry associated with L.

Lemma 6.5.7   *A join-symmetric lattice of finite length is semimodular and strong.*

Proof.   Following Gaskill & Rival [1978], we first show that join symmetry implies semimodularity. Let L be join-symmetric. If L is not semimodular, then there are elements $a, b, c \in L$ such that both a and b cover $a \wedge b$, c covers b, and $a \nleq c < a \vee b$. Let a_m, b_m, c_m be elements of $J(L)$ and minimal satisfying $a_m \leq a$, $b_m \leq b$, $c_m \leq c$ and $a_m \nleq a \wedge b$, $b_m \nleq a \wedge b$, $c_m \nleq b$. Then $a_m \vee (a \wedge b) = a, b_m \vee (a \wedge b) = a$, whence $a_m \vee b_m \vee (a \wedge b) = a \vee b > c_m$. We may choose an antichain K in $J(L)$ such that $K \leq \{a \wedge b\}$ and $(c_m, \{a_m, b_m\} \cup K)$ is a minimal pair. (Notice that if $x < a_m$ then $x \leq a \wedge b$.) Since L is join-symmetric, $(a_m, \{b_m, c_m\} \cup K)$ is a minimal pair. In particular, $a_m \leq b_m \vee c_m \vee \bigvee K \leq c$, whence $a_m \leq a \wedge c = a \wedge b$. As this is impossible, we conclude that L is semimodular.

Next we verify that L is also strong. Suppose on the contrary that there are elements $p \in J(L)$, $x \in L$, such that $x \vee p = x \vee p'$ and $p \nleq x$. Then there must exist an antichain A in $J(L)$ such that for every $a \in A$ we have $a \leq x$ or $x \leq p'$ and, moreover, (p, A) is a minimal pair. Since $p \nleq x$, the antichain A must contain some q with $q < p$. Join symmetry of L implies that $(q, (A - q) \cup p)$ is a minimal pair. But $(A - q) \cup p$ violates (Mp3) with $\tilde{A} = \{q\}$. ∎

Lemma 6.5.8   *A strong semimodular lattice of finite length is join-symmetric.*

Proof.   Suppose L is not join-symmetric. Then there exists a minimal pair (p, A) of L and an element $q \in A$ such that $(q, (A - q) \cup p)$ is not a minimal pair. Let $x = \bigvee(A - q)$. By (Mp3) we have $p \nleq x \vee q'$ and therefore $q \leq x \vee p$ by the strong exchange property (SEP). Hence the conditions (Mp1) and (Mp2) hold for $(q, (A - q) \cup p)$. Note that $(A - q) \cup p$ is an antichain, since otherwise the existence of some $r \in A$ with $r < p$ would imply $p \leq \bigvee(A - r) \vee x$ and hence (by strongness) $p \leq \bigvee(A - x)$, which contradicts (Mp3). Since $(q, (A - q) \cup p)$

violates (Mp3), there exists an antichain $B < A - q$ such that $q \leq (\bigvee B) \vee p$. We claim that $q \not\leq (\bigvee B) \vee p'$. Assume on the contrary that $q \leq (\bigvee B) \vee p'$. Then $q \leq x \vee q'$ implies $p \leq x \vee q \leq x \vee q'$ and hence $p \leq x$, which contradicts (Mp3) for (p, A). Applying (SEP), we now conclude that $p \leq (\bigvee B) \vee q$. Consequently, (p, A) is not a minimal pair unless $A - q$ is empty. But this means $p \leq q$ and $p \leq q'$, that is, $p = q$. Thus we arrived at a contradiction, since $(p, A) = (p, \{p\})$ is not a minimal pair. ■

Lemma 6.5.7 and Lemma 6.5.8, together with the fact that strongness is equivalent to consistency for lattices of finite length, yield (cf. Faigle et al. [1984]):

Theorem 6.5.9 *A lattice of finite length is join-symmetric if and only if it is a consistent semimodular lattice.*

As a consequence we get the following result of Gaskill & Rival [1978].

Corollary 6.5.10 *A lattice L of finite length is modular if and only if both L and its dual L^* are join-symmetric.*

We have seen (cf. Theorems 1.7.2 and 1.7.3) that, for an atomistic lattice L of finite length, semimodularity is equivalent to the condition

(+) pMb holds for all atoms $p \in L$ and for all $b \in L$.

This gives rise to the following question: What does it mean for a lattice L of finite length that

(++) jMb holds for all join-irreducibles $j \in J(L)$ and for all $b \in L$?

We note that if (++) is satisfied, then by Maeda & Maeda [1970] (Lemma 1.2, p. 1) we have $j M^* b$ for all $j \in J(L)$ and for all $b \in L$. Hence condition (iv) of Theorem 3.3.5 is satisfied, and it follows that the lattice is semimodular. If L were not consistent, it would (by Theorem 6.5.1) contain a hexagon sublattice isomorphic to the lattice of Figure 6.18, and it follows that there exists a join-irreducible $j \in J(L)$ and an element $b \in L$ such that jMb does not hold. Thus we have shown

Theorem 6.5.11 *Let L be a lattice of finite length. If jMb holds for all $j \in J(L)$ and for all $b \in L$, then L is a consistent semimodular lattice.*

The converse is not true, that is, in a consistent semimodular lattice of finite length condition (++) does not hold in general (example?).

The following characterization of condition (++) is left as an exercise:

Lemma 6.5.12 *In a lattice L of finite length the following two conditions are equivalent:*

(i) jMb holds for all join-irreducibles $j \in J(L)$ and for all $b \in L$ [i.e., condition $(++)$ holds];

(ii) the intervals $[j \wedge b, j]$ and $[b, b \vee j]$ are isomorphic.

Condition (ii) has also been called the *isomorphism property for join-irreducible elements*. Lattices of finite length having this property have been called generalized matroid lattices (Stern [1978]). Levigion & Schmidt [1995] present a geometric approach to generalized matroid lattices. They investigate hereditary properties of so-called trace geometries and define *generalized exchange geometries* (GEGs) as a common generalization of exchange geometries and projective lattice geometries (introduced by Greferath & Schmidt [1992]; see also Schmidt [1987]). Furthermore, they give constructions for a wide range of GEGs. They also prove that a lattice L of finite length is a generalized matroid lattice if and only if $(L, J(L) \cup \{0\})$ forms a GEG.

Notes

The investigation of cyclically generated lattices seems to go back to Klein-Barmen [1941], who called them "molekulare Verbände." Important classes of cyclically generated modular lattices have been investigated in detail, beginning with the pioneering paper of Baer [1942] and continued by Jónsson & Monk [1969] and others.

Faigle [1986b] also considered closure operators whose lattices of closed sets are not necessarily semimodular. He characterized finite closure spaces with the Steinitz exchange property and discussed the connection between the Steinitz and the Mac Lane exchange property and related exchange properties.

Exercises

1. A semimodular lattice L of finite length is consistent if and only if it does not contain a sublattice of the form shown in Figure 6.18 with $j \in J(L)$ and with five covers indicated by double lines (Faigle et al. [1984], Theorem 2).

2. A lattice of finite length is a strong (and hence consistent) semimodular lattice if and only if it satisfies the strong exchange property (SEP) (Faigle et al. [1984], Theorem 1).

3. Let L be a finite semimodular lattice and $c \in L$. Consider the natural join-preserving map $f_c : L \to [c, 1]$ given by $x \to f_c(x) = x \vee c$. Show that the restriction of f_c to $J_0(L) = J(L) \cup \{0\}$ is an epimorphism onto $J_0([c, 1])$ (the set of all join irreducibles of $[c, 1]$ including c) for every $c \in L$ if and only if L is consistent (Faigle [1980a]).

4. Show that a cyclically generated semimodular lattice of finite length is consistent (strong) if and only if the isomorphism property holds for cycles (see Lemma 6.5.12 for the isomorphism property for join-irreducibles).

5. In a consistent semimodular lattice of finite length, $a \leq b$ implies $r(a) \leq r(b)$ if and only if the lattice is cyclically generated.

References

Baer, R. [1942] A unified theory of projective spaces and finite abelian groups, Trans. Amer. Math. Soc. 52, 283–343.

Dlab, V. [1962] General algebraic dependence relations, Publ. Math. Debrecen 9, 324–355.

Dlab, V. [1966] Algebraic dependence structures, Z. Math. Logik Grundl. Math. 12, 345–377.

Faigle, U. [1980a] Geometries on partially ordered sets, J. Combin. Theory Ser. B 28, 26–51.

Faigle, U. [1986b] Exchange properties of combinatorial closure spaces, Discrete Appl. Math.15, 249–260.

Faigle, U., G. Richter, and M. Stern [1984] Geometric exchange properties in lattices of finite length, Algebra Universalis 19, 355–365.

Gaskill, H. S. and I. Rival [1978] An exchange property for modular lattices, Algebra Universalis 8, 354–356.

Greferath, M. and S. E. Schmidt [1992] A unified approach to projective lattice geometries, Geom. Dedicata 43, 243–264.

Jónsson, B. and G. Monk [1969] Representation of primary arguesian lattices, Pacific J. Math. 30, 95–139.

Klein-Barmen, F. [1941] Molekulare Verbände, Math. Z. 47, 373–394. ·

Levigion, V. and S. E. Schmidt [1995] A geometric approach to generalized matroid lattices, in: *General Algebra and Discrete Mathematics* (ed. K. Denecke and O. Lüders), Heldermann, Berlin, pp. 181–186.

Richter, G. [1991] Strongness in J-lattices, Studia Sci. Math. Hungar. 26, 67–80.

Schmidt, S. E. [1987] Projektive Räume mit geordneter Punktmenge, Mitt. Math. Sem. Gießen 182, 1–77.

Soltan, V. P. [1973] The Jordan form of matrices and its connection to lattice theory (Russian), Mat. Issled. (Kishinev), 1(27), 152–170.

Soltan, V. P. [1975] Jordan elements of lattices and subordinate sums (Russian), Mat. Issled. (Kishinev) 10, 230–237.

Stern, M. [1978] Generalized matroid lattices, in: *Algebraic Methods in Graph Theory*, Colloq. Math. Soc. J. Bolyai, Vol. 25, pp. 727–748.

Stern, M. [1982] Semimodularity in lattices of finite length, Discrete Math. 41, 287–293.

Stern, M. [1990a] Strongness in semimodular lattices, Discrete Math. 82, 79–88.

Stern, M. [1991a] *Semimodular Lattices*, Teubner, Stuttgart.

6.6 Pseudomodular Lattices

Summary. Björner & Lovász [1987] investigated semimodular lattices with so-called pseudointersections and called them pseudomodular. Their main goal is the construction of continuous analogues to geometric lattices. However, as they point out, pseudomodular lattices of finite length are also an object worthwhile to study for its own sake. Here we give some of their results and examples.

Let L be a semimodular lattice of finite length, and denote by $r(x)$ the rank function (height function) of L. For each $x, y \in L$ let

$$P_{x,y} = \{z \le y : r(x \vee z) - r(z) = r(x \vee y) - r(y)\}.$$

Here it would be sufficient to require that $r(x \vee z) - r(z) \le r(x \vee y) - r(y)$, since the reverse inequality always holds by the submodularity of the rank function (see Section 1.9).

It is easy to see that $P_{x,y} \subseteq [x \wedge y, y]$ and that $P_{x,y}$ is a dual order ideal in $[x \wedge y, y]$. If the set $P_{x,y}$ has a unique least element, then we call this the *pseudointersection* of x and y and denote it by $x \rceil y$. A (semimodular) lattice of finite length is called *pseudomodular* if every pair of its elements has a pseudointersection. $x \rceil y$ does not imply $y \rceil x$ (i.e., the existence of a pseudointersection is not a symmetric relation), as the following example shows: Consider three pairwise parallel lines in affine three-space that are not coplanar. Let y be one of these lines, and denote by x the plane spanned by the other two lines. Then in the geometric lattice generated by the points of these lines, $x \rceil y$ exists whereas $y \rceil x$ does not.

The following lemma illustrates the relationship between pseudointersection and the meet operation in the lattice.

Lemma 6.6.1 *Let L be a semimodular lattice and $x, y \in L$. Then the following four conditions are equivalent:*

  (i) *$x M y$, that is, x and y form a modular pair;*
 (ii) *$x \rceil y$ exists and $x \rceil y \leq x$;*
(iii) *$x \rceil y$ exists and $x \rceil y = x \wedge y$;*
 (iv) *$x \wedge y \in P_{x,y}$.*

The proof is straightforward and will be left as an exercise. The preceding lemma also shows that $x \rceil y \neq y \rceil x$. The following result is useful for checking the existence of pseudointersections.

Lemma 6.6.2 *Let L be a semimodular lattice and $x, y \in L$. Then the following three conditions are equivalent:*

  (i) *$x \rceil y$ exists, that is, $P_{x,y}$ has a unique least element;*
 (ii) *$P_{x,y}$ is closed under meets;*
(iii) *if $u, v, z \in P_{x,y}$ and z covers u and v, then $u \wedge v \in P_{x,y}$.*

Proof. The implications (i) $\Rightarrow$ (ii) $\Rightarrow$ (iii) are obvious. To show (iii) $\Rightarrow$ (i) assume that $P_{x,y}$ has minimal elements a, b such that $a \neq b$. Choose a and b so that $a \vee b$ is of minimal height in the lattice. Let u be an element in the interval $[a, a \vee b]$ covered by $a \vee b$, and let v be an element in the interval $[b, a \vee b]$ covered by $a \vee b$. Condition (iii) implies $u \wedge v \in P_{x,y}$. Let c be a minimal element of $P_{x,y}$ below $u \wedge v$. It follows that $a \vee c \leq u < a \vee b$ and $b \vee c \leq v < a \vee b$. Now c is distinct from at least one of a and b. This contradicts the choice of a and b. ∎

The proof of the next assertion is left as an exercise (Exercise 2).

Lemma 6.6.3 *Let L be a geometric lattice and $x, y \in L$. If $x \rceil y$ exists, then it equals the meet of all $z \in L$ such that y covers z and $x \vee y$ covers $x \vee z$.*

The following result characterizes the existence of pseudointersections in terms the nonexistence of certain configurations.

Theorem 6.6.4 *Let L be a semimodular lattice. Then the following five conditions are equivalent:*

(i) *L is pseudomodular.*

(ii) *Let a, b, $c \in L$, and assume that $r(a \vee c) - r(a) = r(b \vee c) - r(b) = r(a \vee b \vee c) - r(a \vee b)$. Then $r((a \vee c) \wedge (b \vee c)) - r(a \wedge b) = r(a \vee c) - r(a)$.*

(iii) *Let x, y, $z \in L$, and assume that x covers $x \wedge z$ and y covers $y \wedge z$. Then $r(x \wedge y) - r(x \wedge y \wedge z) \leq 1$.*

(iv) *Let x, y, z, $u \in L$, and assume that u covers x, y and z, and z covers $x \wedge z$ and $y \wedge z$. Then $r(x \wedge y) - r(x \wedge y \wedge z) \leq 1$.*

(v) *Let x, y, z, $u \in L$ and assume that u covers x and y, $z \leq u$, and z covers $x \wedge z$ and $y \wedge z$. Then $r(x \wedge y) - r(x \wedge y \wedge z) \leq r(u) - r(z)$.*

Proof. (i) $\Rightarrow$ (ii): Let $d = (a \vee c) \wedge (b \vee c)$. Because $a \vee c = a \vee d$ and $a \vee b \vee c = a \vee b \vee d$, we have $a \in P_{d,a \vee b}$. Hence the hypothesis of (ii) implies $r(a \vee d) - r(a) = r(a \vee b \vee d) - r(a \vee b)$. Similarly we have $b \in P_{d,a \vee b}$. Pseudomodularity implies therefore $a \wedge b \in P_{d,a \vee b}$. Thus $r((a \wedge b) \vee d) - r(a \wedge b) = r(a \vee b \vee d) - r(a \wedge b)$. Now the assertion of (ii) follows, since $(a \wedge b) \vee d = d$ and $a \vee b \vee d = a \vee b \vee c$.

(ii) $\Rightarrow$ (iii): We may assume that $z = (x \wedge z) \vee (y \wedge z)$ [otherwise take $(x \wedge z) \vee (y \wedge z)$ for z]. We may also assume that $x \wedge y \nleq z$ (since otherwise $x \wedge y = x \wedge y \wedge z$) and that $x \neq y$. This yields $x \wedge z \neq y \wedge z$ and $x = (x \wedge z) \vee (x \wedge y)$, $y = (y \wedge z) \vee (x \wedge y)$. Moreover $r(z) < r(z \vee (x \wedge y)) \leq r(z \vee x) \leq r(z) + r(x) - r(z \wedge x) = r(z) + 1$. Thus $z \vee (x \wedge y)$ covers z. Putting now $a = x \wedge z$, $b = y \wedge z$, and $c = x \wedge y$ in (ii), we obtain condition (iii).

(iii) $\Rightarrow$ (iv): Obvious.

(iv) $\Rightarrow$ (v): The proof is by induction on $r(u) - r(z)$. Condition (iv) is a special case of (v) when $r(u) - r(z) = 1$. We may assume that $x \wedge y \wedge z \neq x \wedge y$. Denote by p an element of the interval $[x \wedge y \wedge z, x \wedge y]$ covering $x \wedge y \wedge z$. It is clear that $p \nleq z$, whence $p \nleq x \wedge z$ and $p \nleq y \wedge z$. Submodularity implies that $v = z \vee p$ covers z. Similarly, $(z \wedge x) \vee p$ covers $z \wedge x$ and $(z \wedge y) \vee p$ covers $z \wedge y$. Obviously $(z \wedge x) \vee p \leq v \wedge x < v$ and thus $v \wedge x = (z \wedge x) \vee p$. It follows that $v \wedge x$ is covered by v. Similarly, $v \wedge y$ is also covered by v. Applying (iv) with v, $v \wedge x$, $v \wedge y$, and z in place of u, x, y, and z, we obtain

$$r(x \wedge y \wedge v) - r(x \wedge y \wedge z) \leq 1.$$

Applying the induction hypothesis with u, x, y, and v in place of u, x, y, and z, we get

$$r(x \wedge y) - r(x \wedge y \wedge v) \leq r(u) - r(v) = r(u) - r(z) - 1,$$

which proves (v).

(v) $\Rightarrow$ (i): First we verify Lemma 6.6.2(iii). Let u, v, $z \in P_{x,y}$ where z covers both u and v. From the definition of $P_{x,y}$ it follows that $z \vee x$ covers both $u \vee x$ and $v \vee x$, and $z \wedge (u \vee x) = u$, $z \wedge (v \vee x) = v$. Now (v) can be applied with $u \vee x$, $v \vee x$, z, and $z \vee x$ in place of x, y, z, and u, and we obtain $r((u \vee x) \wedge (v \vee x)) - r((u \vee x) \wedge (v \vee x) \wedge z) \leq r(z \vee x) - r(z)$. Because of $(u \vee x) \wedge (v \vee x) \wedge z = u \wedge v$ this implies

$$r((u \wedge v) \vee x) - r(u \wedge v) \leq r((u \vee x) \wedge (v \vee x)) - r(u \wedge v) \leq r(z \vee x) - r(z).$$

This shows that $u \wedge v \in P_{x,y}$. ∎

Property (iv) of the preceding theorem implies the following result of Lindström [1988a], which in turn is a generalization of the Ingleton–Main lemma for full algebraic matroids (cf. Ingleton & Main [1975]):

Corollary 6.6.5   *If a, b and c are three flats such that $r(a) = r(b) = r(c) = n$, $r(a \vee b) = r(a \vee c) = r(b \vee c) = n + 1$, and $r(a \vee b \vee c) = n + 2$, then $a \wedge b = b \wedge c = a \wedge b \wedge c$ and $r(a \wedge b \wedge c) = n - 1$.*

Lindström [1988b] also conjectured that if a, b, and c are three flats in an algebraic matroid such that $r(a) = r(b) = r(c) = n$, $r(a \vee b) = r(a \vee c) = r(b \vee c) = n + k$, and $r(a \vee b \vee c) = n + 2k$, then $a \wedge b = b \wedge c = a \wedge b \wedge c$ and $r(a \wedge b \wedge c) = n - k$. This conjecture is a consequence of condition (ii) of the preceding theorem.

Modular lattices of finite length and semimodular lattices of length at most 3 are pseudomodular. Several classes of geometric lattices have been shown to be pseudomodular.

For example, let F and K be algebraically closed fields and $F \subset K$. Then the algebraically closed subfields of K containing F form a geometric lattice $L(F, K)$ (cf. Mac Lane [1938]). Lattices of this kind are also called full algebraic matroid lattices. Dress & Lovász [1987] proved that $L(F, K)$ has pseudointersections. A simple description of the operation ⌉ is the following one (see Björner & Lovász [1987]): Let X and Y be two algebraically closed fields with $F \subset X$ and $Y \subset K$. Let $\{x_1, \ldots, x_m\}$ be a transcendence basis of X over F. Consider the ideal I of all polynomials over Y in m variables that are satisfied by $(x_1, \ldots, x_m)$ and a basis $q_1, \ldots, q_N$ of this ideal. We may assume that each q_i has at least one coefficient that is equal to 1. Then the algebraically closed subfield T of Y generated by the coefficients of $q_1, \ldots, q_N$ is the pseudointersection of X and Y.

Partition lattices are pseudomodular (Exercise 3). The fact that partition lattices are pseudomodular can be restated so that the Dilworth truncation of a Boolean lattice (cf. Section 6.2) is pseudomodular. Björner & Lovász [1987] pose the problem of finding a broader class of lattices whose Dilworth truncations are pseudomodular.

Björner & Lovász [1987] also prove the pseudomodularity of a large class of transversal matroids. (For information on transversal matroids see Brualdi [1987].)

Their result implies, in particular, the pseudomodularity of full transversal matroids. Some special antimatroids are pseudomodular, for example, antimatroids with Carathéodory number 2 (cf. Björner & Lovász [1987], cf. also Notes to Section 7.3).

Let K be a field of prime characteristic p, and assume that K is finitely generated over K^p, the field of pth powers of the elements of K, and distinct from K^p. Then K is a matroid on an infinite set. Mac Lane [1938] observed that the matroid lattice of fields k, $K^p \subseteq k \subseteq K$, is not modular in general. By a result of Teichmüller [1936] it can be shown that this lattice is modular if K is generated by at most two elements, transcendentals over K^p. The following result is due to Lindström [1990].

Theorem 6.6.6 *Let p be a prime, and K finitely generated over K^p(K is an inseparable extension of exponent 1 over K^p). Then the matroid lattice of fields k, $K^p \subseteq k \subseteq K$, is pseudomodular.*

Proof. We show that condition (iii) of Theorem 6.6.4 is satisfied. Let u, v, w be three fields such that u covers $u \wedge w$ and v covers $v \wedge w$ and $r(u \wedge v) - r(u \wedge v \wedge w) \geq 2$. We shall derive a contradiction. We may assume $w = (u \wedge w) \vee (v \wedge w)$. Otherwise put $w_1 = (u \wedge w) \vee (v \wedge w)$. Then $u \wedge w = u \wedge w_1$ and $v \wedge w = v \wedge w_1$, and we may replace w by w_1 in the proof.

From $r(u \wedge v) - r(u \wedge v \wedge w) \geq 2$ it follows that there exist elements x, y of the field $u \wedge v$ that are independent over the subfield $u \wedge v \wedge w$. Since u covers $u \wedge w$, we obtain that x and y are dependent over $u \wedge w$. Thus $x = P(y)$ holds for a polynomial $P(Y)$ with coefficients from the field $u \wedge w$. Similarly, there is a polynomial $Q(Y)$ over $v \wedge w$ such that $x = Q(y)$. We choose degrees of P and Q less than p. Then the polynomials are unique. If $P(Y) = Q(Y)$, then the coefficients of the polynomial belong to the field $u \wedge v \wedge w$, and x [$= P(y)$] and y will be dependent over $u \wedge v \wedge w$, contradicting our assumption. Thus $P(Y) - Q(Y)$ will be a nonzero polynomial of degree less than p with y a zero. It follows that y belongs to the field $(u \wedge w) \vee (v \wedge w) = w$ and (by assumption) also to $u \wedge v$. Thus y is in $u \wedge v \wedge w$, contrary to the assumptions. Hence the lattice is pseudomodular. ∎

The following theorem of Björner & Lovász [1987] generalizes a well-known result for modular lattices (see Birkhoff [1967], pp. 73–74).

Theorem 6.6.7 *Let L be a pseudomodular lattice, and $a_1, \ldots, a_k$ elements of L such that $r(a_1) + \cdots + r(a_k) = r(a_1 \vee \cdots \vee a_k)$. Then the sublattice generated by the intervals $[0, a_i]$ is isomorphic to the direct product of these intervals.*

Proof. It suffices to consider the case $k = 2$. Note that the condition of submodularity and the hypothesis that $r(a_1) + r(a_2) = r(a_1 \vee a_2)$ yield that $r(x_1) + r(x_2) =$

$r(x_1 \vee x_2)$ for all $x_i \le a_i$. Let L' be the sublattice generated by the intervals $[0, a_i]$, and define the mapping $\varphi(x_1, x_2) = x_1 \vee x_2$. It is easily seen that this is an injection of $[0, a_1] \times [0, a_2]$ and that this injection is join-preserving. We show that it is also meet-preserving, which implies that the mapping is bijective.

Let $x_i \vee y_i \in [0, a_i]$, and set $p = (x_1 \vee x_2) \wedge (y_1 \vee y_2)$, $q = (x_1 \vee y_1) \vee (x_2 \vee y_2)$. Then $p \ge q$, and we show that equality holds. To see this, we prove $r(p) \le r(q)$, which yields that p and q have the same rank. Obviously, $r(q) = r(x_1 \wedge y_1) + r(x_2 \wedge y_2)$.

Let now $a = x_1$, $b = y_1$ and $c = x_2 \vee y_2 \vee p$ in Theorem 6.6.4 (ii). Then trivially $a \vee b \vee c = x_1 \vee y_1 \vee x_2 \vee y_2$, and thus $r(a \vee b \vee c) = r(x_1 \vee y_1) \vee r(x_2 \vee y_2)$. In a similar way we obtain

$$r(a \vee b) = r(x_1 \vee y_1), \qquad r(a \vee c) = r(x_1) \vee r(x_2 \vee y_2),$$
$$r(b \vee c) = r(y_1) \vee r(x_2 \vee y_2).$$

It follows that a, b, and c satisfy the assumptions in Theorem 6.6.4(ii), and thus by pseudomodularity we have $r(c) \le r(a \wedge b) + r(a \vee c) - r(c)$. Substituting yields $r(x_2 \vee y_2 \vee p) \le r(x_1 \wedge y_1) + r(x_2 \vee y_2)$. Interchanging the subscripts, we get

$$r(x_1 \vee y_1 \vee p) \le r(x_2 \wedge y_2) + r(x_1 \vee y_1).$$

Now submodularity implies

$$r(p) \le r(p \vee x_1 \vee y_1) + r(p \vee x_2 \vee y_2) - r(x_1 \vee y_1 \vee x_2 \vee y_2)$$
$$\le r(x_1 \wedge y_1) + r(x_2 \wedge y_2) = r(q),$$

which proves the theorem. ∎

This theorem does not hold in a general semimodular lattice of finite length (see Exercise 4).

The preceding result can be used to construct "stretch embeddings" for various classes of matroids. Let $L_1, L_2, \ldots$ be a sequence of pseudomodular geometric lattices such that L_n has rank n. Assume that for each $n, m \ge 1$ such that $m \mid n$, there exist in L_n n/m elements $a_1, \ldots, a_{n/m}$ of rank m such that $a_1 \vee \cdots \vee a_{n/m} = 1$ and $[0, a_i] \cong L_m$. We call these elements the representatives of L_m in L_n. It is now easy to define a *stretch embedding* of L_m in L_n, that is, a lattice embedding

$$\varphi = \varphi_m^n : L_m \to L_n \qquad \text{such that} \quad r(\varphi(x)) = \frac{n}{m} r(x) \quad \text{for each } x \in L_m.$$

To see this, let $\varphi_i : L_m \to [0, a_i]$ $(i = 1, \ldots, n/m)$ be an arbitrary isomorphism, and define

$$\varphi(x) = \varphi_1(x) \vee \cdots \vee \varphi_{n/m}(x).$$

Theorem 6.6.7 implies that this is indeed a stretch embedding. Björner [1987] introduced a similar construction under the hypothesis that the elements $a_1, \ldots, a_{n/m}$ are modular elements. To construct the "continuous limit" of the above sequence of lattices, one has to assume that the mappings φ_m^n form a directed system, that is, if $k \mid m$ and $m \mid n$ then $\varphi_m^n \circ \varphi_k^m = \varphi_k^n$. This can be done by compatibly choosing the representatives, constructing the direct limit $L_{(\infty)}$ and its metric completion L_∞. For partition lattices this was accomplished by Björner [1987], who introduced thereby continuous analogues to finite partition lattices.

Notes

For a general background on matroid theory see the books Welsh [1976] and Oxley [1993].

Lindström [1988b] surveys algebraic and nonalgebraic matroids and deals in particular with successive generalizations of the Ingleton–Main lemma.

There are some standard operations on semimodular lattices that preserve pseudomodularity, such as forming the direct product, truncation, and forming the principal extension. The latter two constructions are best known for geometric lattices (see Brylawski [1986]). Björner & Lovász [1987] proved that any principal extension of a pseudomodular lattice is again pseudomodular. They point out that by this result every matroid that can be obtained by principal extensions from Boolean lattices is pseudomodular. In particular, full transversal matroids (whose definition is not given here) are pseudomodular.

Pseudomodular lattices were used by Dress & Lovász [1987] for the investigation of combinatorial properties of algebraic matroids. Dress et al. [1994] give local conditions characterizing modular sublattices of a geometric pseudomodular lattice. As an application they derive a result of Hochstättler & Kern [1989] implying that Lovász's min–max formula for matchings in projective geometries remains valid for geometric pseudomodular lattices.

Exercises

1. Show that assertion (iii) in Lemma 6.6.2 holds automatically if (u, v) is a modular pair.
2. Let L be a geometric lattice and $x, y \in L$. If $x \rceil y$ exists, then it equals the meet of all $z \in L$ such that y covers z and $x \vee y$ covers $x \vee z$ (Björner & Lovász [1987], Lemma 1.3).
3. Show that finite partition lattices are pseudomodular (Björner & Lovász [1987], Example 4). [Hint: Use Lemma 6.6.2 (iii) and Exercise 1.]
4. Give examples showing that Theorem 6.6.7 does not hold for semimodular lattices or geometric lattices in general. (See Björner & Lovász [1987] for an example.)

References

Birkhoff, G. [1967] *Lattice Theory*, Amer. Math. Soc. Colloquium Publications, Vol. 25, Providence, R.I. (reprinted 1984).

Björner, A. [1987] Continuous partition lattice, Proc. Natl. Acad. Sci. U.S.A. 84, 6327–6329.

Björner, A. and L. Lovász [1987] Pseudomodular lattices and continuous matroids, Acta Sci. Math. (Szeged) 51, 295–308.

Brualdi, R.A. [1987] Transversal matroids, in: White [1987], pp. 72–97.

Brylawski, T. [1986] Constructions, in: White [1986], pp. 127–223.

Dress, A. and L. Lovász [1987] On some combinatorial properties of algebraic matroids, Combinatorica 7, 39–48.

Dress, A., W. Hochstättler, and W. Kern [1994] Modular substructures in pseudomodular lattices, Math. Scand. 74, 9–16.

Hochstättler, W. and W. Kern [1989] Matroid matching in pseudomodular lattices, Combinatorica 9, 145–152.

Ingleton, A. W. and R. A. Main [1975] Non-algebraic matroids exist, Bull. London Math. Soc. 7, 144–146.

Lindström, B. [1988a] A generalization of the Ingleton–Main lemma and a class of non-algebraic matroids, Combinatorica 8, 87–90.

Lindström, B. [1988b] Matroids, algebraic and non-algebraic, in: *Algebraic, Extremal and Metric Combinatorics, 1986,* London Math. Soc. Lecture Notes Ser. 131 (ed. M. M. Deza, P. Frankl, and I. G. Rosenberg), pp. 166–174.

Lindström, B. [1990] p-Independence implies pseudomodularity, Eur. J. Combin. 11, 489–490.

Mac Lane, S. [1938] A lattice formulation for transcendence degrees and p-bases, Duke Math. J. 4, 455–468.

Oxley, J. G. [1993] *Matroid Theory*, Oxford Univ. Press.

Teichmüller, O. [1936] p-Algebren, Deutsche Math. 1, 362–388.

Welsh, D. [1976] *Matroid Theory*, Academic Press, London.

White, N. L. (ed.) [1986] *Theory of Matroids*, Encyclopedia of Mathematics and its Applications, Vol. 26, Cambridge Univ. Press, Cambridge.

White, N. L. (ed.) [1987] *Combinatorial Geometries,* Encyclopedia of Mathematics and its Applications, Vol. 29, Cambridge Univ. Press, Cambridge.

7

Local Distributivity

7.1 The Characterization of Dilworth and Crawley

Summary. In this section we briefly mention characterizations of strongly atomic algebraic lattices having unique irredundant meet decompositions. These characterizations – which are due to Dilworth and Crawley – are similar to the corresponding characterizations in the finite case (for which see Section 7.2), though different techniques of proof are required. One of these equivalent conditions is that the lattice is upper locally distributive (join-distributive).

In this section all lattices are assumed to be algebraic and strongly atomic. For algebraic strongly atomic lattices the properties of (*upper*) *local distributivity* and (*upper*) *local modularity* can be defined in complete analogy to the finite length case. From these definitions it is clear that a locally distributive lattice is locally modular. By definition it is also immediate that any locally distributive or locally modular lattice satisfies Birkhoff's condition (Bi) (weak semimodularity). Hence Theorem 1.7.1 implies

Corollary 7.1.1 *A locally distributive or locally modular, algebraic, strongly atomic lattice is semimodular.*

Upper local distributivity for strongly atomic algebraic lattices has been characterized in the following way.

Theorem 7.1.2 *Let L be an algebraic strongly atomic lattice. Then the following three conditions are equivalent:*

(i) *L is upper locally distributive;*

(ii) *each element of L has a unique irredundant meet decomposition;*

(iii) (a) *L is semimodular and*

(b) *if $a, p, x, y \in L$, $x, y \geq a$, and $p \succ a$, then $p \wedge (x \vee y) = (p \wedge x) \vee (p \wedge y)$.*

267

This result is due to Dilworth & Crawley [1960] (cf. also Bogart et al. [1990], pp. 145–166; for a proof we also refer to Crawley & Dilworth [1973], Theorem 7.4).

We remark that condition (i) and condition (ii) are just conditions (f') and (e'), respectively, in Theorem 7.2.30 below. Condition (iii)(b) of the preceding theorem implies, in particular, that every atom of L has a pseudocomplement. As in the finite case (see Chameini-Nembua & Monjardet [1992], Theorem 3.3), it can be shown that then every element has a pseudocomplement, that is, the lattice is pseudocomplemented (cf. also Theorem 1.4.2). More precisely, condition (iii)(b) implies that any interval of L is pseudocomplemented. This is the strong form of pseudocomplementation mentioned in Section 1.4. Note also that condition (iii)(b) is equivalent to meet semidistributivity for strongly atomic algebraic lattices (for finite lattices this was observed in Duquenne [1991], Theorem 17, condition (4)).

A part of the proof of Theorem 7.1.2 was isolated in the following result of Dilworth & Crawley [1960], Lemma 6.4.

Lemma 7.1.3 *Let L be an upper locally distributive strongly atomic algebraic lattice. Then each element of L has a unique irredundant decomposition into meet irreducibles.*

An examination of the proof of the preceding lemma shows that the only conditions required for the sufficiency argument are Birkhoff's condition (Bi) and special conditions on the independence of covering elements (cf. Dilworth & Crawley [1960], Corollary to Theorem 6.1; see also Crawley & Dilworth [1973], p. 53). More precisely, we have

Corollary 7.1.4 *An algebraic, strongly atomic lattice is upper locally distributive if and only if for every set of four distinct elements a, p_1, p_2, $p_3 \in L$ for which $p_1, p_2, p_3 \succ a$, the following two conditions hold: (i) If p_i, $p_k \succ a$ with $p_i \neq p_k$, then the sublattice $[a, p_i \vee p_k]$ is a four-element Boolean lattice, and (ii) the sublattice $[a, p_1 \vee p_2 \vee p_3]$ is an eight-element Boolean lattice.*

Freese & Nation (Clarification to "Congruence lattices of a semilattice," 1993, 1995, unpublished comments[1] on Freese & Nation [1973]) pointed out that if in the preceding corollary condition (i) is omitted, then the requirement that every three distinct covering elements generate a cover-preserving Boolean sublattice of order eight is weaker than upper local distributivity (which is shown by N_5). We obtain the following characterization (Dilworth & Crawley [1960], Corollary to Theorem 6.1; see also Crawley & Dilworth [1973], p. 53):

Corollary 7.1.5 *An algebraic, strongly atomic lattice is upper locally distributive if and only if it is upper semimodular and every modular sublattice is distributive.*

This characterization is analogous to that given by Dilworth [1940], Theorem 1.1, for the finite case [see also condition (h') in Theorem 7.2.30].

---

[1] Available on the web at http://www.math.hawaii.edu/~ralph/papers.html.

Notes

Crawley and Dilworth define local distributivity only for strongly atomic (algebraic) lattices. This definition can be extended to an arbitrary complete lattice L in the following obvious way: If $a \in L$, let a^+ be the join of the covers of a; if a has no covers, then set $a^+ = a$. L is locally distributive if $[a, a^+]$ is distributive for each $a \in L$.

Exercises

1. Using the extended version of local distributivity (see preceding Notes), give an example of an algebraic locally distributive lattice that is not semimodular (Freese & Nation, 1993, unpublished).

2. Using the preceding extended version of local distributivity, prove the following result: Let L be a semimodular algebraic lattice such that each interval of L is pseudocomplemented. Then L is locally distributive (Freese & Nation, 1993, unpublished).

References

Bogart, K. P., R. Freese, and J. P. S. Kung (eds.) [1990] *The Dilworth Theorems. Selected Papers of Robert P. Dilworth*, Birkhäuser, Boston.

Chameni-Nembua, C. and B. Monjardet [1992] Les treillis pseudocomplémentés finis, Eur. J. Combin. 13, 89–107.

Crawley, P. and R. P. Dilworth [1973] *Algebraic Theory of Lattices*, Prentice-Hall, Englewood Cliffs, N.J.

Dilworth, R. P. [1940] Lattices with unique irreducible decompositions, Ann. of Math. 41, 771–777.

Dilworth, R. P. [1990a] Background to Chapter 3, in: Bogart et al. [1990], pp. 88–92.

Dilworth, R. P. and P. Crawley [1960] Decomposition theory for lattices without chain conditions, Trans. Amer. Math. Soc. 96, 1–22.

Duquenne, V. [1991] The core of finite lattices, Discrete Math. 88, 133–147.

Freese, R. and J. B. Nation [1973] Congruence lattices of semilattices, Pacific J. Math. 49, 51–58.

7.2 Avann's Characterization Theorem

Summary. We present here the characterizations of local distributivity given by Avann [1961a] for finite lattices. The results and proofs will be given in terms of lower semimodular lattices and lower locally distributive lattices (meet-distributive lattices), that is, we adopt here Avann's approach, which is dual to that of Dilworth and Crawley. If not indicated otherwise, the results below are from Avann [1961a] and [1964b].

Let here L always denote a finite lattice. As before, we denote by $J(L)$ its poset of join-irreducibles ($\neq 0$), and we use $J(x)$ as shorthand for $J([0, x]) = \{j \in J(L) : j \leq x\}$.

Aside from the rank function $r(a)$, we consider two other numerical functions defined on a finite lattice: the *join-order function* $\tau(a) = |J(a)|$, that is, the order or cardinality of $J(a)$, and the *join-excess function* $v(a) = \tau(a) - r(a)$. The join-order function is also called *join-rank function* by some authors (see e.g. Markowsky [1992]).

The lattice S_7^* (the dual to the centered hexagon) is the prototype of a lower locally distributive (LLD) lattice. It is a lower semimodular lattice in which every element has a unique irredundant join decomposition. It is readily checked that $v(x) = \tau(x) - r(x) = 0$ holds for all elements x of the lattice. Later (see Theorem 7.2.27) it will be shown that $v(x) \equiv 0$ is one of eight equivalent conditions for a graded lattice to be lower locally distributive.

Let us first list some general properties of the three numerical functions $r(x)$, $v(x)$, and $\tau(x)$ and of the sets $J(x)$. The following statement is an immediate consequence of the definition of rank.

Lemma 7.2.1   *If $a \geq b$ in L, then $r(a) - r(b)$ is not less than the maximum length of a connected chain from b to a. In particular, if also $r(a) = r(b) + 1$ then $a \succ b$.*

Some elementary facts are noted in

Lemma 7.2.2

  (i) $a > b$ if and only if $J(a) \supset J(b)$;
 (ii) $J(a \wedge b) = J(a) \cap J(b)$;
(iii) $J(a \vee b) \supseteq J(a) \cup J(b)$ and $J(a \vee b) - J(a) \supseteq J(b) - J(a \wedge b)$;
 (iv) $J(a \vee b) \supset J(a) \cup J(b)$ if and only if $\tau(a \vee b) + \tau(a \wedge b) > \tau(a) + \tau(b)$.

The proofs of both lemmas are left as exercises. Lemma 7.2.2 (ii) implies that for any finite lattice L, the map $x \to J(x)$ is a meet-preserving injection of L into the lattice of order ideals of L. The standard proof of the fundamental theorem on finite distributive lattices (cf. Corollary 1.3.2) shows that this map is a bijective isomorphism if and only if L is distributive.

Corollary 7.2.3   *In any connected chain $a_1 \prec a_2 \prec \cdots \prec a_n \prec \cdots$ of a finite lattice L, the functions $\tau(x)$ and $r(x)$ are strictly increasing, whereas the function $v(x)$ is nondecreasing.*

Proof.   We apply Lemma 7.2.1 and Lemma 7.2.2 to $r(x)$ and $\tau(x)$, respectively. Let $c \succ a$ be an arbitrary covering in L with $r(c) = t > s = r(a)$. From an ascending connected chain $0 = c_0 \prec c_1 \cdots \prec c_t = c$ of maximum length select a subchain $c_{j_1} < c_{j_2} < \cdots < c_{j_k}, k < t$, consisting of all c_{j_i} having the following property: $x_i \leq c_{j_i}$ and $x_i \nleq c_{j_i - 1}$ hold for some $x_i \in J(a)$ $(i = 1, \ldots, k)$. From $x_1 \vee \cdots \vee x_{i-1} \leq c_{j_i - 1} \leq c_{j_i - 1}$ it follows that $x_i \nleq x_1 \vee \cdots \vee x_{i-1}$. Hence $0 < x_1 < x_1 \vee x_2 < \cdots < x_1 \vee \cdots \vee x_k \leq a$ is a strictly ascending chain and $k \leq s$.

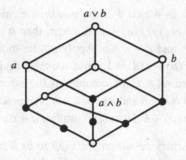

Figure 7.1

Each of the $t - k$ remaining c_m must contain at least one join-irreducible $y_m \notin J(a)$ such that $y_m \leq c_m$ and $y_m \not\leq c_{m-1}$. Thus $\tau(c) - \tau(a) \geq t - k \geq t - s$, which implies the assertion. ∎

The lattice in Figure 7.1 (cf. Avann [1964b]) illustrates the preceding corollary (the elements of $J(L)$ being indicated by solid dots). With the notation of Figure 7.1 we have the following values for $\nu(x) = \tau(x) - r(x)$: $\nu(a \vee b) = \tau(a \vee b) - r(a \vee b) = 6 - 5 = 1, \nu(a \wedge b) = \tau(a \wedge b) - r(a \wedge b) = 2 - 2 = 0, \nu(a) = \tau(a) - r(a) = 4 - 3 = 1,$ $\nu(b) = \tau(b) - r(b) = 4 - 3 = 1,$ and $\nu(x) = \tau(x) - r(x) = 0$ for all other values of x. In particular we have $\nu(a \vee b) + \nu(a \wedge b) = 1 + 0 < 1 + 1 = \nu(a) + \nu(b)$, that is, $\nu(x \vee y) + \nu(x \wedge y) \geq \nu(x) + \nu(y)$ does not hold for all elements x, y. In other words, $\nu(x)$ is not a supermodular (lower semimodular) function.

If $\lambda(a)$ is a numerical function on a finite lattice, then we denote by $\lambda^*(a)$ the dual function. The following result is immediate from Lemma 7.2.2 and its dual.

Corollary 7.2.4   *The function $\tau(a)$ and its dual $\tau^*(a)$ are supermodular.*

We have seen that a *sufficient* condition for the rank function $r(x)$ to be submodular or supermodular on a finite lattice is that the lattice itself is upper semimodular or lower semimodular, respectively (cf. Theorem 1.9.7 and its dual). We shall have a look now at the validity of the converses in the next two results.

Theorem 7.2.5   *A necessary condition for the rank function $r(x)$ to be submodular on a finite lattice is that the lattice itself is semimodular.*

Sketch of proof.   One first shows by induction on the M-*closed subsets* that the Jordan–Dedekind chain condition (JD) holds in the lattice L (the M-closure of any subset S of a poset P is defined as the set $\bar{S}$ of all t such that $t \leq s$ for one or more $s \in S$; cf. Birkhoff [1948], p. 11). The assertion holds trivially in the M-closed subset $\{0\}$. Assume it holds for the M-closed subset K of L, and let c be minimal in $L - K, c > a, a \in K$. Let $c \succ \cdots \succ b \succ d \succ \cdots \succ 0$ be a connected chain of maximal length in which d is the maximal element for which $d \leq a$. If $d = a$, then $b = c$ and $r(c) = r(a) + 1$. If $d < a$, then $b < c$, so that $b \in K$. We have moreover $c = a \vee b$ and $d = a \wedge b$. Applying the induction hypothesis, we get $1 \leq r(c) - r(a) \leq r(b) - r(d) = 1$. Thus we have again $r(c) = r(a) + 1$ and (JD) holds in the M-closed set $K \cup \{c\}$, completing the induction. Let now $a, b \in L$ such that $a \wedge b \prec a$ and $a \wedge b \prec b$ and let $c = a \vee b$. The preceding inequality again holds and hence $r(c) = r(a) + 1$. Now (JD) implies $a \prec c$ and $b \prec c$. ∎

Theorem 7.2.6 *A necessary condition for $r(x)$ to be a supermodular function on a finite graded lattice is that the lattice itself is lower semimodular.*

Proof. Let L be a finite graded lattice, $a, b \in L$, and let $r(a)$ be a supermodular function on L. If $a, b \prec a \vee b$, then $r(a \vee b) + r(a \wedge b) \geq r(a) + r(b)$ yields $1 = r(a \vee b) - r(a) \geq r(b) - r(a \wedge b) \geq 1$. It follows that equality subsists and we get $a \wedge b \prec b$ and [because of (JD)] $a \wedge b \prec a$. ∎

The pentagon (cf. Figure 7.2) shows that (JD) cannot be dropped from the hypothesis of the theorem: $r(x)$ is supermodular, but the lattice is not lower semimodular.

Theorem 7.2.7 *In a finite graded lattice, the function $v(x)$ and its dual $v^*(x)$ are supermodular.*

Proof. We proceed by induction over the M-closed subsets of a finite graded lattice L and first note that trivially $v[a, b] = v(a \vee b) + v(a \wedge b) - v(a) - v(b) = 0$ for $a = b = 0$. We assume that $v[a, b] = 0$ whenever $a, b, a \vee b$, and $a \wedge b$ are all in the M-closed subset K. Consider the M-closed subset $K \cup \{c\}$ where c is minimal in $L - K$. We need only show that $v[a, b] \geq 0$ for those $a, b \in K$ for which $a \vee b = c$. We distinguish two cases.

Case 1. If $a, b \prec c$ then $\tau[a, b] = \tau(a \vee b) + \tau(a \wedge b) - \tau(a) - \tau(b) \geq 0$ by Corollary 7.2.4. Because $r(c) - 1 = r(a) = r(b) \geq r(a \wedge b) + 1$, we have $r[a, b] = (a \vee b) + r(a \wedge b) - r(a) - r(b) \leq 0$. Now subtraction yields the desired result.

Case 2. If $a, b < c$, let $a \leq a_1 \prec c$ and $b \leq b_1 \prec c$, and let $d_1 = a_1 \wedge b_1$. Then by Corollary 7.2.3 we have $v(a_1) - v(a \vee d_1) \geq 0$, $v(b_1) - v(b \vee d_1) \geq 0$, and

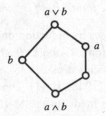

Figure 7.2

$v(d_1) - v[(a \wedge d_1) \vee (b \wedge d_1)] \geq 0$. Also $v[a_1, d_1] \geq 0$ by case 1. Since the elements involved are all in K, the induction hypothesis yields $v[a, d_1] \geq 0$, $v[b, d_1] \geq 0$, and $v[a \wedge d_1, b \wedge d_1] \geq 0$. Summing the seven inequalities, the last four in expanded form, yields the desired results and completes the induction. ∎

The lattice of Figure 7.1 shows that the Jordan–Dedekind chain condition is needed in the hypothesis of the theorem: this lattice does not satisfy (JD), and $v(x)$ is not a supermodular function.

In what follows we give some results concerning increments in $v(x)$. First we shall see that in a graded lattice the failure of lower semimodularity yields that $v(x)$ is increasing. This is an immediate consequence of

Theorem 7.2.8 *If, in a finite graded lattice, $a \vee b \succ a$ and b is not an upper cover of $a \wedge b$, then $v(a \vee b) > v(a)$.*

Proof. The function $\tau(x)$ is supermodular (cf. Corollary 7.2.4). This and the fact that $\tau(x)$ is strictly increasing in chains (see Corollary 7.2.3) yield $\tau(a \vee b) - \tau(a) \geq \tau(b) - \tau(a \wedge b) > 1$. Subtracting $r(a \vee b) - r(a) = 1$ yields $(\tau(a \vee b) - r(a \vee b)) - (\tau(a) - r(a)) = v(a \vee b) - v(a) > 0$. ∎

Corollary 7.2.9 *In a finite graded lattice the failure of lower semimodularity, that is, the existence of elements a, b such that $a \vee b$ covers a and b but $a \wedge b$ is not covered by a and b, implies $v(a)$, $v(b) < v(a \vee b)$.*

In what follows we briefly write v for $v(1)$.

Corollary 7.2.10 *If a finite graded lattice is not lower semimodular, then $v > 0$.*

Proof. If a graded lattice is not lower semimodular, there exist elements a and b both covered by $a \vee b$ but with a or b not covering $a \wedge b$. The preceding corollary implies $v(a \vee b) > v(a)$ and thus $v > 0$ [since $v(x)$ is not decreasing in a connected ascending chain – see Corollary 7.2.3]. ∎

Corollary 7.2.11 *If a finite upper semimodular lattice is not modular, then $v > 0$.*

The proof is immediate by the preceding corollary.

Besides the failure of lower semimodularity, there exists also another possibility for $v(x)$ to increase in a finite graded lattice:

Theorem 7.2.12 *Let $c > a_i > d$ $(i = 1, 2, 3)$ be the elements of a nondistributive modular sublattice of order 5 (i.e. a sublattice of type M_3) in a finite graded lattice. Then $v(c) > v(a_i)$ $(i = 1, 2, 3)$.*

Proof. The sets $J(a_1) - J(d)$, $J(a_2) - J(d)$, $J(a_3) - J(d)$ are nonempty and pairwise disjoint by Lemma 7.2.2(i) and (ii). Let $c \succ e_1 \geq a_1$. There exists an $x_2 \in J(a_2) - J(d)$ such that $x_2 \not\leq e_1$; otherwise we had $J(a_2) \subseteq J(e)_1$, $c = a_1 \vee a_2 \leq e_1$, a contradiction. Similarly there exists $x_3 \in J(a_3) - J(d)$ such that $x_3 \not\leq e_1$

and $x_2 \nleq e_1$ and $x_3 \neq x_2$. Thus $\tau(c) > \tau(e_1) + 1$ and subtracting $r(c) = r(e_1) + 1$ yields $\nu(c) > \nu(e_1) \geq \nu(a_1)$ by Corollary 7.2.3. Similarly $\nu(c) > \nu(a_2), \nu(a_3)$. ∎

Corollary 7.2.13 *If a finite graded lattice contains a nondistributive modular sublattice of order 5 (i.e. a sublattice of type M_3), then $\nu > 1$.*

For $c \succ a$ in a finite graded lattice, we have already shown that each of the conditions (a), (b), (c) of the following theorem separately implies $\nu(c) > \nu(a)$ (by Theorem 7.2.8, Theorem 7.2.7, and Theorem 7.2.12, respectively). The following result is a partial converse.

Theorem 7.2.14 *In a finite graded lattice let $c \succ a$ imply $\nu(c) > \nu(a)$. Then at least one of the following conditions holds:*

(a) *The lattice is not lower semimodular, that is, there exist elements a, b covered by an element c such that $a \wedge b = d$ is not covered by a and b;*
(b) *there exists a proper lower transpose $[d, b]$ of $[a, c]$ such that $\nu(b) > \nu(d)$;*
(c) *there exists a modular nondistributive cover-preserving sublattice of order 5 (i.e. a cover-preserving M_3): $c \succ a$, $b_1, b_2 \succ d$.*

Proof. Suppose first that c is join-irreducible, that is, $c \succ a$ only. For every join-irreducible $x < c$ we have $x \leq a$ so that $J(c) = J(a) \cup \{c\}$ and therefore $\tau(c) = \tau(a) + 1$. Subtraction of $r(c) = r(a) + 1$ leads to a contradiction. Hence there exists $b \prec c$, $b \neq a$. By the Jordan–Dedekind chain condition (JD), b is not an upper cover of $a \wedge b = d$ if and only if a is not an upper cover of d, in which case we obtain condition (a).

Consider now $b \succ d$ and $a \succ d$ with $r(b) = r(d) + 1$. If $\tau(b) > \tau(d) + 1$, then we obtain condition (b).

Finally let $b = b_1$, $d = d_1$, and $\tau(b_1) = \tau(d_1) + 1$ so that $J(b_1) - J(d_1) = x_1$ only. Since $c = a \vee b_1$ we get by Lemma 7.2.2(iii) that $x_1 \in J(c) - J(a)$.

Since $\tau(c) > \tau(a) + 1$, there exists $x_2 \in J(c) - J(a)$, $x_2 \neq x_1$. There exists b_2, necessarily distinct from a and b_1, such that $x_2 \leq b_2 \prec c$. As with $d = d_1$, we can restrict ourselves to $b_2 \succ d_2 = a \wedge b_2$ along with $\tau(b_2) = \tau(d_2) + 1$. Because $x_2 \nleq a$, we have $x_2 \nleq d_2$ and $J(b_2) - J(d_2) = x_2$ only. Suppose $d_1 \neq d_2$ is possible. Then d_1, d_2, and $d_{12} = b_1 \wedge b_2$ are all distinct. Let $d = d_1 \wedge d_2 \wedge d_{12}$. Since $[d_i, b_i]$ is an upper transpose of $[d, d_{12}]$, Corollary 7.2.4 yields $1 - \tau(b_i) - \tau(d_i) \geq \tau(d_{12}) - \tau(d) > 0$ $(i = 1, 2)$. It follows that $J(d_{12}) - J(d)$ is of order 1. But by Lemma 7.2.2(iii) we have $x_i = J(b_i) - J(d_i) \supseteq J(d_{12}) - J(d)$ $(i = 1, 2)$ for a contradiction. Hence $d_1 = d_2$. In this case $d_1 = d_2 = d_{12} = d$ and also $d \prec a$, yielding (c). ∎

The remark preceding Theorem 7.2.14 together with the nondecreasing property of $\nu(x)$ in an ascending chain (cf. Corollary 7.2.3) yields some corollaries, which involve the following conditions in a finite graded lattice L.

(a) $\nu > 0$;
(a*) $\nu = \nu^* > 0$;
(b) L is not lower semimodular;
(c) there exists a cover-preserving sublattice of type M_3;
(c*) there exists a sublattice of type M_3;
(d) L is not distributive.

Corollary 7.2.15 *If $\nu > 0$ in a finite graded lattice, then the lattice is not lower semimodular or there exists a cover-preserving sublattice of type M_3.*

Proof. In a finite graded lattice with $\nu > 0$, let c be a minimal element such that $\nu(c) > 0$. For $a \prec c$ it then follows that $\nu(a) = 0 < \nu(c)$. Thus the assertion follows from Theorem 7.2.14 [condition (b) of this theorem is excluded]. ∎

The proofs of the following three corollaries are left as exercises.

Corollary 7.2.16 *If a finite graded lattice contains no sublattice of type M_3, then $\nu > 0$ holds if and only if the lattice is not lower semimodular.*

Corollary 7.2.17 *In a finite lower semimodular lattice the preceding conditions (a), (c), and (c*) are equivalent and imply (d).*

Corollary 7.2.18 *In a finite modular lattice the following four conditions are equivalent:*

(i) $\nu > 0$;
(ii) *there exists a cover-preserving sublattice of type M_3;*
(iii) *there exists a sublattice of type M_3;*
(iv) *the lattice is not distributive.*

The preceding corollary was first proved by Ward [1939]. The foregoing results yield

Theorem 7.2.19 *In a finite graded lattice L the following four conditions are equivalent:*

(a) $\nu = 0$;
(b) $\nu(x) = 0$ *holds for every $x \in L$, that is, $\nu(x) \equiv 0$;*
(c) L *is lower semimodular with $\nu = 0$;*
(d) L *is lower semimodular and has no sublattice isomorphic to M_3.*

Proof. Corollary 7.2.3 and the contrapositives of Corollaries 7.2.10 and 7.2.11 suffice for the equivalences of (a) and (b), (a) and (c), and (c) and (d), respectively. ∎

The preceding results (in particular Theorem 7.2.19) will now be related to the principal results of Dilworth [1940].

Theorem 7.2.20   *In a lattice L, if $c \succ c_i$ with $\tau(c) - \tau(c_i) = 1$ $(i = 1, \ldots, t)$ and $d = c_1 \wedge \cdots \wedge c_t$, then the c_i are (meet-)independent, that is, they generate under c an interval $[d, c]$ that is a Boolean sublattice B^r. Moreover, all elements of $[d, c]$ have the same join excess as c.*

Proof.   The t sets of join-irreducibles $J(c) - J(c_i)$ each consist of one element x_i $(i = 1, \ldots, t)$. The 2^t sets of join-irreducibles $J(c) - S_j$ where S_j ranges over all the subsets $\{x_1, \ldots, x_t\}$ are ordered as a Boolean sublattice B_1^t. By Lemma 7.2.2(ii) and (i) they form respectively the sets of join-irreducibles contained in 2^t distinct elements of L contained in c, which are also ordered as a Boolean sublattice B_2^t of L. Let $a \succ b$ be an arbitrary covering in the sublattice B_2^t of L. To show that $a \succ b$ in L itself, note that $1 = |J(a) - J(b)| = \tau(a) - \tau(b) = (r(a) - r(b)) + (\nu(a) - \nu(b)) = 1 + 0$ by Corollary 7.2.3. Hence $r(a) - r(b) = 1$, which implies $a \succ b$ in L and also $\nu(a) - \nu(b) = 0$. The latter yields $\nu(a) = \nu(b)$ for all $a, b \in B_2^t$. Finally $c \geq a \geq d$ in L implies $J(c) \supseteq J(a) \supseteq J(d)$ by Lemma 7.2.2(i). Thus $J(a) = J(c) - S \in B_1^t$ for a subset S of $\{x_1, \ldots, x_t\}$ and $a \in B_2^t$. This means that B_2^t is an interval $[d, c]$ of L.   ∎

In Theorem 7.2.20 and in the following corollary we do not require that the c_i be all the elements covered by c.

Corollary 7.2.21   *If in a graded lattice $c \succ c_i$ $(i = 1, \ldots, t)$ and the c_i have the same join excess as c, then the c_i are (meet-)independent and satisfy the statement of Theorem 7.2.20.*

Proof.   For each c_i we have $\tau(c) - \tau(c_i) - 1 = (r(c) - r(c_i)) + (\nu(c) - \nu(c_i)) - 1 = 1 + 0 - 1 = 0$, which is the hypothesis of Theorem 7.2.20.   ∎

The pentagon N_5 shows that the property of being graded cannot be omitted in Corollary 7.2.21.

Theorem 7.2.22   *In a lattice L let $c_1, \ldots, c_t$ be all the elements covered by c, and let $d = c_1 \wedge \cdots \wedge c_t$. Then the following conditions (a), (b), (c) are equivalent and imply (d). If L is a graded lattice, all conditions are equivalent:*

(a)   *c has a unique irredundant join-representation.*
(b)   *$\tau(c) - \tau(c_i) = 1$ for each c_i.*
(c)   *$r(c) - r(c_i) = 1$ and $\nu(c) - \tau(c_i) = 0$ for each c_i.*
(d)   *The c_i are (meet-)independent under c, that is, the c_i generate under c an interval $[d, c]$ of L that is a Boolean lattice B^t. Moreover, all the elements of $[d, c]$ have the same join excess as c.*

Proof. Suppose (a) holds, and assume that (b) is false, that is, for some $c_i = a$ we have $\tau(c) - \tau(a) > 1$. Let $a = x_1 \vee \cdots \vee x_s$ be an irredundant representation of a as the join of join-irreducible elements. Then c contains at least two join-irreducibles x_{11} and x_{12} not contained in a and thus distinct from $x_1, \ldots, x_s$. Two irredundant representations of c, obtained from the representations $c = x_1 \vee \cdots \vee x_s \vee x_{11}$ and $c = x_1 \vee \cdots \vee x_s \vee x_{12}$ by omitting superfluous join-irreducibles, must still retain x_{11} and x_{12} respectively. They are therefore distinct, which contradicts (a).

Let now (b) hold. By hypothesis and by Corollary 7.2.3, we have $1 = \tau(c) - \tau(c_i) \geq r(c) - r(c_i) \geq 1$. Thus for each i equality subsists, and by subtraction we obtain $\nu(c) - \tau(c_i) = 0$. Hence (b) implies (c).

Condition (c) implies (b) by addition.

Assuming now (b), we shall prove (d) and (a). Theorem 7.2.20 yields (d).

Let $x_i = J(c) - J(c_i)$ $(i = 1, \ldots, t)$, and let $x_{t+1}, \ldots, x_s$ be the remaining join-irreducibles contained in c. We have $J(c) - \{x_i, \ldots, x_t\} = J(d) = \{x_{t+1}, \ldots, x_s\}$ so that $d = x_{t+1} \vee \cdots \vee x_s$. For $i = 1, \ldots, t$ we have that $x_1 \vee \cdots \vee x_{i-1} \vee x_{i+1} \cdots \vee x_s = c_i \prec c$. It follows that every redundant representation of c as the join of join-irreducibles must contain all of $x_1, \ldots, x_t$. But $c \geq x_1 \vee \cdots \vee x_t \not\leq c_i$ $(i = 1, \ldots, t)$ requires that $c = x_1 \vee \cdots \vee x_t$ be a join-irreducible representation of c. Hence this representation is irredundant and unique, which proves (a). Thus we have shown that (a), (b), and (c) are equivalent and imply (d). If L is a graded lattice, then (d) implies both parts of (c), making all four conditions equivalent. ∎

We leave it as an exercise to give an example showing that condition (d) in the preceding theorem does not imply (c) if the lattice is not graded. The preceding theorem should also be compared with Theorem 1.7.4.

Theorem 7.2.23 *Let L be a finite nonmodular upper semimodular lattice, and c be a minimal element covering two distinct elements a and b such that $a \wedge b$ is not covered by a and b (note that such elements exist because the lattice is not lower semimodular). Let $a \wedge b = a_0 \prec a_1 \prec \cdots \prec a_t = a$ and $a \wedge b = b_0 \prec b_1 \prec \cdots \prec b_t = b$ be two maximal chains (observe that $t \geq 2$). Then these two chains generate a cover-preserving sublattice L_0 of the type shown in Figure 7.3.*

Proof. Since the lattice L is modular below c (due to the minimality property), the proof is a straightforward modification of the corresponding proof when the whole lattice L is modular (see Birkhoff [1948], p.71, or Grätzer [1978], Theorem 14, p.169). ∎

This yields again the result that a finite semimodular lattice is nonmodular if and only if it contains no cover-preserving S_7 (cf. Theorem 3.1.10).

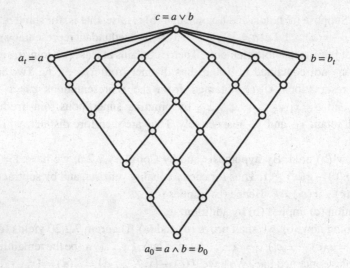

Figure 7.3

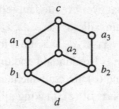

Figure 7.4

Corollary 7.2.24 *Let $v > 0$ in an upper semimodular lattice, and let c be minimal for which $v(c) > 0$. Then c is the greatest element of a cover-preserving M_3 or of a cover-preserving S_7.*

Proof.   In Theorem 7.2.14 condition (b) is excluded. ∎

Theorem 7.2.25 *Let L be a graded lattice having a sublattice isomorphic to S_7. Then (with the notation of Figure 7.4) $v(c) > v(a_2)$.*

Proof.   Let $c \succ c_2 \geq a_2$ in L. Then there exists a join-irreducible $x_1 \in J(a_1) - J(d)$ such that $x_2 \not\leq c_2$ (otherwise $a_1 \leq c_2$ and $c = a_1 \vee a_2 \leq c_2$, a contradiction). Similarly, there exists a join-irreducible $x_3 \in J(a_3) - J(d)$ such that $x_3 \not\leq c_2$.

Now $J(a_1) \cap J(a_3) = J(d)$ implies $x_1 \neq x_2$ with $x_1, x_2 \in J(c) - J(c_2)$. Subtracting $r(c) - r(c_2) = 1$ from $\tau(c) - \tau(c_2) = 1$, we thus get $v(c) > v(c_2) \geq v(a_2)$. ∎

Corollary 7.2.26 *If a lower semimodular lattice has a sublattice isomorphic to S_7, then $v > 0$.*

By means of Theorems 7.2.22, 7.2.23, and 7.2.25 we can now extend Theorem 7.2.19 to obtain Avann's latticial characterization of lower locally distributive (meet-distributive) lattices.

Theorem 7.2.27 *In a finite lattice L the following conditions are equivalent:*

(a) *L is graded (by its rank function) with $\nu = 0$ $[= \nu(1) = \tau(1) - r(1)]$, that is, $r(1) = |J(L)|$;*

(b) *L is graded with $\nu(a) = 0$ for all $a \in L$;*

(c) *L is a lower semimodular lattice with $\nu = 0$;*

(d) *L is a lower semimodular lattice having no cover-preserving sublattice isomorphic to M_3;*

(e) *every $a \in L$ has a unique irredundant join decomposition;*

(f) *L is a lower locally distributive (meet-distributive) lattice;*

(g) *L is lower semimodular, and every upper semimodular sublattice of L is distributive;*

(h) *L is lower semimodular, and every modular sublattice of L is distributive.*

Proof. The equivalence of (a), (b), (c), (d) is a reformulation of Theorem 7.2.19. These imply condition (c) of Theorem 7.2.22. Hence conditions (a) and (d) of Theorem 7.2.22 yield (e) and (f). Next assume (f), which implies lower semimodularity. If (g) is false, then there exists (by Exercise 2) a nondistributive upper semimodular sublattice L_0 for which $\nu_0 > 0$. Then (by Corollary 7.2.24), L_0 and hence L each contain a nondistributive modular sublattice of order 5 (i.e. an M_3-sublattice) or a nonmodular upper semimodular sublattice of order 7 (i.e. an S_7-sublattice). It follows (by Corollary 7.2.13 or Corollary 7.2.26) that $\nu > 0$ in L. Then Corollary 7.2.17 leads to a contradiction of (f). Thus (f) implies (g). Now (g) implies (h), since any modular lattice is lower semimodular. Finally (h) implies (d) trivially. Hence all the conditions are equivalent. ■

Every cover-preserving upper semimodular sublattice of a finite lower semimodular lattice is modular. This should be compared with the stronger condition (g) in Theorem 7.2.27. This and other distributive-like properties of Theorem 7.2.27 motivated Avann to use the name "lower semidistributive lattice" for a finite lower semimodular lattice in which all modular sublattices are distributive.

As immediate consequences we note two corollaries. For the first one see also Monjardet [1990] [dual of condition (O)].

Corollary 7.2.28 *In a finite lattice L, the following condition is equivalent to any of the conditions of Theorem 7.2.27:*

(k) *for each $x, y \in L$ with $x \prec y$ there exists a unique join-irreducible j such that $j \leq y$ and $j \not\leq x$.*

Proof.   Follows from the obvious implications (e) $\Rightarrow$ (k) and (k) $\Rightarrow$ (b).   ■

For the next corollary see also Greene & Markowsky [1974], Markowsky [1980], and Monjardet [1990] [dual of condition (M)].

Corollary 7.2.29   *In a finite lattice L, the following condition is equivalent to any of the conditions of Theorem 7.2.27:*

(m)   *L is graded, and there exists a meet-preserving and rank-preserving embedding of L into a distributive lattice.*

The equivalence of condition (m) with anyone of the properties (a)–(h) of Theorem 7.2.27 appears (in dual form) explicitly first in Greene & Markowsky [1974], Theorem 2, but it can also be deduced from some of the equivalent conditions of Avann's characterization. We leave this deduction as an exercise. For the sake of convenience and for later use we also give the dual of Theorem 7.2.27:

Theorem 7.2.30   *In a finite lattice L the following conditions are equivalent:*

(a′)   *L is graded and $r(1) = |M(L)|$;*
(b′)   *L is graded, and for each $x \in L$, $r(x) = \{m \in M(L) : m \not\geq x\}$;*
(c′)   *L is upper semimodular and $r(1) = |M(L)|$;*
(d′)   *L is upper semimodular and has no covering-preserving sublattice isomorphic to M_3;*
(e′)   *every element of L has a unique irredundant meet decomposition;*
(f′)   *L is an upper locally distributive (join-distributive) lattice;*
(g′)   *L is upper semimodular and every lower semimodular sublattice of L is distributive;*
(h′)   *L is upper semimodular and every modular sublattice of L is distributive.*

The main result of Dilworth [1940], Theorem 1.1, is the proof of the equivalence of (e′) and (h′) and that these conditions imply the statements of (b′) and (f′). Dilworth showed this by a somewhat different approach. More generally, Dilworth & Crawley [1960] proved the equivalence of (e′), (f′), and (h′) as well as of a further condition for strongly atomic algebraic lattices – see Theorem 7.1.2 [(e′) is condition (ii) of Theorem 7.1.2, and (f′) is condition (i) of Theorem 7.1.2; the equivalence of (f′) and (h′) is the content of Corollary 7.1.5]. Further distributive properties of semimodular lattices were studied by Avann [1961b].

Notes

In a finite nonmodular upper semimodular lattice or a finite nonmodular lower semimodular lattice there is no restrictive inequality between ν and ν^*. In a finite modular lattice, however, one has $\tau = \tau^*$ (this follows from Dilworth's covering theorem – cf. Theorem 6.1.9) and therefore (after subtracting $r = r^*$) one gets $v = v^*$.

The length of every finite lattice is bounded above by the minimum of the number of meet-irreducibles (the meet rank) and the number of join-irreducibles (the join rank) that it has. Markowsky [1992] investigated lattices for which length = join rank or length = meet rank and calls these lattices *p-extremal*. If a *p*-extremal lattice is graded, then it is either lower locally distributive (join rank = length; cf. Theorem 7.2.27) or upper locally distributive (meet rank = length; cf. Theorem 7.2.30) or distributive (join rank = meet rank = length; cf. Exercise 2). Markowsky showed that even in the absence of the Jordan–Dedekind chain condition, *p*-extremal lattices still enjoy many interesting properties.

A finite upper semimodular lattice that is cover-preserving embeddable in a Boolean lattice cannot have a covering sublattice M_3 and is therefore already upper locally distributive (cf. Theorem 7.2.30). This has been generalized from upper semimodular lattices to upper balanced lattices by Wild [1992], Theorem 3.

Exercises

1. Prove that, in a finite modular lattice, $J(c) \supset J(a) \cup J(b)$ for $c = a \vee b$ implies $v(a) < v(c)$ and $v(b) < v(c)$, but not conversely (Avann [1961a], Corollary 4.4.5).

2. In a finite lattice L the following conditions are equivalent:
   (a) L is distributive;
   (b) L is a graded lattice and $v = 0 = v^*$;
   (c) L is a graded lattice and $v(a) = 0 = v^*(a)$ for all $a \in L$;
   (d) L is an upper semimodular lattice and $v = 0$;
   (e) L is modular and $v = 0$;
   (f) L is modular, and there exists no sublattice of type M_3;
   (g) $\tau(a)$ is a modular function (i.e. both submodular and supermodular).
   (For a proof see Avann [1961a], Theorem 4.6; some of the above equivalences were previously known.)

3. Show that a finite semimodular lattice is nondistributive if and only if $v > 0$ (Avann [1961a], Corollary 4.6). (Hint: Use the preceding exercise.)

4. Prove that a lower semimodular sublattice of a lower locally distributive lattice is itself lower locally distributive (Avann [1961a], Theorem 5.6).

5. Show a finite lower locally distributive lattice is balanced if and only if it is distributive (Stern [1992b]).

References

Avann, S. P. [1961a] Application of the join-irreducible excess function to semimodular lattices, Math. Annalen 142, 345–354.

Avann, S. P. [1961b] Distributive properties in semimodular lattices, Math. Z. 76, 283–287.

Avann, S. P. [1964b] Increases of the join-excess function in a lattice, Math. Annalen 54, 420–426.

Birkhoff, G. [1948] *Lattice Theory* (2nd edition), Amer. Math. Soc. Colloquium Publications, Vol. 25, New York.

Bogart, K. P., R. Freese, and J. P. S. Kung (eds.) [1990] *The Dilworth Theorems. Selected Papers of Robert P. Dilworth*, Birkhäuser, Boston.

Dilworth, R. P. [1940] Lattices with unique irreducible decompositions, Ann. of Math. 41, 771–777.

Dilworth, R. P. and P. Crawley [1960] Decomposition theory for lattices without chain conditions, Trans. Amer. Math. Soc. 96, 1–22.

Grätzer, G. [1978] *General Lattice Theory*, Birkhäuser, Basel.

Greene, C. and G. Markowsky [1974] A combinatorial test for local distributivity, Research Report RC4129, IBM T.J. Watson Research Center, Yorktown Heights, N.Y.

Markowsky, G. [1980] The representation of posets and lattices by sets, Algebra Universalis 11, 173–192.

Markowsky, G. [1992] Primes, irreducibles and extremal lattices, Order 9, 265–290.

Monjardet, B. [1990] The consequences of Dilworth's work for lattices with unique irreducible decompositions, in: Bogart et al. [1990], pp. 192–200.

Stern, M. [1992b] On meet-distributive lattices, Studia Sci. Math. Hungar. 27, 279–286.

Ward, M. [1939] A characterization of Dedekind structures, Bull. Amer. Math. Soc. 45, 448–451.

Wild, M. [1992] Cover-preserving order embeddings into Boolean lattices, Order 9, 209–232.

7.3 Meet-Distributive Lattices and Convexity

Summary. In Section 7.2 we discussed Avann's latticial characterizations of lower locally distributive lattices (meet-distributive lattices). There are also more combinatorial approaches to lower locally distributive lattices. One of these approaches is via abstract convexity and was developed jointly by Edelman and Jamison after they had separately developed many of the key ideas. Here we draw on Edelman & Jamison [1985] and Edelman [1986].

Let E be a finite subset of $\mathbb{R}^n$, and for subsets $A \subseteq E$ let $\mathrm{cl}(A) = E \cap \mathrm{cvx}(A)$, where $\mathrm{cvx}(A)$ denotes the *convex hull* of A in the sense of Euclidean geometry. The *convex-hull operator* is a closure operator that has the so-called antiexchange property:

(AEP) If $x \neq y$ and $y \in \mathrm{cl}(A \cup x)$ but $x, y \notin \mathrm{cl}(A)$, then $x \notin \mathrm{cl}(A \cup y)$.

The antiexchange property in its lattice-theoretic form was given in the Notes to Section 2.5. A closure operator having the antiexchange property is also called an *antiexchange closure*.

A *convex geometry* is a pair (E, cl) where E is a finite set and cl is a closure operator on E having the antiexchange property. Thus convex geometries combinatorially abstract the notion of the convex hull of a set of points in Euclidean space in the same way as matroids abstract the idea of linear independence (linear hull). For the characterization of convex geometries to be given here we need some more concepts.

Let E be a finite set equipped with a closure operator cl, and let A be a subset of E. A point $p \in A$ is called an *extreme point* of A if $p \notin \mathrm{cl}(A - p)$. By $\mathrm{ex}(A)$ we denote the set of extreme points of A. Extreme points may or may not exist. Note that $\mathrm{ex}(A)$ is contained in every basis of A. Let $p \in E$. A *copoint* C attached at p is a maximal convex set in $E - p$. There may be more than one copoint attached at a point. For the following characterization of convex geometries and its proof we refer to Edelman & Jamison [1985], Theorem 2.1.

Theorem 7.3.1 *Let E be a finite set equipped with a closure operator cl. Then the following conditions are equivalent:*

(a) *The closure operator cl has the antiexchange property (AEP), that is, (E, cl) is a convex geometry.*
(b) *For every convex set K, there exists a point $p \in E$ such that $K \cup p$ is convex.*
(c) *For every point p and C a copoint attached at p, $C \cup p$ is convex.*
(d) *Every subset $A \subseteq E$ has a unique basis.*
(e) *For every convex set K, $K = \mathrm{cl}(\mathrm{ex}(K))$.*
(f) *For every convex set K and $p \notin K$, $p \in \mathrm{ex}(\mathrm{cl}(K \cup p))$.*

A consequence of Avann's characterization (cf. Theorem 7.2.27) is that we can identify a lower locally distributive (meet-distributive) lattice L with a particular collection of subsets of $J(L)$, which forms a convex geometry. In fact, we have the following representation theorem due to Edelman [1980], which shows that convex geometries are essentially lower locally distributive lattices.

Theorem 7.3.2 *A finite lattice is lower locally distributive (meet-distributive) if and only if it is isomorphic to the lattice of closed sets of a convex geometry.*

Sketch of proof. Let (E, cl) be a convex geometry. In the lattice L of its closed subsets (flats) the set of join-irreducibles is $J(L) = \{\mathrm{cl}(p) : p \in E\}$. Applying Theorem 7.3.1(d) and Avann's latticial characterization (Theorem 7.2.27) shows that L is lower locally distributive. Suppose now L is lower locally distributive. By Corollary 7.2.29, L can be embedded in a Boolean lattice of the same rank so that meets are preserved. This ensures that for $x \in L$, $J(x) = \{y \in J(L) : y \leq x\}$ defines a closure operator cl. It follows from Theorem 7.2.27 and from Theorem 7.3.1(d) that $(J(L), \mathrm{cl})$ is a convex geometry. ∎

There is an abundance of examples for convex geometries, some of which may be found in Edelman [1986] and Edelman & Jamison [1985]. In what follows we present some of these.

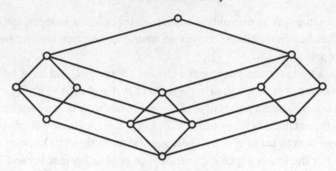

Figure 7.5

Let P be a finite poset. If A is a subset of P then $\mathrm{cl}(A) = \{p : p \le a$ for some $a \in A\}$ defines a closure operator. Under this closure the closed sets are the order ideals of A, and hence we shall call this the *order-ideal closure*. The order-ideal closure of a poset has the antiexchange property. It follows that Theorem 7.3.2 can be viewed as a generalization of the fundamental theorem for finite distributive lattices (Corollary 1.3.2).

Let P be again a finite poset, and A a subset of P. Define $\mathrm{cl}\,(A) = \{p : a_1 \le p \le a_2$ for some $a_1, a_2 \in A\}$. This is the *interval closure*, which is a closure operator having the antiexchange property. The closed subsets with respect to this closure operator are the order-convex subsets. (A subset A of a poset P is said to be *order-convex* if $a, b \in A$ and $a \le b$ imply $[a, b] \subseteq A$.)

By $\mathrm{Co}(P)$ we denote the (lower locally distributive) lattice of convex subsets of the poset P. Edelman [1980] assumes P to be a finite poset, while Birkhoff & Bennett [1985] admit arbitrary posets P. The lattice $\mathrm{Co}(P)$ is atomistic (every convex subset of P is the join – in fact, the union – of its convex one-element subsets) and algebraic. It follows that $\mathrm{Co}(P)$ is a strongly atomic algebraic lattice. If P is an antichain, then $\mathrm{Co}(P) \cong 2^{|P|}$, the complete Boolean lattice of all subsets of P. At the opposite extreme, if $P \cong n$ is a finite chain, then $\mathrm{Co}(P)$ coincides with the planar lattice denoted by $\mathrm{Int}(C_n)$. It has already been indicated (cf. Section 1.3) that $\mathrm{Co}(3) = \mathrm{Int}(3)$ is isomorphic with S_7^*; for $\mathrm{Co}(4) = \mathrm{Int}(4)$ see Figure 1.27 (Section 1.7). A further specific instance is $\mathrm{Co}(2^2)$, which is shown in Figure 7.5.

Notes

It has been observed that there is a one-to-one correspondence between convex geometries and antimatroids. The two concepts are equivalent through a certain duality (see Björner & Ziegler [1992], Proposition 8.7.3). This duality between convex geometries and antimatroids has the effect of switching from lower locally distributive (meet-distributive) lattices to their lattice-theoretic duals, that is, to upper locally distributive lattices (join-distributive lattices, locally free lattices). Sometimes the term "antimatroid" is also used synonymously with "convex geometry."

In a series of papers M. K. Bennett and G. Birkhoff investigated lattices of convex sets and their generalizations. The main concept appearing in Bennett & Birkhoff [1983], [1985] is that of a *convexity lattice*. Convexity lattices generalize the properties of lattices of convex sets. Properties of convexity lattices are also studied by Libkin [1992], [1995]. In particular, Libkin [1995] investigates the dual of the concept of n-distributivity (cf. Huhn [1972]) in certain convexity lattices.

The *Carathéodory number* of a convex geometry (or of the associated antimatroid) is defined as the least integer k with the following property: whenever an element p is contained in the convex hull of a set G, then it is also contained in the convex hull of some subset $G' \subset G$ with $|G'| \leq k$. Korte & Lovász [1984] proved that if the Carathéodory number of an antimatroid is 1, then its lattice of flats is distributive and hence modular. On the other hand, it is easy to see that for all other kinds of antimatroids the lattice of flats is nonmodular. Björner & Lovász [1987] showed that if an antimatroid has Carathéodory number at most 2, then the lattice of flats is pseudomodular (see Section 6.6 for pseudomodular lattices). In particular, the dual of $Co(P)$ is a pseudomodular lattice (see Björner & Lovász [1987], Example 6a).

Birkhoff & Bennett [1985] had already shown that the lattice $Co(P)$ of all convex subsets of a poset P has "Carathéodory rank" (as they call it) 2, which implies that the lattice is biatomic (see also Bennett [1987]). Birkhoff & Bennett [1985] also proved that $Co(P)$ satisfies what they call the "Altwegg condition" (a condition suggested by axiom Z_6 of Altwegg [1950], p. 150). Conversely they show that a complete atomistic directly indecomposable lattice L is isomorphic to $Co(P)$ for some poset P if L is join-semidistributive, has Carathéodory rank 2, and satisfies Altwegg's condition.

Although the order convexity mentioned above is not to be confused with the *affine convexity* discussed in Bennett [1974], [1977] and Bennett & Birkhoff [1983], [1985], the two notions of convexity share several properties (e.g., the associated lattices are algebraic and biatomic). Moreover, for any ordered division ring D, $Co(D)$ is the same in both interpretations. However, for $n > 1$, the lattices $Co(D^n)$ defined by *order* betweenness are very different from those defined by *affine* betweenness.

Lovász & Saks [1993] present an approach to communication complexity based on convex geometries and the lattices associated with them. For a representation of locally distributive lattices cf. also Behrendt [1991].

References

Altwegg, M. [1950] Zur Axiomatik der teilweise geordneten Mengen, Comment. Math. Helv. 24, 149–155.

Behrendt, G. [1991] Representation of locally distributive lattices, Portugal. Math. 48, 351–355.

Bennett, M. K. [1974] On generating affine geometries, Algebra Universalis 4, 207–219.

Bennett, M. K. [1977] Lattices of convex sets, Trans. Amer. Math. Soc. 234, 279–288.

Bennett, M. K. [1987] Biatomic lattices, Algebra Universalis 24, 60–73.

Bennett, M. K. and G. Birkhoff [1983] A Peano axiom for convexity lattices, in: *Calcutta Math. Soc. Diamond Jubilee Commemorative Volume*, pp. 33–43.

Bennett, M. K. and G. Birkhoff [1985] Convexity lattices, Algebra Universalis 20, 1–26.

Birkhoff, G. and M. K. Bennett [1985] The convexity lattice of a poset, Order 2, 223–242.

Björner, A. and L. Lovász [1987] Pseudomodular lattices and continuous matroids, Acta Sci. Math. (Szeged) 51, 295–308.

Björner, A. and G. M. Ziegler [1992] Introduction to greedoids, in: White [1992], pp. 284–357.

Edelman, P. H. [1980] Meet-distributive lattices and the anti-exchange closure, Algebra Universalis 10, 290–299.

Edelman, P. H. [1986] Abstract convexity and meet-distributive lattices, in: *Contemporary Mathematics*, Vol. 57 (ed. I. Rival), pp. 127–150.

Edelman, P. H. and R. E. Jamison [1985] The theory of convex geometries, Geom. Dedicata 19, 247–270.

Edelman, P. H. and M. E. Saks [1988] Combinatorial representation and convex dimension of convex geometries, Order 5, 23–32.

Huhn, A. [1972] Schwach distributive Verbände, Acta Sci. Math. (Szeged) 33, 297–305.

Korte, B. and L. Lovász [1984] Shelling structures, convexity and a happy end. In: *Graph Theory and Combinatorics, Proc. Cambridge Combin. Conf. in Honor of Paul Erdős*, (ed. B. Bollobás), Academic Press, London, pp. 219–232.

Korte, B., L. Lovász, and R. Schrader [1991] *Greedoids*, Springer-Verlag, Berlin.

Libkin, L. O. [1992] Parallel axiom in convexity lattices, Periodica Math. Hungar. 24, 1–12.

Libkin, L. O. [1995] n-distributivity, dimension and Carathéodory's theorem, Algebra Universalis 34, 72–95.

Lovász, L. and M. Saks [1993] Communication complexity and combinatorial lattice theory, J. Comput. System Sci. 47, 322–349.

White, N. L. (ed.) [1992] *Matroid Applications*, Encyclopedia of Mathematics and Its Applications, Vol. 40, Cambridge Univ. Press, Cambridge.

7.4 Other Characterizations

Summary. Monjardet [1990] gives a thorough survey of papers dealing with upper locally distributive (join-distributive) lattices and lower locally distributive (meet-distributive) lattices, mostly in the finite case. He gives the key ideas of proofs of many characterizations of this class of lattices and lists altogether almost 50 references. Since Dilworth [1940], upper and lower locally distributive lattices have been rediscovered many times in different guises. In this section we mention Avann's "upper splitting property" and Crapo's combinatorial approach to locally free selectors, which is equivalent to a part of the theory of greedoids. Here we draw on Monjardet [1990] and Crapo [1984].

First we consider the upper splitting property. We start with some notation. For simplification we denote by a/b an ordered pair (a, b) of elements of a lattice L satisfying $b \le a$. We call a/b a *quotient* of L; c/d is called a *subquotient* of a/b if $b \le d \le c \le a$. We call a/b a *proper quotient* if $b < a$. If $b \prec a$, then a/b is called a *prime quotient*. We write $a/b = c/d$ if $a = c$ and $b = d$. If $c = a \vee d$ and $b = a \wedge d$, then we write $a/b \nearrow c/d$ and say that c/d is *perspective upward* to

a/b. If $a = b \vee c$ and $d = b \wedge c$, then we write $a/b \searrow c/d$ and say that c/d is *perspective downward* to a/b.

Let us now restrict upward perspectivity to the set of prime quotients $x \prec y$ of L. This set is partially ordered by the upward perspectivity relation $\nearrow$. We say that a set C of prime quotients of L is a *projective covering class* if $(C, \nearrow)$ is a connected component of this poset. By $\overline{c/d}$ we denote the projective covering class of the prime quotient c/d.

It follows from Crapo [1984], Theorem 31 and Theorem 32, that an upper locally distributive lattice L has the property: L is upper semimodular, and every projective covering class has a greatest element m^*/m with $m \in M(L)$ and m^* denoting the unique upper cover of m. Conversely, this condition implies upper locally distributivity [(ULD) for short] (see e.g. Duquenne [1991]). We leave both results as exercises. Avann [1968] had already shown a more comprehensive result, which we are going now to describe. Following Avann, we introduce the upper splitting property (USP):

(USP) A finite lattice L is said to be *upper splitting* if every projective covering class of L satisfies the following three conditions:
   (a) it has a greatest element m^*/m [$m \in M(L)$];
   (b) its minimal elements are the quotients j_α/j'_α with $j_\alpha \in J(L)$;
   (c) L is the disjoint union of $\{x \in L : x \le m\}$ and $\{y \in L : y \ge j_\alpha\}$ for at least one j_α.

The fact that m is indeed meet-irreducible and that the j_α are join-irreducible follows easily and is left as an exercise (see also Avann [1968], Lemma 2.3).

Avann [1968], Theorem 6.1, proved that (USP) is equivalent to any of the conditions of Theorem 7.2.30 and hence (USP) characterizes (ULD):

Theorem 7.4.1 *In a finite lattice* (USP) *and* (ULD) *are equivalent and imply upper semimodularity [condition* (Sm)*].*

This result implies a generalization of the natural bijection between the sets $M(L)$ and $J(L)$ in a finite distributive lattice L, which was mentioned in Section 1.3.

In Figure 7.6 we illustrate property (c) of (USP) by means of a representation of S_7 as a disjoint union of the corresponding subsets: $S_7 = (m] + [j_1) \cup [j_2)$ (with $+$ denoting here the disjoint union and $\cup$ the set theoretic union).

A covering pair $x \prec y$ (i.e. a prime quotient y/x) of a finite lattice is said to be meet-irreducible if and only if x is a meet-irreducible element and thus y is the unique element covering x. Let $s \prec t$ be a covering pair in an upper locally distributive lattice L. The unique meet-irreducible covering pair of L up to which $s \prec t$ is perspective is also called the *generator* of $s \prec t$, briefly denoted by $g(s \prec t)$. The following results and examples are due to Crapo [1984].

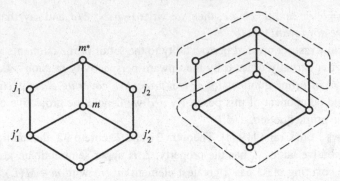

Figure 7.6

Theorem 7.4.2   *Let C be a maximal chain $0 = z_0 \prec z_1 \prec \cdots \prec z_n = 1$ in an upper locally distributive lattice L, and for $i = 1, \ldots, n$ let $x_1 = g(z_{i-1} \prec z_i)$ be the sequence of generators of covering pairs in L. Then the string $\alpha = x_1 \ldots x_n$ is a linear order on $M(L)$.*

Proof.   Let C be such a maximal chain in L. For each meet-irreducible element $x \in L$, let z_{i-1} be the last element of the chain C such that $z_{i-1} \leq x$. Then z_i covers z_{i-1} and $z_i \not\leq x$, and thus $z_i \vee x$ covers x. Since x is meet-irreducible, $x \prec z_i \vee x$ is a meet-irreducible pair, and since z_i/z_{i-1} is perspective upward to $(z_i \vee x)/x$, we have $x = g(z_{i-1} \prec z_i)$. Hence $x = x_i$, and every meet-irreducible element appears somewhere in the string α. If two distinct links $x_{i-1} \prec x_i$ and $x_{j-1} \prec x_j$ in the chain C would have the same generator, these two covering pairs would be perspective upward to the same meet-irreducible covering pair $s \prec t$. However, this leads to the contradiction (for $i < j$) that $s \vee x_i = t$ while $x_i \leq x_{j-1} \leq s$. It follows that the map from covering pairs in C into the set $M(L)$ of meet-irreducible elements of L is both one-to-one and onto.   ∎

Selectors were introduced in Section 6.4. A selector Λ on a finite set E is called *locally free* (see Crapo [1984]) if the following two axioms are satisfied:

 (i) For any word $\alpha \in \Lambda$ and distinct letters $x, y \in E$, if αx and αy are words, so is $\alpha x y$.
(ii) If $\alpha, \alpha' \in \Lambda$ are words using the same letters, and for some string β, $\alpha \beta \in \Lambda$, then $\alpha' \beta \in \Lambda$.

Locally free selectors have been extensively investigated and – similarly to upper locally distributive lattices – there exist many different names for them, for example, antimatroids, alternative precedence structures, upper interval greedoids, and antiexchange greedoids. The last two names indicate that locally free selectors (antimatroids) form – like matroids – a subclass of greedoids. For more details on the theory of greedoids we refer to the monograph Korte et al. [1991] and to the comprehensive introduction by Björner & Ziegler [1992].

We now give two examples of locally free selectors.

Consider first a finite poset P, and remove its elements one by one, in such a way that at each stage the element to be removed is minimal among the elements that remain. Such a linear order $x_1, \ldots, x_n$ of the elements of P will be referred to as a *shelling order* for the poset P. Equivalently, the string $\alpha = x_1 \ldots x_n$ has the property that every initial segment of α is an order ideal of P. Also equivalently, α is a linear extension on P (i.e., a linear order which dominates the partial order on $P : x \leq y$ in P implies x comes before y in the linear order α). It is also clear that α is the unique generating sequence of join-irreducibles for a maximal chain in the distributive lattice of order ideals of P.

Theorem 7.4.3   *For any finite poset P, the set Λ of initial segments of its shelling orders form a locally free selector.*

This amounts to a restatement of the fundamental theorem for finite distributive lattices (cf. Corollary 1.3.2). We also note that Theorem 7.4.3 is a special case of Theorem 6.4.6 concerning Faigle geometries. Let us moreover remark that the above notion of shelling order has nothing to do with the concepts of shelling and shellability as used in Section 4.4.

As a specific example of the preceding theorem consider the poset given in Figure 7.7. The locally free selector Λ arising from the initial segments of the shelling orders of the poset in Figure 7.7 is given by the following table:

λ	Λ
0	$\emptyset$
1	a, d
2	ab, ac, ad, da
3	$abc, abd, acb, adb, adc, dab, dac$
4	$abcd, abdc, acbd, acdb, adbc, adcb, dabc, dacb$
5	$abcde, abdce, acbde, acdbe, adbce, adcbe, dabce, dacbe$

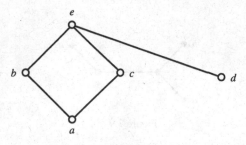

Figure 7.7

The lattice of partial alphabets (all of which are closed) of this selector is the distributive lattice of Figure 7.8.

The second example is provided by the shelling orders for a tree. A *tree* is a connected graph with no cycles. A vertex $b \in X$ in a tree T with vertex set X is *terminal* if and only if it has valency equal to 1 (or 0) in T. (The only case in which a terminal vertex b has valency 0 is when $X = \{b\}$ and the tree has no edges.)

A shelling order for a tree T on an n-element vertex set X is a total order $\alpha = x_1 \ldots x_n$ of X such that for $i = 1, \ldots, n - 1$ the vertices in every final segment $x_i \ldots x_n$, together with the T-edges that join them, form a tree T_i with the vertex set x_i terminal in T_i. Crapo [1984], Theorem 24, proved the following result (for the notion of basis of a selector we refer to Section 6.4):

Theorem 7.4.4   *The shelling orders for a tree on a vertex set X form the bases of a locally free selector on X.*

As a concrete example consider the tree of Figure 7.9. The locally free selector Λ formed by the shelling orders for that tree is given by the following table (all

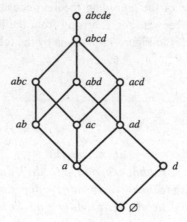

Figure 7.8

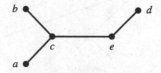

Figure 7.9

partial alphabets are closed):

λ	Λ
0	∅
1	a, b, d
2	ab, ad, ba, bd, da, db, de
3	abc, abd, adb, ade, bac, bad, bda, bde, dae, dba, dbe, dea, deb
4	abce, abcd, abdc, abde, adbc, adbe, adeb, adec, bace, bacd, badc, bade, bdac, bdae, bdea, bdec, dabc, dabe, daeb, daec, dbac, dbae, dbea, dbec, deab, deac, deba, debc
5	abced, abcde, abdce, abdec, adbce, adbec, adebc,adecb, baced, bacde, badce, badec, bdace, bdaec, bdeac, bdeca, dabce, dabec, daebc, daecb, dbace, dbaec, dbeac, dbeca, deabc, deacb, debac, debca

The upper locally distributive (join-distributive, locally free) lattice of partial alphabets of this selector is shown in Figure 7.10.

Crapo [1984], Theorem 34, proved the following representation theorem:

Theorem 7.4.5   *For any finite upper locally distributive lattice L, the sets of strings of generators for maximal chains in L form the bases of a locally free selector Λ on the set $M(L)$ of meet-irreducible elements of L. The lattice of closed partial alphabets (flats) of Λ is isomorphic to L.*

Sketch of proof.   Let z be any element of L, and consider a maximal chain $C : 0 = z_0 < z_1 < \cdots < z_n = 1$ in L containing z, for example, $z = z_i$. Let $\alpha = x_1 \ldots x_n$

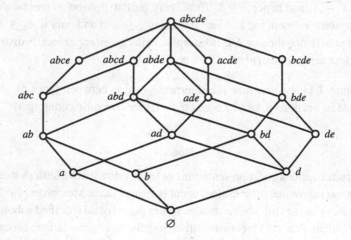

Figure 7.10

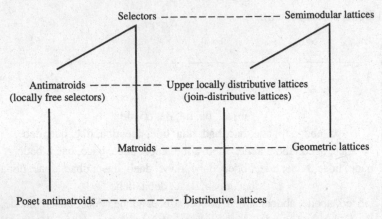

Figure 7.11

be the corresponding string of generators for covering pairs in C. For any covering pair $s \prec t$ and any meet-irreducible $x \in M(L)$, the quotient t/s is perspective upward to $(x \vee t)/x$ if and only if $s \leq x, t \nleq x$. It follows that the initial segment of $x_1 \ldots x_i$ of α is a linear order of the set $A_z = \{x \in M(L) : s \nleq x\}$ of meet-irreducibles not greater than or equal to z. Hence any such A_z, for $z \in L$, is a partial alphabet. If $A \in X$ is a partial alphabet, let $z = \bigwedge(X - A)$ denote the meet of all meet-irreducibles not in A. We show that $A = A_z$. For any element x in the partial alphabet A we can find a maximal chain $0 = z_0 < z_1 < \cdots < z_k = w$ such that $g(z_{i-1} \prec z_i) \in A$ for $i = 1, \ldots, k$ and $g(z_{k-1} \prec z_k) = x$. Hence $A_w \subseteq A$. Since $w = \bigwedge(X - A_w)$, we have $w \leq \bigwedge(X - A) = z$. From $w \nleq x$ we get $z \nleq x$ and $x \in A_z$. Conversely, assume $x \in A_z$, so $z \nleq x$. From $z \in \bigwedge(X - A)$ we get $x \notin \bigwedge(X - A)$ and hence $x \in A$. Thus every partial alphabet is represented as a set A_z for some element $z \in L$. Since moreover $z \leq w$ if and only if $A_z \leq A_w$, the lattice of partial alphabets of Λ is isomorphic to the given upper locally distributive (locally free semimodular) lattice L.   ■

In Figure 7.11 we visualize the interrelationships between types of selectors mentioned in Sections 6.4 and 7.5 and their lattice-theoretic counterparts.

Notes

The characterizations and representations of locally distributive lattices mentioned in this chapter show that these lattices occur in many places. Monjardet [1985] gives a survey mentioning still other instances where the set of all specified subobjects of a mathematical object is a (nondistributive) locally distributive lattice, for example, "separating" sets of a graph (see Pym & Perfect [1970], Polat [1976], Sabidussi [1976], and Halin [1993]).

Exercises

1. Show that a finite upper locally distributive lattice has the following property: it is (upper) semimodular, and every projective covering class has a greatest element m^*/m [m being a meet irreducible ($\neq 1$) and m^* denoting the unique upper cover of m] (Crapo [1984]).

2. Show that the property in the preceding exercise implies upper locally distributivity (Duquenne [1991]).

3. Use the fact that the upper splitting property characterizes upper locally distributivity to show that finite upper locally distributive lattices are (meet-)pseudo-complemented. (Hint: the pseudocomplement $g(x)$ of an element x is the unique meet-irreducible element that is $\not\geq x$.) (Chameni-Nembua & Monjardet [1992].)

References

Avann, S. P. [1968] Locally atomic upper locally distributive lattices, Math. Annalen 175, 320–336.

Björner, A. and G. M. Ziegler [1992] Introduction to greedoids, in: White [1992], pp. 284–357.

Bogart, K. P., R. Freese, and J. P. S. Kung (eds.) [1990] *The Dilworth Theorems. Selected Papers of Robert P. Dilworth*, Birkhäuser, Boston.

Chameni-Nembua, C. and B. Monjardet [1992] Les treillis pseudocomplémentés finis, Eur. J. Combin. 13, 89–107.

Crapo, H. H. [1984] Selectors: a theory of formal languages, semimodular lattices, and branching and shelling processes, Adv. in Math. 54, 233–277.

Dilworth, R. P. [1940] Lattices with unique irreducible decompositions, Ann. of Math. 41, 771–777.

Duquenne, V. [1991] The core of finite lattices, Discrete Math. 88, 133–147.

Halin, R. [1993] Lattices related to separation graphs, in: *Finite and Infinite Combinatorics in Sets and Logic* (ed. N. W. Sauer et al.), Kluwer Academic, pp. 153–167.

Korte, B., L. Lovász, and R. Schrader [1991] *Greedoids*, Springer-Verlag, Berlin.

Monjardet, B. [1985] A use for frequently rediscovering a concept, Order 1, 415–417.

Monjardet, B. [1990] The consequences of Dilworth's work for lattices with unique irreducible decompositions, in: Bogart et al. [1990], pp. 192–200.

Polat, N. [1976] Treillis de séparation des graphs, Canad. J. Math. 28, 725–752.

Pym, J. S. and H. Perfect [1970] Submodular functions and independence structures, J. Math. Anal. Appl. 30, 1–31.

Sabidussi, G. [1976] Weak separation lattices of graphs, Canad. J. Math. 28, 691–724.

White, N. L. (ed.) [1992] *Matroid Applications*, Encyclopedia of Mathematics and Its Applications, Vol. 40, Cambridge Univ. Press, Cambridge.

8

Local Modularity

8.1 The Kurosh–Ore Replacement Property

Summary. The Kurosh–Ore theorem for finite decompositions in modular lattices was given in Section 1.6. In particular we have seen that modularity implies the Kurosh–Ore replacement property for meet decompositions ($\wedge$-KORP). On the other hand, the lattice N_5 shows that the $\wedge$-KORP (together with its dual, the $\vee$-KORP) does not even imply semimodularity. Thus the question arose how to characterize the Kurosh–Ore replacement property in general and, in particular, in the semimodular case. The pertinent fundamental results are due to Dilworth and Crawley. In this section we have a brief look at the characterization of the $\wedge$-KORP for strongly atomic algebraic lattices. In the following section we turn to the semimodular case.

The equivalence of the $\wedge$-KORP, Crawley's condition (Cr*), and dual consistency for lattices of finite length was mentioned in Theorem 4.5.1.

From Section 1.8 we recall the definition of completely meet-irreducible elements and the fact that in an algebraic lattice every element is a meet of completely meet-irreducible elements, that is, infinite meet decompositions exist (cf. Theorem 1.8.1). We also recall that if an algebraic lattice is strongly atomic, then any meet-irreducible element is completely meet-irreducible, that is, the two concepts are identical. Moreover, Crawley [1961] proved the existence of irredundant meet decompositions (cf. Theorem 1.8.2).

In Section 1.8 the $\wedge$-KORP was defined for complete lattices. Crawley's condition (Cr*) (defined in Section 4.5) can also be formulated for strongly atomic algebraic lattices. Finally let us state that we may define consistency for strongly dually atomic dually algebraic lattices as Kung [1985] did for lattices of finite length (cf. Section 4.5). In a dual way we define dual consistency for strongly atomic algebraic lattices. With these reformulations of the $\wedge$-KORP, Crawley's condition (Cr*), and dual consistency for strongly atomic algebraic lattices, we have, in complete analogy to the finite-length case (see Theorem 4.5.1), the following result (cf. Crawley [1961]).

Theorem 8.1.1　*Let L be a strongly atomic algebraic lattice. Then the following three conditions are equivalent:*

(i)　*L has the* $\wedge$-*KORP;*
(ii)　*L satisfies* (Cr*);
(iii)　*L is dually consistent.*

For a proof of the equivalence of (i) and (ii) we also refer to Crawley & Dilworth [1973], Theorem 7.5 [the equivalence of condition (iii) is implicit in the proof]. The claims of the first two exercises below are the essential steps used by Crawley & Dilworth [1973] in the proof of the equivalence of (i) and (ii).

It is easy to see that (Cr*) holds in a lattice L of finite length if and only if it holds in $[a, a^+]$ for every $a \in L$ (recall that by a^+ we mean the join of the upper covers of a). If L is a general strongly atomic algebraic lattice and (Cr*) holds in L, then (Cr*) also holds in every interval of L; in particular, (Cr*) holds then in $[a, a^+]$ for every $a \in L$. The converse is not true: Crawley [1961] provided the following example of a lattice L that does not satisfy (Cr*) although $[a, a^+]$ satisfies (Cr*) for every $a \in L$.

Consider first the lattice L_1 consisting of two infinite chains $a_1 < a_2 < \cdots < a_i < \cdots$ and $b_1 < b_2 < \cdots < b_i < \cdots$ such that the interval $[a_i, b_{i+1}]$ is isomorphic with the lattice of Figure 8.1 for each $i = 1, 2, 3, \ldots$. The lattice of Figure 8.1 has the property $p_{i,1} \vee a_{i+1} = b_{i+1}$ and $p_{i,k} \vee a_{i+1} = p_{i+1,k-1}$ for $k \geq 2$. Therefore, in L_1, for every i and k we have $p_{i,k} \vee a_{i+k} = b_{i+k}$. Consider now $L = \mathbf{I}(L_1)$, the lattice of ideals of L_1. Every nonprincipal ideal of L_1 contains the ideal $A = (a_1, a_2, a_3, \ldots)$ [by $(x_1, x_2, \ldots)$ we denote the ideal generated by the set $\{x_1, x_2, \ldots\}$]. Suppose B is an ideal of L_1 with $B > A$. Then B must contain the element $p_{i,k}$ for some i and k. Thus B contains $p_{i,k} \vee a_{i+k} = b_{i+k}$, and hence

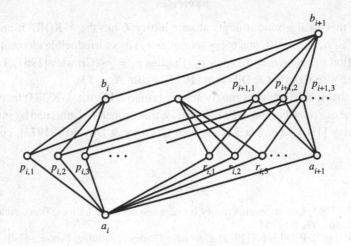

Figure 8.1

$B \geq (b_1, b_2, b_3, \ldots) = L_1$. Therefore A and L_1 are the only nonprincipal ideals of L, and hence it follows that L is a strongly atomic algebraic lattice. The only ideals X covered by more than two elements of L are the ideals (a_i). Hence if $X \neq (a_i)$, then $[X, X^+]$ satisfies (Cr*). Moreover $[(a_i), (a_i)^+] \cong [a_i, b_{i+1}]$, and it is obvious from Figure 8.1 that $[(a_i), (a_i)^+]$ also satisfies (Cr*) for each i. However, L does not satisfy (Cr*), since $[A, A \vee (b_1)^+] = L_1 \succ A$ and $[(b_1) \wedge A, (b_1)^+] = [(a_1), (b_1)^+] = (b_1)$ does not cover (a_1).

The lattice $\mathbf{I}(L_1)$ is not semimodular. It can be shown that if a strongly atomic algebraic lattice is semimodular, then it has the $\wedge$-KORP [and thus satisfies (Cr*)] if and only if the lattice is upper locally modular (cf. Theorem 8.2.1 below). Similarly, we have already mentioned (see Section 7.2) that a strongly atomic algebraic lattice has unique meet decompositions if and only if the lattice is upper locally distributive (cf. Theorem 7.1.2). Thus the uniqueness of meet decompositions (in a general strongly atomic algebraic lattice) and the $\wedge$-KORP (in a semimodular strongly atomic algebraic lattice) are local properties in the sense that they are determined by the intervals $[a, a^+]$.

Notes

Walendziak [1994b] pointed out that Crawley's characterization [Theorem 8.1.1(i), (ii)] still holds if "algebraic" is replaced by "upper continuous." In fact, Walendziak uses the dual formulation, that is, he characterizes consistency in lower continuous strongly coatomic lattices by means of condition (Cr), the dual to (Cr*). Richter [1982] proved that a strongly coatomic lower continuous lattice has the $\vee$-KORP if and only if it is consistent.

Exercises

1. Prove that an algebraic strongly atomic lattice L has the $\wedge$-KORP if and only if, for all a, p_1, $p_2 \in L$ and every (completely) meet irreducible element q, the condition $p_1, p_2 \succ a = q \wedge (p_1 \vee p_2)$ implies $p_1 = p_2$ (Crawley [1961], Lemma 3.1; see also Crawley & Dilworth [1973], claim A, p. 53).

2. Show that if an algebraic strongly atomic lattice L has the $\wedge$-KORP, then every interval of L also has the $\wedge$-KORP, that is, this property is inherited by intervals (Crawley [1961], Lemma 3.2; see also Crawley & Dilworth [1973], claim B, p. 54).

References

Crawley, P. [1961] Decomposition theory for non-semimodular lattices, Trans. Amer. Math. Soc. 99, 246–254.

Crawley, P. and R. P. Dilworth [1973] *Algebraic Theory of Lattices*, Prentice-Hall, Englewood Cliffs, N.J.

Kung, J. P. S. [1985] Matchings and Radon transforms in lattices I. Consistent lattices, Order 2, 105–112.

Richter, G. [1982] The Kuroš–Ore theorem, finite and infinite decompositions, Studia Sci. Math. Hungar. 17, 243–250.

Walendziak, A. [1994b] On consistent lattices, Acta Sci. Math. (Szeged) 59, 49–52.

8.2 Dually Consistent Semimodular Lattices

Summary. Crawley's characterization of the Kurosh–Ore replacement property for meet decompositions ($\wedge$-KORP) or dual consistency in strongly atomic algebraic lattices was given in the preceding section (Theorem 8.1.1). If the lattices in question are moreover semimodular, it can be shown that the $\wedge$-KORP is equivalent with upper local modularity.

If every interval of a semimodular lattice is of finite length, then meet decompositions and covering elements exist. It can then be proved that the number of components is unique for each element if and only if the intervals generated by the elements covering a lattice element are modular, that is, the lattice is upper locally modular. Dilworth [1941b] extended this result to lattices satisfying the ACC. Some 20 years later, Dilworth and Crawley investigated these questions for lattices without a chain condition. Motivated by examples from abelian groups, vector-space lattices, and lattices of congruence relations, Dilworth & Crawley [1960], Theorem 7.1, treated these questions for lattices without a chain condition and proved the following result.

Theorem 8.2.1 *An upper semimodular strongly atomic algebraic lattice is dually consistent (has the $\wedge$-KORP) if and only if it is locally modular.*

For a proof see also Crawley & Dilworth [1973], Theorem 7.6, pp. 54–55. The lattice shown in Figure 1.27 (Section 1.7) is a dually consistent upper semimodular lattice. For more on the duals of these lattices (i.e. consistent lower semimodular lattices) see also Sections 8.3 and 8.4.

The Kurosh–Ore property for join decomposition ($\vee$-KOP, for short) was mentioned in connection with the Kurosh–Ore theorem for modular lattices (Theorem 1.6.2). A complete lattice is said to have the Kurosh–Ore property for meet decompositions ($\wedge$-KOP, for short) if each of its elements has a finite irredundant meet decomposition and whenever an element b has two irredundant meet decompositions

$$b = m_1 \wedge \cdots \wedge m_n = \bigwedge T,$$

one has $|T| = n$ (i.e., for each b, the number of meet-irreducible elements in any irredundant meet decomposition of b is unique).

When an algebraic strongly atomic lattice satisfies the DCC or the ACC, then for any of its elements all the irredundant meet-decompositions are finite. Consequently, if the lattice has the $\wedge$-KORP, then it also has the $\wedge$-KOP. In a semimodular

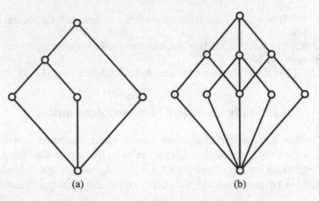

Figure 8.2

algebraic lattice satisfying the DCC, the converse also holds (Dilworth [1941a]).
Thus we have

Theorem 8.2.2   *A semimodular algebraic lattice satisfying the* DCC *has the* ∧-
KOP *if and only if it is locally modular.*

For a proof see also Crawley & Dilworth [1973], Theorem 7.7, p. 55. Dilworth
& Crawley [1960], Section 4, give an example of an algebraic, strongly atomic,
modular lattice in which the least element has two irredundant meet decompositions of different cardinalities. This example shows that the restriction to finite
irredundant meet decompositions is essential.

In particular, we get from Theorem 8.2.1 and Theorem 8.2.2 that in an upper
semimodular lattice of finite length the ∧-KOP is equivalent to the ∧-KORP. The
lattice of Figure 8.2(a) (cf. Crawley & Dilworth [1973], Chapter 7, p. 56) shows
that upper semimodularity cannot be dropped: this lattice has the ∧-KOP, but it
does not have the ∧-KORP.

Crawley & Dilworth [1973], p. 56, remark that no necessary and sufficient
conditions are known for an arbitrary finite lattice to have the ∧-KOP.

Notes

Richter [1982] and Walendziak [1990a] investigate the relationship of consistency
and the KORP [properties (V) and (V1) in their setting] in several classes of lattices that are not necessarily of finite length. Replacing "algebraic" by "upper
continuous" in Crawley & Dilworth [1973], Theorem 7.7, p. 55, Walendziak
[1990b] proves that in a semimodular upper continuous lattice L satisfying the
DCC, the following four conditions are equivalent: (i) L is locally modular; (ii) L
has the ∧-KORP; (iii) L has the ∧-KOP; (iv) L satisfies (Cr*). In this result the
DCC cannot be replaced by "strongly atomic": Consider for example the lattice

2^X of all subsets of an infinite set X. 2^X does not have the $\wedge$-KOP, but it satisfies the preceding conditions (i), (ii), and (iv).

Dilworth & Crawley [1960] also consider a replacement property stronger than $\wedge$-KORP. Crawley [1961], Theorem 3.3, defines doubly replaceable irredundant meet decompositions and proves that every element of a semimodular algebraic strongly atomic lattice L has doubly replaceable irredundant meet decompositions if and only if, for every $a \in L$, $[a, a^+]$ is a direct product of modular lattices of finite length. Some of these results are extended by Walendziak [1991].

Exercises

1. Any upper semimodular semicomplemented lattice of finite length with $\vee$-KORP is atomistic (Reeg & Weiß [1991], Theorem 6.8.1, p. 83).
2. A locally modular lattice of finite length is consistent (or strong, or balanced) if and only if it is modular.
3. Let L be an upper continuous strongly atomic lattice that is dually consistent. Then L has the $\wedge$-KORP (Walendziak [1990b]).
4. The lattice of Figure 8.2(b) shows that a finite lattice in which every join decomposition of any element contains at most two elements need not have a planar diagram. However, if a finite lattice L has a planar diagram, then L has the $\vee$-KOP and every irredundant join decomposition of an element of L contains at most two elements (Richter [1982]).
5. Let L be a semimodular, algebraic, and strongly dually atomic lattice satisfying the DCC. If L has both the $\wedge$-KORP and the $\vee$-KORP, then L is modular (Stern [1996c]).

References

Bogart, K. P., Freese, R., and Kung, J. P. S. (eds.) [1990] *The Dilworth Theorems. Selected Papers of Robert P. Dilworth*, Birkhäuser, Boston.

Crawley, P. [1961] Decomposition theory for non-semimodular lattices, Trans. Amer. Math. Society 99, 246–254.

Crawley, P. and R. P. Dilworth [1973] *Algebraic Theory of Lattices*, Prentice-Hall, Englewood Cliffs, N.J.

Dilworth, R. P. [1941a] The arithmetical theory of Birkhoff lattices, Duke Math. J. 8, 286–299.

Dilworth, R. P. [1941b] Ideals in Birkhoff lattices, Trans. Amer. Math. Soc. 49, 325–353.

Dilworth, R. P. and P. Crawley [1960] Decomposition theory for lattices without chain conditions, Trans. Amer. Math. Soc. 96, 1–22.

Reeg, S. and W. Weiß [1991] Properties of finite lattices, Diplomarbeit, Technische Hochschule Darmstadt.

Richter, G. [1982] The Kuroš–Ore theorem, finite and infinite decompositions, Studia Sci. Math. Hungar. 17, 243–250.

Stern, M. [1996c] A converse to the Kurosh–Ore theorem, Acta Math. Hungar. 70, 177–184.

Walendziak, A. [1990a] Meet-decompositions in complete lattices, Periodica Math. Hungar. 21, 219–222.

Walendziak, A. [1990b] The Kurosh–Ore property, replaceable irredundant decompositions, Demonstratio Math. 23, 549–556.

Walendziak, A. [1991] Lattices with doubly replaceable decompositions, Annales Soc. Math. Polonae Ser. I: Comment. Math. 30, 465–472.

8.3 Lattices of Subnormal Subgroups

Summary. The lattice of Figure 1.27 (Section 1.7) is dually consistent and semimodular. We have already mentioned that its dual (which is consistent and lower semimodular) is isomorphic to the lattice of all subnormal subgroups of the dihedral group D_4. In this section we shall have a closer look at finite, consistent, and lower semimodular lattices and, in particular, at the lattice of all subnormal subgroups of a finite group.

Let us first recall some relevant group-theoretical concepts. We don't restrict ourselves to finite groups from the beginning on, although later, in connection with lattices of subnormal subgroups, all groups will assumed to be finite. For more details on groups and subgroup lattices see the following books: Suzuki [1956], Hall [1959], Zassenhaus [1958], Zappa [1965], Rose [1978], Robinson [1982], Lennox & Stonehewer [1987], and Schmidt [1994]. Here we follow the presentation of Hall [1959].

We shall consider a chain of subgroups of a group G with identity element e,

$$G = A_0 \supseteq A_1 \supseteq A_2 \supseteq \cdots \supseteq A_r = \{e\}, \qquad (1)$$

where each A_i is a normal subgroup of A_{i-1}, which will be denoted by

$$A_i \lhd A_{i-1}, \qquad i = 1, \ldots, r. \qquad (2)$$

The groups A_i are called *subnormal* (or *subinvariant*) *subgroups* of G. With the chain (1) we associate the sequence of factor groups A_{i-1}/A_i, $i = 1, \ldots, r$. If every A_i is a normal subgroup of G, we call (1) a *normal chain* or *normal series*. From $A_i \lhd A_{i-1}$, $i = 1, \ldots, r$, it does not in general follow that $A_i \lhd G$. Hence the requirements for a normal series are stronger than (2). If we assume only (2), the series will be called *subnormal* (or *subinvariant*). A normal series in which every A_i is a maximal normal subgroup of G contained in A_{i-1} will be called a *principal* or *chief series*. In lattice terminology: if the inclusions in (1) are coverings, a normal series is called a principal or chief series. A subnormal series in which every A_i is a maximal normal subgroup of A_{i-1} will be called a *composition series*. In lattice terminology: if the inclusions in (1) are coverings, a subnormal series is called a composition series.

A group G is called *solvable* if it has a composition series

$$G = A_0 \supseteq A_1 \supseteq A_2 \supseteq \cdots \supseteq A_r = \{e\}$$

in which every factor A_{i-1}/A_i, $i = 1, \ldots, r$, is abelian. There are two properties of groups qualitatively stronger than solvability, which are of considerable importance. These are supersolvability and nilpotence.

A group G is *supersolvable* if it has a finite normal series

$$G = A_0 \supseteq A_1 \supseteq A_2 \supseteq \cdots \supseteq A_r = \{e\}$$

in which each factor A_{i-1}/A_i, $i = 1, \ldots, r$, is cyclic. A group G is *nilpotent* if it has a finite normal series

$$G = A_0 \supseteq A_1 \supseteq A_2 \supseteq \cdots \supseteq A_r = \{e\}$$

in which each factor A_{i-1}/A_i, $i = 1, \ldots, r$, is in the center of G/A_i. Finally, we recall that a *p-group* is a group in which every element except the identity has order a power of a prime number p.

For both supersolvable and nilpotent groups the factors A_{i-1}/A_i are abelian, and hence these groups are solvable. For general groups the interrelationships are indicated in Figure 8.3(a). Moreover, it can be shown that a finitely generated nilpotent group is supersolvable (cf. Hall [1959], Theorem 10.2.4, p. 152); the interrelationships in this case are shown in Figure 8.3(b).

For a comprehensive treatment of subnormal subgroups of groups we refer to Lennox & Stonehewer [1987]. For general remarks on Wielandt's contributions to subnormality see Isaacs [1994].

If H is a subnormal subgroup of a group G, we also write $H \lhd\lhd G$. Wielandt introduced subnormality with the motivation that in groups of finite composition length, and in particular in finite groups, the subnormal subgroups are precisely the subgroups that appear in composition series. The corresponding factors are of great importance in describing the structure of a group. Since subnormality generalizes normality, it is natural to try to find theorems about subnormal subgroups by generalizing facts known for normal subgroups.

Let us first indicate a property of normal subgroups that need *not* hold for subnormal subgroups: If $H, K \lhd G$, then $HK = KH$, that is, H and K are permutable and HK is a subgroup. It is immediate that this property of normal subgroups need

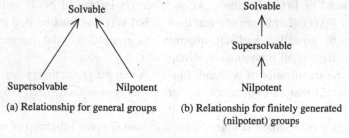

(a) Relationship for general groups	(b) Relationship for finitely generated (nilpotent) groups

Figure 8.3

not hold for subnormal subgroups, since every subgroup of a finite p-group is subnormal (see Theorem 8.3.2 below) and yet it is not true that every two subgroups
are permutable.

On the other hand, it is trivially true, for instance, that if H and K are normal
in some group, then $H \cap K$ and $\langle H, K \rangle$ are also normal. It is also easy to see that
under the weaker assumption that $H, K \lhd\lhd G$ we get $H \cap K \lhd\lhd G$. The following
join theorem is due to Wielandt [1939].

Theorem 8.3.1   *If a group G has finite composition length and $H, K \lhd\lhd G$, then*
$\langle H, K \rangle \lhd\lhd G$.

For proofs we refer to Lennox & Stonehewer [1987], Rose [1978], or Zappa
[1965]. The result remains valid (as Wielandt remarked) if the assumption that G
has finite composition length is weakened to the maximal condition on the set of
subnormal subgroups. For general groups, however, the assertion of the preceding
theorem is false; a counterexample can be found in Zassenhaus [1958] (Exercise 23
on p. 235). Dean & Kruse [1966] present a considerably simplified version of this
example.

It follows from Wielandt's join theorem that, in a group with composition series,
the subnormal subgroups form a sublattice of the lattice of subgroups. Let us recall
some notation.

By $L(G)$ we denote the set of all subgroups of a group G. With respect to
set inclusion $L(G)$ forms a lattice, which we denote by $\mathbf{L}(G) = \langle L(G), \subseteq \rangle$. For
$H, K \in L(G)$ the lattice operations are defined by $H \wedge K = H \cap K$ and $H \vee K =
\langle H, K \rangle$, that is, the meet of two subgroups is their set-theoretic intersection, while
the join is the subgroup generated by the two subgroups. Similarly we denote by
$N(G)$ the set of all normal subgroups of a group G and get the lattice $\mathbf{N}(G) =
\langle N(G), \subseteq \rangle = \langle N(G), \wedge, \vee \rangle$. Dedekind proved that, for an arbitrary group G, the
normal subgroups form a modular sublattice of $\mathbf{L}(G)$ (see Section 1.6). By $W(G)$
we denote the set of all subnormal subgroups of a group G with finite composition
length. The *lattice of subnormal subgroups* will be denoted by $\mathbf{W}(G) = \langle W(G),
\subseteq \rangle = \langle W(G), \wedge, \vee \rangle = \langle W(G), \cap, \vee \rangle$. Somewhat sloppily, we shall also write
$L(G)$ instead of $\mathbf{L}(G)$. Similarly, we write $N(G)$ instead of $\mathbf{N}(G)$ and $W(G)$
instead of $\mathbf{W}(G)$. [Let us remark that the symbol $\mathbf{W}(G)$ is usually used in group
theory for the so-called Wielandt subgroup of a group G, i.e. the intersection of
the normalizers of all subnormal subgroups of G.]

While the significance of Wielandt's join theorem for group theory was immediate, it seems that its importance for lattice theory was recognized only much
later.

Let us give now some examples of finite groups G *every* subgroup of which is
subnormal, that is, the lattice $W(G)$ of subnormal subgroups of G coincides with
the lattice $L(G)$ of all subgroups of G.

The first case we consider is that of a group each of whose subgroups is normal; such groups are called *Dedekind groups* (Dedekind [1897]). Clearly abelian groups are Dedekind groups, and the structure of nonabelian Dedekind groups (*Hamiltonian groups*; see Zassenhaus [1958]) is given by the following result: A nonabelian group has each of its subgroups normal if and only if it is the direct product of a quaternion group of order 8 and a periodic abelian group with no elements of order 4.

The next example are finite nilpotent groups. We have the following result (see Rose [1978], Theorem 11.3, p. 266).

Theorem 8.3.2   *For a finite group G the following statements are equivalent:*

(i)   *G is nilpotent;*
(ii)   *every subgroup of G is subnormal in G;*
(iii)   *every maximal subgroup of G is normal in G;*
(iv) $G' \leq \Phi(G)$ *[$\Phi(G)$ is the Frattini subgroup of G, i.e., the intersection of all maximal subgroups];*
(v)   *every Sylow subgroup of G is normal in G;*
(vi)   *G is a direct product of groups of prime power orders.*

Let us now turn to lattices of subnormal subgroups. The following results and examples are due to Gragg & Kung [1992].

Theorem 8.3.3   *The lattice of subnormal subgroups $W(G)$ of a finite group G is lower semimodular and consistent (relative to local properties).*

Proof.   If $x \vee y \succ x$ in $W(G)$, then x is a maximal subnormal subgroup and hence a normal subgroup of $x \vee y$ (any maximal subnormal subgroup is normal – see Lennox & Stonehewer [1987]). By the second isomorphism theorem of group theory (see Rose [1978], Theorem 3.40, p. 56) the interval $[x \wedge y, y]$ is isomorphic to $[x, x \vee y]$, and thus $y \succ x \wedge y$. Hence $W(G)$ is lower semimodular. For the definition of consistency relative to local properties see Section 4.5. We introduce some more notation: if $y \leq x$, then we denote by $\langle y : x \rangle$ the meet of x and all the elements in $[y, 1]$ that are covered by x. Let now j and x be elements of $W(G)$. Since the element $\langle x : x \vee j \rangle$ is the intersection of all maximal subnormal subgroups covered by $x \vee j$, it is a normal subgroup of $x \vee j$. Again, by the second isomorphism theorem we get $[\langle x : x \vee j \rangle, x \vee j] \cong [j \wedge \langle x : x \vee j \rangle, j]$. Hence, if a coatomistic lattice is an upper subinterval of $[\langle x : x \vee j \rangle, x \vee j]$, then it is also an upper subinterval of $[j \wedge \langle x : x \vee j \rangle, j]$. This implies that $W(G)$ is consistent (relative to local properties). ∎

Corollary 8.3.4   *The subgroup lattice of a finite nilpotent group is consistent and lower semimodular. In particular, the subgroup lattice of a finite p-group is consistent and lower semimodular.*

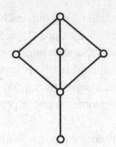

Figure 8.4

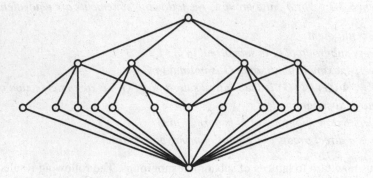

Figure 8.5

Proof.   This follows from the fact that every subgroup of a finite nilpotent group
is subnormal (cf. Theorem 8.3.2) and from the nilpotency of finite p-groups.   ∎

In the following three examples the whole group G is a p-group and thus
$W(G) = L(G)$ (in all three groups, the Frattini subgroup is also in the center).

Figure 8.4 shows the subgroup lattice of the quaternion group of order 8. The
lattice is modular.

The subgroup lattice of the group D_4 of the symmetries of a square is the dual
to the lattice of Figure 1.27 (Section 1.7). Figure 8.5 shows the subgroup lattice
of the group Unp(3, 3) of all 3×3 unipotent matrices, that is, upper triangular
matrices with 1's on the diagonal, over the finite field GF(3).

Notes

Since the subgroup lattice of a finite p-group is lower semimodular, it is upper
semimodular if and only if it is modular. This happens exactly if the group is quasi-
Hamiltonian, that is, if any two subgroups are permutable (see Suzuki [1956],
p. 10). For related results see Richter [1983].

In Section 7.4 we defined upward and downward perspectivity of quotients. If
a quotient a/b is upward perspective or downward perspective to a quotient c/d,

then we say that a/b is *perspective to* c/d and write $a/b \sim c/d$. If for some natural number n there exist $a/b = e_0/f_0, e_1/f_1, \ldots, e_n/f_n$ such that $e_i/f_i \sim e_{i+1}/f_{i+1}$ for $i = 0, \ldots, n-1$, then we say that a/b is *projective to* c/d and write $a/b \approx c/d$. Thus projectivity is the transitive extension of perspectivity. Tamaschke [1960] considers for lattices of finite length the following condition:

(QS) $x \succ x \wedge y$ implies that all prime quotients p/q with $x \vee y \geq p \succ q \geq y$ are projective to $x/(x \wedge y)$

and the dual condition

(QS*) $x \vee y \succ y$ implies that all prime quotients p/q with $x \geq p \succ q \geq x \wedge y$ are projective to $(x \vee y)/y$.

Any upper semimodular lattice of finite length satisfies (QS) (but not conversely), and any lower semimodular lattice of finite length satisfies (QS*) (but not conversely). Lattices with (QS) and (QS*) have also been called *quasi upper semimodular* and *quasi lower semimodular*, respectively (cf. e.g. Zappa [1965], pp. 122–123). Tamaschke calls a lattice L of finite length *submodular* if both (QS) and (QS*) hold in L and investigates the properties of submodular lattices in several papers (see e.g. Tamaschke [1960], [1961], [1962]). Among other results, he proved that $W(G)$ is submodular. Tamaschke [1961] also noted that $W(G)$ is lower semimodular.

A group G is called submodular if its lattice of subgroups $L(G)$ is of finite length and satisfies the conditions (QS) and (QS*). Napolitani [1973] characterized submodularity for finite groups. Leone & Maj [1982], [1982–3] investigated submodularity for minimal finite subgroups and for factor groups of a finite solvable group that is not submodular. Note that other authors use the concept *submodular* in a different sense.

Another approach to the lattice-theoretic treatment of normal and subnormal subgroups of a group (via lattices endowed with a normality relation) can be found in Zassenhaus [1958], Appendix B. From the numerous papers dealing with this approach and related ones let us mention here only the survey Barbilian [1946] and Dean & Kruse [1966]. The latter paper introduces a relation ⊲ between the elements of a lattice that generalizes the relation of normality between subgroups [in the lattice $L(G)$ of subgroups of a group G].

Exercises

1. Show that the lattice of subnormal subgroups of a finite group is submodular (Tamaschke [1960]).
2. Give an example of a submodular lattice of finite length that is not isomorphic to the lattice of subnormal subgroups of a group (Tamaschke [1960]).
3. Show that a homomorphic image of a submodular lattice is submodular.

References

Barbilian, D. [1946] Metrisch-konkave Verbände, Disquisitiones Math. Phys. (Bucureşti), V (1–4), 1–63.

Dean, R. A. and R. L. Kruse [1966] A normality relation for lattices, J. Algebra 3, 277–290.

Dedekind, R. [1897] Über Zerlegungen von Zahlen durch ihren größten gemeinsamen Teiler, in *Festschrift der Techn. Univ. Braunschweig*, pp. 1–40; *Gesammelte Werke*, Vol. 2, pp. 103–147.

Gragg, K. and J. P. S. Kung [1992] Consistent dually semimodular lattices, J. Combin. Theory Ser. A 60, 246–263.

Hall, M. [1959] *The Theory of Groups*, Macmillan, New York.

Isaacs, I. M. [1994] Helmut Wielandt on subnormality, in: Wielandt [1994], pp. 299–306.

Lennox , J. C. and S. E. Stonehewer [1987] *Subnormal Subgroups of Groups*, Clarendon, Oxford.

Leone, A. and M. Maj [1982] Gruppi finiti minimali non submodulari, Ricerche Mat. 31, 377–388.

Leone, A. and M. Maj [1982–3] Gruppi finiti non submodulari a quozienti propri submodulari, Rend. Accad. Sci. Fis. e Mat. Ser. IV XLX, 185–193.

Napolitani, F. [1973] Submodularità nei gruppi finiti, Rend. Sem. Mat. Univ. Padova 50, 355–363.

Richter, G. [1983] Applications of some lattice theoretic results on group theory, in: *Proc. Klagenfurt Conf. Universal Algebra, June 10–13, 1982*, Hölder-Pichler-Tempsky, Wien, pp. 305–317.

Robinson, D. J. S. [1982] *A Course in the Theory of Groups*, Springer-Verlag, New York.

Rose, J. S. [1978] *A Course in Group Theory*, Cambridge Univ. Press, Cambridge.

Schmidt, R. [1994] *Subgroup Lattices of Groups*, de Gruyter, Berlin.

Suzuki, M. [1956] *Structure of a Group and the Structure of Its Lattice of Subgroups*, Ergebnisse, Vol. 10, Springer-Verlag, Berlin.

Tamaschke, O. [1960] Submodulare Verbände, Math. Z. 74, 186–190.

Tamaschke, O. [1961] Die Kongruenzrelationen im Verband der zugänglichen Subnormalteiler, Math. Z. 75, 115–126.

Tamaschke, O. [1962] Verbandstheoretische Methoden in der Theorie der subnormalen Untergruppen, Archiv d. Math. 13, 313–330.

Wielandt, H. [1939] Eine Verallgemeinerung der invarianten Untergruppen, Math. Z. 45, 209–244.

Wielandt, H. [1994] *Mathematical Works Vol. 1: Group Theory*, (ed. B. Huppert and H. Schneider), de Gruyter, Berlin.

Zappa, G. [1965] *Fondamenti di Teoria dei Gruppi*, Vol. I, Edizioni Cremonese, Roma.

Zassenhaus, H. [1958] *The Theory of Groups* (2nd edition), New York.

8.4 Breadth and Reach

Summary. Following Gragg & Kung [1992], we consider a sufficient condition for finite lower semimodular lattices to be consistent. Then we apply this condition to obtain a matching result of Rival [1976b] for finite upper semimodular lattices of breadth at most 2.

Let us first recall some notions and results. The *breadth* of a finite lattice L is the least positive integer m such that any join $\bigvee_{i=1}^{n} x_i$, $x_i \in L$, $n \geq m$, is always a join of m of the x_i's. This definition of breadth is order-dual to that of Birkhoff [1967], p. 99. We leave it as an exercise to show that the property of breadth 2 is self-dual. Lattices of breadth at most two are those lattices which do not contain a sublattice isomorphic to the eight-element Boolean lattice. If a finite lattice is dismantlable (see Section 5.6), then it has breadth at most 2 (cf. Kelly & Rival [1974]).

In a semimodular lattice L of finite length, breadth and reach are related by the inequality

$$\text{breadth}(L) \geq \text{reach}(L).$$

For the notions of reach and coreach see Section 6.1. For consistency relative to a property see Section 4.5. The following result is due to Gragg & Kung [1992], Theorem 4.1.

Theorem 8.4.1 *Let $c \geq 2$, and L be a lower semimodular lattice having coreach c. Then L is consistent relative to the property of having coreach at most $c - 1$.*

Proof. Let coreach$(j) \leq c - 1$. Suppose that for some x, $x \vee j$ has coreach greater than $c - 1$ in the interval $[x, 1]$. In this case $j \not\leq x$. Let $u_1, u_2, \ldots, u_c$ be elements in $[x, 1]$ covered by $x \vee j$ such that $r(\bigwedge u_i) = r(x \vee j) - c$. Since $j \not\leq x$, there exists an element y such that $x \vee j$ covers y and $x \not\leq y$. Because $x \not\leq y$ but $x \leq \bigwedge u_i$, we have $y \wedge (\bigwedge u_i) \neq \bigwedge u_i$. Thus $r(x \vee j) - (c + 1) \geq r(y \wedge (\bigwedge u_i)) \geq r((x \vee j)_+)$ and coreach$(x \vee j) \geq c + 1$, a contradiction. ∎

Corollary 8.4.2 *A lower semimodular lattice having coreach 2 is consistent.*

The lattice of subgroups of the group 3×3 unipotent matrices (see Figure 8.5, Section 8.3) over a prime field is a lower semimodular lattice with coreach 2.

Rival [1976b] investigated upper semimodular lattices having breadth 2. The order duals of his examples are lower semimodular lattices having coreach 2.

Note that there is an abundance of finite nonmodular upper semimodular lattices of breadth two. The examples of Figure 8.6 (cf. Rival [1976b]) indicate that there are finite, nonmodular, semimodular lattices of breadth 2 and of arbitrary finite width [Figure 8.6 (a)] and length [Figure 8.6 (b)]. For the definition of width see the Notes below.

We have seen (cf. Corollary 8.4.2) that a lower semimodular lattice L having coreach 2 is consistent. By a result of Kung (see Theorem 6.1.12), consistency implies that for such a lattice there exists a matching between $J_c(1, L) = J(1, L)$ and $M(1, L)$ and hence $|J(1, L)| \leq |M(1, L)|$. Dualizing this, we get that, in a finite upper semimodular lattice L of breadth at most 2, one has

$$|M(1, L)| \leq |J(1, L)|.$$

This is one part of the following matching result due to Rival [1976b], Theorem 8.

Theorem 8.4.3 *Let L be a finite upper semimodular lattice of breadth at most 2. Then $|M(1, L)| \leq |J(1, L)|$, and equality occurs if and only if L is modular.*

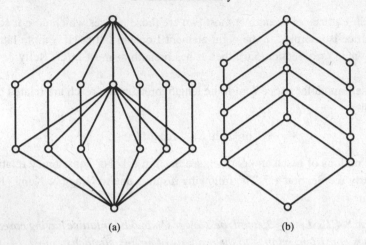

(a)   (b)

Figure 8.6

One direction of the other part of this result (i.e. the preceding inequality is an equality if and only if L is modular) is immediate: if L is modular, then equality follows from the above inequality and from the reverse inequality (or as a special case of Dilworth's covering theorem – see Theorem 6.1.9). The other direction [i.e., if L is a finite upper semimodular lattice of breadth 2 satisfying $|M(1, L)| = |J(1, L)|$, then L is modular] can be proved directly (as in Rival [1976b]) or in the broader context of *boundary lattices* as treated in Gragg & Kung [1992]. We shall indicate the latter approach in Section 8.5. For a generalization of Rival's matching result see Kung [1986b], Theorem 3.2.2.

Notes

For a lattice L we define the width by $w(L) = \sup\{|A| : A$ is a finite antichain of $L\}$. Using Exercise 1, the proof of the next result (cf. Rival [1976b], Theorem 9) can be carried through in essentially the same way as in the proof of the main result of Wille [1976]: If L is a finite semimodular lattice of breadth at most 2, then $w(J(L)) = w(D(L)) \leq w(M(L)) = w(J(L) \cup M(L))$.

This theorem, together with its dual, yields the corresponding result for finite modular lattice of breadth at most 2 due to Lea [1974] and Wille [1976] (who provided a short combinatorial proof of a slightly generalized version of Lea's result).

Exercises

1. Let L be a semimodular lattice of finite length and of breadth at most 2. If b is a maximal join irreducible element in L, then b is doubly irreducible in L (Rival [1976b], Proposition 5).

2. A coatomistic and semicomplemented lattice of breadth 2 and of finite length is complemented. (Reeg & Weiß [1991], Theorem 6.6.2).

3. An upper semimodular and complemented lattice of breadth 2 and of finite length is of length 2, atomistic, and modular (Reeg & Weiß [1991], Theorem 6.6.3).

4. An upper semimodular and lower balanced lattice of breadth 2 and of finite length is modular (Reeg & Weiß [1991], Theorem 6.6.4).

References

Birkhoff, G. [1967] *Lattice Theory*, Amer. Math. Soc. Colloquium Publications, Vol. 25, Providence, R.I. (reprinted 1984).

Gragg, K. and J. P. S. Kung [1992] Consistent dually semimodular lattices, J. Combin. Theory Ser. A 60, 246–263.

Kelly, D. and I. Rival [1974] Crowns, fences and dismantlable lattices, Canad. J. Math. 26, 1257–1271.

Kung, J. P. S. [1986b] Radon transforms in combinatorics and lattice theory, in: *Combinatorics and Ordered Sets*, Contemporary Mathematics, Vol. 57 (ed. I. Rival), pp. 33–74.

Lea, J. W. [1974] Sublattices generated by chains in modular topological lattices, Duke Math. J. 41, 241–246.

Reeg, S. and W. Weiß [1991] Properties of finite lattices, Diplomarbeit, Technische Hochschule Darmstadt.

Rival, I. [1976b] Combinatorial inequalities for semimodular lattices of breadth two, Algebra Universalis 6, 303–311.

Wille, R. [1976] On the width of sets of irreducibles in finite modular lattices, Algebra Universalis 6, 257–258.

8.5 Boundary Lattices

Summary. In this section we have a look at a combinatorial characterization of modularity among finite consistent lower semimodular lattices due to Gragg & Kung [1992].

We recall that in a finite lattice L the existence of a matching from a subset J into a subset M implies the numerical inequality $|J| \leq |M|$ (see Section 6.1). This suggests the following extremal problem: Let J and M be subsets of a finite lattice L. For which pairs of subsets J and M does equality $|J| = |M|$ imply that L is modular?

Some results on finite geometric lattices along this line were mentioned at the end of Section 6.1. The following result is due to Gragg & Kung [1992].

Theorem 8.5.1 *Let L be a finite, consistent, and lower semimodular lattice. Then the number of join-irreducibles is less than or equal to the number of meet-irreducibles, with equality if and only if L is modular.*

Sketch of proof. Consistency implies $|J(1, L)| \leq |M(1, L)|$ (cf. Corollary 6.1.13). If L is modular, then by Dilworth's covering theorem (Theorem 6.1.9), $|J(1, L)| =$

$|M(1, L)|$ (alternatively, this also follows from the fact that if L is modular, then its dual is likewise consistent and hence the reverse inequality also holds).

We indicate now why equality also implies modularity. The proof is carried through by induction on the rank of L. Since all lower semimodular lattices of rank 1 or 2 are modular, the assertion holds for $r(L) \leq 2$. Let now L be a consistent lower semimodular lattice of rank ≥ 3 such that

$(+)$ $|J(1, L)| = |M(1, L)|$, .

and assume that L is not modular. We show that this yields a contradiction.

A finite lattice L is said to be a *boundary lattice* if $(+)$ holds. More generally, Gragg & Kung [1992] define a boundary lattice as a finite lattice in which $|J| = |M|$ holds for a pair of concordant sets J and M (see Section 6.1). It can be shown that for many concordant sets, upper intervals of boundary lattices are also boundary lattices (Gragg & Kung [1992], Lemma 7.2). In particular, for a consistent lattice satisfying $(+)$ the following statement holds: Let $J[x]$ be the set of consistent join irreducibles in the interval $[x, 1]$, and $M[x]$ the set of meet irreducibles of L that are $\geq x$. Then $|J[x]| = |M[x]|$.

By this result and by our assumptions, we may suppose that L is a finite lower semimodular but nonmodular lattice in which every proper upper interval is modular.

The dual of Theorem 3.1.10 guarantees that a lower semimodular but nonmodular lattice has a cover-preserving sublattice isomorphic to S_7^*. Since every proper upper interval of L is modular, such a S_7^* must be contained in an interval $[0, x \vee y]$. It follows that x and y may be chosen to be atoms. (For a detailed independent proof see Gragg & Kung [1992], Lemma 7.4.)

Choose now such a sublattice S_7^* with its least element coinciding with 0 and label it as in Figure 8.7. There exists a meet-irreducible m such that $y \leq m$ but $y_1 \not\leq m$ (since an element is the meet of all meet-irreducibles above it). Note that $y \vee z \geq y_1$ implies $z \not\leq m$. Similarly $x \not\leq m$ and $x_1 \not\leq m$. It follows that $m \wedge x_1 = 0$.

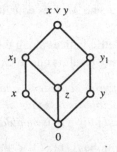

Figure 8.7

For an element a in L and an element b in $M(1, L)$ Gragg & Kung define a value $\gamma(a, b)$ and then consider the x, m entry $\sum_{z \le x \wedge m} \mu(z, x)\gamma(z, m)$ in the inverse of the incidence matrix $[\zeta(a, b)]$ between $J(1, L)$ and $M(1, L)$. Because $m \wedge x = 0$, there is only one term in the sum, namely $\mu(0, x)\gamma(0, m)$. Then $\gamma(0, m)$ is computed using two different contact functions (see Section 6.1 for the definition of a contact function). Since $\mu(0, x) \ne 0$, the x, m entry $\mu(0, x)\gamma(0, m)$ in the right inverse of the incidence matrix $[\zeta(a, b)]$ obtained using the two contact functions yields two different values. This contradicts the fact that $[\zeta(a, b)]$ is a square matrix and has a unique inverse. ∎

If the assumption of lower semimodularity is dropped, the assertion is obviously false: the pentagon is consistent, and the number of join-irreducibles equals the number of meet-irreducibles, but it is not lower semimodular.

The preceding result implies the second part of Theorem 8 in Rival [1976b]:

Corollary 8.5.2 *A finite semimodular lattice of breadth at most 2 with the same number of join-irreducibles and meet-irreducibles is modular.*

Proof. The dual of the assertion follows from the Theorem 8.5.1 and from the fact that a lower semimodular lattice having coreach 2 is consistent (see Corallary 8.4.2). ∎

Notes

The arguments used in the proof of Theorem 8.5.1 can also be used for an alternative proof of the corresponding result for geometric lattices obtained by Dowling & Wilson [1975] (see Section 6.1).

References

Dowling, T. A. and R. M. Wilson [1975] Whitney number inequalities for geometric lattices, Proc. Amer. Math. Soc. 47, 504–512.

Gragg, K. and J. P. S. Kung [1992] Consistent dually semimodular lattices, J. Combin. Theory Ser. A 60, 246–263.

Race, D. M. [1986] Consistency in lattices, Ph.D. thesis, North Texas State University, Denton, Tex.

Rival, I. [1976b] Combinatorial inequalities for semimodular lattices of breadth two, Algebra Universalis 6, 303–311.

9

Congruence Semimodularity

9.1 Semilattices

Summary. Algebras whose congruence lattices satisfy nontrivial lattice identities have been thoroughly studied during the last two decades. In particular, varieties have been investigated in which the congruence lattice of all algebras is distributive or modular. Although weaker than modularity, semimodularity is still an interesting property in this connection. Ore [1942] proved that sets have semimodular congruence lattices (in fact, these are matroid lattices). Here we describe some properties of congruence lattices of semilattices that were shown to be semimodular by Hall [1971].

An algebra $S = \langle S; \cdot \rangle$ with a binary operation $\cdot$ is called a *semigroup* if this operation is associative, that is, $x \cdot (y \cdot z) = (x \cdot y) \cdot z$ holds for all $x, y, z \in S$. For simplification we briefly write ab for $a \cdot b$. We shall identify S with S if there is no danger of confusion. A *semilattice* is a semigroup satisfying $xy = yx$ (commutativity) and $x^2 = x$ (idempotency) for all $x, y \in S$. We may impose a partial order on S by defining $x \leq y$ if $xy = x$. Under this ordering, any two elements $x, y \in S$ have a greatest lower bound, namely their product xy. Hence S is a (meet) semilattice.

An algebra A is said to be *locally finite* if every subalgebra of A generated by finitely many elements is finite. A variety V is said to be *locally finite* if every algebra in V is locally finite, or equivalently, every finitely generated algebra in V is finite. Semilattices form a locally finite variety. We denote this variety by Semilattices.

Congruence relations on a semilattice themselves form an algebraic lattice. The lattice $\mathrm{Con}(S)$ of congruence relations on a semilattice S has a number of further properties, which have been investigated by several authors. We recall here some of these results. While the lattice $\mathrm{Con}(L)$ of congruence relations on a lattice L is distributive (Funayama & Nakayama [1942]), which is also expressed by saying that the variety of lattices is congruence-distributive (see Section 1.3), the lattice $\mathrm{Con}(S)$ of congruence relations on a semilattice S is not even modular in general (i.e., the variety Semilattices is not congruence-modular), as the following example shows.

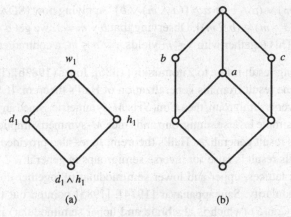

Figure 9.1

The centered hexagon S_7 [cf. Figure 9.1(b)] is isomorphic to the congruence lattice of $Sl_4 = E_2 \times E_2$ [cf. Figure 9.1(a)], where $E_2 = \langle \{0, 1\}, \wedge \rangle$ denotes the two-element (meet) semilattice. The lattice S_7 is upper semimodular and meet-semidistributive [condition (SD$\wedge$); see Section 1.3]. Both of the latter properties were shown to hold for the lattice $\mathrm{Con}(S)$ of congruence relations on an arbitrary semilattice S: Papert [1964] proved that $\mathrm{Con}(S)$ is coatomistic (cf. also Schmidt [1969]) and satisfies (SD$\wedge$). Hall [1971] proved that, for a semilattice S, the congruence lattice $\mathrm{Con}(S)$ is upper semimodular (see also Freese & Nation [1973]). The fact that, for a semilattice S, the lattice $\mathrm{Con}(S)$ is upper semimodular and satisfies (SD$\wedge$) is also expressed by saying that the variety Semilattices is *congruence semimodular* (CSM, for short) and *congruence meet-semidistributive*.

The abovementioned result of Papert [1964] can be used to sharpen the result of Hall [1971] in that "semimodular" can be replaced by the stronger "M-symmetric."

Theorem 9.1.1 *A bounded coatomistic lattice satisfying condition* (SD$\wedge$) *(meet semidistributivity) is M-symmetric.*

Proof. Let L be a bounded coatomistic lattice, and assume that $b \, M \, a$ holds in L, that is,

$(+)$ $x \leq a \;\Rightarrow\; x \vee (b \wedge a) = (x \vee b) \wedge a.$

We show that $a \, M \, b$ also holds, that is, $y \leq b \;\Rightarrow\; y \vee (a \wedge b) = (y \vee a) \wedge b$. We always have $y \vee (a \wedge b) \leq (y \vee a) \wedge b$. Suppose now that equality does not hold, that is, for some $y \leq b$ we have $y \vee (a \wedge b) < (y \vee a) \wedge b$. Since L is coatomistic, there exists a dual atom m such that $m \geq y \vee (a \wedge b)$ but $m \not\geq (y \vee a) \wedge b$. The latter relation implies $b \not\leq m$ and $y \vee a \not\leq m$. Putting $x = a \wedge m$ in $(+)$, we get

$a \wedge m = (a \wedge m) \vee (a \wedge b) = a \wedge [(a \wedge m) \vee b]$. Applying now (SD∧), we obtain $a \wedge m = a \wedge [(a \wedge m) \vee (b \vee m)]$. Observing that $b \vee m = 1$, we get $a \wedge m = a$ and hence $a \leq m$. This together with $y \leq m$ yields $y \vee a \leq m$, a contradiction. ∎

The preceding result is due to Žitomirskii [1986]. Jones [1983b], Theorem 3.3, derived the same result from his generalization of Hall's theorem: If S is a hyper-semisimple inverse semigroup, then $\mathrm{Con}(S)$ is M-symmetric. Since any semilattice is a hypersemisimple inverse semigroup and since M-symmetry implies semimodularity, Jones's result generalizes Hall's theorem. Jones also provided an example showing that his result is false for inverse semigroups in general.

In algebraic lattices, upper and lower semimodularity together do not in general imply modularity. Sankappanavar [1974], [1985] pointed out that if, for a semilattice S, $\mathrm{Con}(S)$ (which is algebraic and upper semimodular) is also lower semimodular, then it is not only modular but distributive. Before indicating a proof of this, we still need some preparations in which the semilattice Sl_4 [cf. Figure 9.1(a)] plays a significant role. As already noted, its congruence lattice is isomorphic to the lattice of Figure 9.1(b), that is, it is isomorphic to S_7.

The following two lemmas are due to Sankappanavar [1985].

Lemma 9.1.2 *Let S be a (meet) semilattice, and for $A \subseteq S$ denote by A^u the set of upper bounds of $A \subseteq S$. Assume $d, h, w \in S$ are such that d and h are incomparable and $w \in \{d, h\}^u$. Then there exists an epimorphism $\alpha : S \to Sl_4$ such that $\alpha(d) = d_1, \alpha(h) = h_1,$ and $\alpha(w) = w_1$.*

Sketch of proof. Define $\alpha : S \to Sl_4$ as follows:

$$\alpha(a) = d_1 \wedge h_1 \qquad \text{if} \quad a \not\geq d \wedge h, \tag{1}$$

$$\alpha(a) = d_1 \wedge h_1 \qquad \text{if} \quad a \geq d \wedge h, \quad a \not\geq d, \quad a \not\geq h, \tag{2a}$$

$$\alpha(a) = w_1 \qquad\qquad \text{if} \quad a \geq d \text{ and } a \geq h, \tag{2b}$$

$$\alpha(a) = d_1 \qquad\qquad \text{if} \quad a \geq d \text{ and } a \not\geq h, \tag{2c}$$

$$\alpha(a) = h_1 \qquad\qquad \text{if} \quad a \not\geq d \text{ and } a \geq h. \tag{2d}$$

It is clear that the condition $a \geq d \wedge h$ is equivalent to the disjunction of conditions (2a)–(2d) and thus α is defined on S. We leave it as an exercise to show that α is indeed a homomorphism. ∎

Lemma 9.1.3 *Let $d, h, w \in S$ as in Lemma 9.1.2. Then there exists an interval in $\mathrm{Con}(S)$ isomorphic to the lattice given in Figure 9.1(b).*

Proof. Let $\alpha : S \to Sl_4$ be the epimorphism defined in Lemma 9.1.2, and let $\theta = \ker \alpha$, where $\ker \alpha = \{(a, b) \in S \times S : \alpha(a) = \alpha(b)\}$. It is well known (cf. Burris

& Sankappanavar [1981], Theorem 6.20) that $\text{Con}(Sl_4) \cong [\theta, \nabla]$ (where ∇ denotes the greatest congruence of S). The proof is finished by observing that $\text{Con}(Sl_4)$ is isomorphic to the lattice of Figure 9.1(b). ∎

Sankappanavar [1985], Theorem 2.3, gives the following characterization of lower semimodularity for the congruence lattice of a semilattice.

Theorem 9.1.4 *Let S be a semilattice. Then the following conditions are equivalent:*

(1) $\text{Con}(S)$ *is lower semimodular;*
(2) *for all $x, y, z \in S$, $(x \leq z$ and $y \leq z)$ imply $(x \leq y$ or $y \leq x)$;*
(3) $\text{Con}(S)$ *is modular;*
(4) $\text{Con}(S)$ *is distributive;*
(5) $\text{Con}(S)$ *is a relatively pseudocomplemented lattice;*
(6) $\text{Con}(S)$ *is meet-semidistributive;*
(7) $\text{Con}(S)$ *is join-semidistributive;*
(8) Sl_4 *is not embeddable as a semilattice in S;*
(9) *every interval in the dual of $\text{Con}(S)$ is pseudocomplemented.*

Sketch of proof. $(1) \Rightarrow (2)$: If (2) is false, then there exist elements $d, h, w \in S$ as in Lemma 9.1.2. It follows from Lemma 9.1.3 that $\text{Con}(S)$ contains an interval that is not lower semimodular. This implies that $\text{Con}(S)$ is not lower semimodular.

In a similar way the implications $(7) \Rightarrow (2)$, $(8) \Rightarrow (2)$, and $(9) \Rightarrow (2)$ can be verified. The equivalences $(2) \Rightarrow (3) \Rightarrow (4)$ are well known (cf. e.g. Varlet [1965]), and the remaining implications are obvious. ∎

Semilattices whose congruence lattices are distributive (i.e. congruence-distributive semilattices) were characterized by Dean & Oehmke [1964] (see also Hamilton [1974]). Varlet [1965] proved that modularity of $\text{Con}(S)$ is equivalent to distributivity for any semilattice S. Papert [1964] provided another characterization of distributivity for the congruence lattice of a semilattice. Other specific properties of the congruence lattice of a semilattice have also been characterized, for example, the property of being relatively pseudocomplemented or Boolean. We refer to Mitsch [1983] for further details.

Let us now turn to congruence lattices of pseudocomplemented semilattices. The investigation of this topic was initiated by Sankappanavar [1979], [1982]. Let S be a meet semilattice S with least element 0 and with a unary operation g such that for $a \in S$, $g(a)$ is the pseudocomplement of a in S. Let us view this semilattice as an algebra $\mathbf{S} = \langle S; \wedge, g, 0 \rangle$. We shall denote by PCS the class of all pseudocomplemented semilattices, whereas PCS will be used as shorthand for "pseudocomplemented semilattice." As done before, we will also identify an algebra with its universe (i.e., S with S) when there is no danger of confusion; for example, the PCS $\langle S; \wedge, g, 0 \rangle$ is called "the PCS S."

Balbes & Horn [1970] have shown that the class PCS is indeed a variety whose defining identities, in addition to the three semilattice identities (commutativity, idempotency, and associativity), are:

(i) $x \wedge 0 = 0$,

(ii) $x \wedge g(x \wedge y) = x \wedge g(y)$,

(iii) $x \wedge g(0) = x$,

(iv) $g(g(0)) = 0$,

where $g(0)$ is the greatest element of the PCS.

From a characterization given by Frink [1941] for Boolean algebras (by a set of identities that are of the same similarity type as that of PCS and that include the defining identities of PCS) it follows that the class of Boolean algebras, BA, is a subvariety of PCS. Jones [1972] proved that BA viewed in this way is the only nontrivial proper subvariety of PCS.

By $\mathrm{Con}_p(S)$ we denote the lattice of all congruence relations on a PCS viewed as an algebra $\langle S; \wedge, g, 0 \rangle$. Examples of pseudocomplemented (meet) semilattices (viewed in the above-described way as algebras) and their corresponding congruence lattices are given in Figure 9.2(a), (b) and Figure 9.3(a), (b).

Congruence lattices of PCSs show a behavior different from that of congruence lattices of semilattices. For example, $\mathrm{Con}_p(P_5)$ [cf. Figure 9.2(b)] is semimodular but not coatomistic, while $\mathrm{Con}_p(P_6)$ [cf. Figure 9.3(b)] is neither semimodular nor coatomistic. However, congruence lattices of PCSs also have a number of interesting properties, which were investigated in detail by Sankappanavar [1979],

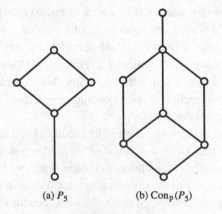

(a) P_5 (b) $\mathrm{Con}_p(P_5)$

Figure 9.2

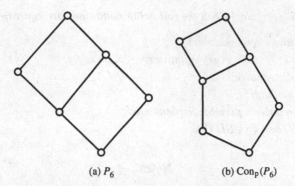

(a) P_6 (b) $\mathrm{Con_p}(P_6)$

Figure 9.3

[1982]. For example, $\mathrm{Con_p}(S)$ is algebraic, and it is a complete sublattice of $\mathrm{Con}(S)$. Hence $\mathrm{Con_p}(S) \subseteq \mathrm{Con}(S)$.

It is well known that the congruence lattice of a Boolean algebra determines that Boolean algebra up to isomorphism. However, nonisomorphic PCSs may have isomorphic congruence lattices (cf. Sankappanavar [1979]).

Let us also note that $\mathrm{Con_p}(S)$ inherits some further properties from $\mathrm{Con}(S)$. For example, it follows from Papert [1964] that $\mathrm{Con_p}(S)$ also has the property that each interval is pseudocomplemented. This implies again that M_3 cannot be embedded as a sublattice into the congruence lattice of a PCS (cf. Varlet [1965]).

We have stated above that $\mathrm{Con_p}(S)$ is not coatomistic in general. On the other hand, it can be shown that $\mathrm{Con_p}(S)$ is coatomic, that is, every element $(\neq 1)$ is contained in some coatom.

We have seen that $\mathrm{Con_p}(P_5)$ [Figure 9.2(b)] is semimodular, while $\mathrm{Con_p}(P_6)$ [Figure 9.3(b)] is not. It is easily verified that all PCSs having at most five elements have semimodular congruence lattices. The special role of P_5 [Figure 9.2(a)] and P_6 [Figure 9.3(a)] is also seen from the following two results of Sankappanavar [1982] characterizing upper and lower semimodularity for PCSs:

Theorem 9.1.5   *For any PCS S the following conditions are equivalent:*

(1) *$\mathrm{Con}(S)$ satisfies Birkhoff's condition* (Bi);
(2) *for all $x, y \in S$, $x < g(g(y))$ implies $g(x) = g(y)$ or $x \leq y$;*
(3) *$\mathrm{Con}(S)$ is upper semimodular, that is, S is congruence-semimodular* (CSM *for short*);
(4) *$P_6 \notin SH(S)$.*

For the definition of the operators S and H we refer to Section 1.2. In fact, Sankappanavar [1982] gives a further equivalent condition and shows that if $\mathrm{Con}(S)$ is of finite length, then each of the above conditions is equivalent to "$\mathrm{Con}(S)$ is a graded lattice." By condition (3) in the preceding result the variety PCS is CSM.

Theorem 9.1.6   *For any* PCS *S the following conditions are equivalent:*

(1) $\mathrm{Con}(S)$ *is lower semimodular;*

(2) *for all* $x, y \in S$, $x < g(g(y))$ *implies* $y \leq x$ *or* $x \leq y$;

(3) $\mathrm{Con}(S)$ *is distributive;*

(4) $\mathrm{Con}(S)$ *is modular;*

(5) $\mathrm{Con}(S)$ *is relatively pseudocomplemented;*

(6) $P_6 \notin SH\ (S)$ *and* $P_5 \notin SH\ (S)$.

Notes

It was mentioned that the class of Boolean algebras is the only nontrivial subvariety of PCS. The class of congruence lattices of Boolean algebras, being distributive, satisfies every nontrivial lattice identity. The class of congruence lattices of PCSs, on the other hand, does not satisfy any nontrivial lattice identity. This result can be derived from the corresponding result for congruence lattices of semilattices which was proved by Freese & Nation [1973], who solved thereby Problem 6 of Schmidt [1969].

Petrich [1987] raised the problem of characterizing those semilattices Y whose congruence lattices $\mathrm{Con}(Y)$ are atomistic, that is, every congruence is the join of the minimal proper congruences it contains. For example, locally finite trees have this property. Auinger [1990] shows that the converse also holds: for a semilattice Y it is necessary for it to be a locally finite tree in order for $\mathrm{Con}(Y)$ to be atomistic.

Congruence lattices of semilattices were also investigated by Burris & Sankappanavar [1975].

Exercises

1. Show that the map α as defined in the proof of Lemma 9.1.2 is indeed a homomorphism.

2. Give an example of a semilattice S with $\mathrm{Con}_p(S) \subset \mathrm{Con}(S)$ (Sankappanavar [1979]).

3. Give an example of a lattice T and a semilattice S with $\mathrm{Con}_p(T) = \mathrm{Con}_p(S)$ (Sankappanavar [1979]).

References

Auinger, K. [1990] Atomistic congruence lattices of semilattices, Semigroup Forum 40, 115–117.

Balbes, R. and A. Horn [1970] Stone lattices, Duke Math. J. 37, 537–546.

Burris, S. and H. P. Sankappanavar [1975] Lattice-theoretic decision problems in universal algebra, Algebra Universalis 5, 163–177.

Burris, S. and H. P. Sankappanavar [1981] *A Course in Universal Algebra*, Springer-Verlag, New York.

Dean, R. A. and R. H. Oehmke [1964] Idempotent semigroups with distributive right congruence lattices, Pacific J. Math. 14, 1187–1209.

Freese, R. and J. B. Nation [1973] Congruence lattices of semilattices, Pacific J. Math. 49, 51–58.

Frink, O. [1941] Representation of Boolean algebras, Bull. Amer. Math. Soc. 47, 755–756.

Funayama, N. and T. Nakayama [1942] On the distributivity of a lattice of lattice congruences, Proc. Imp. Acad. Tokyo 18, 553–554.

Hall, T. E. [1971] On the lattice of congruences on a semilattice, J. Austral. Math. Soc. 12, 456–460.

Hamilton, H. B. [1974] Semilattices whose structure lattice is distributive, Semigroup Forum 8, 245–254.

Jones, J. T. [1972] Pseudocomplemented semilattices, Ph.D. Dissertation, UCLA.

Jones, P. R. [1983b] On congruence lattices of regular semigroups, J. Algebra 82, 18–39.

Mitsch, H. [1983] Semigroups and their lattices of congruences, Semigroup Forum 26, 1–63.

Ore, O. [1942] Theory of equivalence relations, Duke Math. J. 9, 573–627.

Papert, D. [1964] Congruence relations in semi-lattices, J. London Math. Soc. 39, 723–729.

Petrich, M. [1987] Congruences of completely regular semigroups, Working paper II, University of Vienna, June 1987.

Sankappanavar, H. P. [1974] A study of congruence lattices of pseudocomplemented semilattices, Ph.D. thesis, Univ. of Waterloo.

Sankappanavar, H. P. [1979] Congruence lattices of pseudocomplemented semilattices, Algebra Universalis 9, 304–316.

Sankappanavar, H. P. [1982] Congruence-semimodular and congruence distributive pseudocomplemented semilattices, Algebra Universalis 14, 68–81.

Sankappanavar, H. P. [1985] Congruence-distributivity and join-irreducible congruences on a semilattice, Math. Japonica 30, 495–502.

Schmidt, E. T. [1969] *Kongruenzrelationen algebraischer Strukturen*, VEB Deutscher Verlag der Wissenschaften, Berlin.

Varlet, J. C. [1965] Congruence dans les demi-lattis, Bull. Soc. Roy. Liège 34, 231–240.

Žitomirskii, G. I. [1986] Some remarks on properties of lattices which are generated by dual atoms (Russian), Uporyad. Množestva i Reshotki 9, 16–18.

9.2 Semigroups

Summary. The varieties Sets and Semilattices are congruence-semimodular, but not congruence-modular. Semilattices are special semigroups. It can be shown that congruence modularity has little relevance in the variety of semigroups, since any congruence-modular variety of semigroups is in fact a variety of groups (Freese & Nation [1973]). This suggests that congruence semimodularity is an important concept for semigroups and that results obtained for semilattices might be extended to other varieties of semigroups.

Let us first recall that the varieties of semigroups form a lattice under inclusion, which we denote by L_{sg} (the *lattice of semigroup varieties*). The lattice L_{sg} has been investigated by many authors (see Evans [1971] for a detailed survey). The least element of L_{sg} is the variety of all one-element semigroups, called the trivial variety and denoted by T; its greatest element is the variety of all semigroups. For the following result see Evans [1971].

Theorem 9.2.1   *The atoms of L_{sg} are the following varieties:*

the variety RZ *of the right zero semigroups defined by the identity $xy = y$;*
the variety LZ *of the left zero semigroups defined by the identity $xy = x$;*
the variety N *of the null semigroups defined by the identity $xy = zt$;*
the variety Semilattices, *that is, the semigroups defined by $xy = yx$, $x^2 = x$;*
the variety G_p *of the groups of exponent p (for each prime p) defined by $x^{p+1} = x$, $x^p = y^p$.*

The following properties (i)–(iii) are significant properties of the atoms of L_{sg} (cf. Jones [1988], Theorem 1.2).

(i) Each atom of L_{sg} is congruence-semimodular (CSM for short):

Semigroup varieties consisting of groups are congruence modular and thus CSM; in Section 9.1 we have indicated that the variety Semilattices is CSM (which was proved by Hall [1971]). On any right zero, left zero, or null semigroup every equivalence relation is a congruence and thus $\mathrm{Con}(A) = \Pi(A)$ is semimodular.

(ii) The only atoms of L_{sg} whose congruence lattices satisfy a nontrivial lattice identity are the group varieties.

The class of partition lattices satisfies no nontrivial lattice identity (Whitman [1946]), eliminating RZ, LZ, and N as in (i). For Semilattices this was proved by Freese & Nation [1973].

(iii) A variety of semigroups is congruence modular if and only if it consists of groups.

Semigroup varieties consisting of groups are congruence modular. Let now K be a congruence modular variety of semigroups. Then K contains none of RZ, LZ, N, and Semilattices, from which it follows that K consists entirely of groups.

In the literature there are several results on congruence semimodularity for special types of semigroups (different from semilattices). Let us mention some of these results.

Lallement [1967] proved that every completely simple semigroup is CSM. A proof also appears in Howie [1976], where further details on completely simple semigroups may be found. The class CS of such semigroups is not a variety, but contains the varieties RZ, LZ, N, and the G_p's. A semigroup S with zero is *nil* if for each x in S there exists $n > 0$ such that $x^n = 0$. T. E. Hall (unpublished) has shown that such semigroups are CSM (see Jones [1988]). A nil variety satisfies an identity of the form $x^n = 0$ and is therefore CSM. A semigroup S with zero is *nilpotent* if $S^n = \{0\}$ for some $n > 0$. Nilpotent varieties of semigroups satisfy an identity of the form $x_1 \cdots x_n = 0$ and are thus nil.

Let S be a semigroup, and K be an ideal of S. Then S is an *ideal extension* of K by a nil semigroup Q if $S/K \cong Q$. If Q is nilpotent, then S is an ideal extension of K by a nilpotent semigroup Q. We say briefly that S is an ideal nil extension of K and an ideal nilpotent extension of K, respectively. Trueman [1983] has shown that any finite semigroup that is an ideal extension of a group by a nil semigroup is CSM. Jones [1983b] proved that any finite combinatorial inverse semigroup is CSM.

Jones [1988] unified and generalized these results from a varietal point of view, and in doing so he demonstrated that significant varieties of semigroups are congruence-semimodular. Let us first have a brief look at his results in the regular case and then in the irregular case.

An equation $\sigma = \tau$ is *regular* if a variable appears in σ if and only if it appears in τ. A variety of algebras is *regular* if only regular equations hold in it. A variety of algebras is *irregular* if it is not regular. It is clear that the variety of semilattices is regular and that a variety of semigroups is regular if and only if it contains the variety of semilattices. Jones [1988], Theorem 2.1, proved the following characterization.

Theorem 9.2.2 *A regular variety V of semigroups is congruence semimodular if and only if it contains no nontrivial completely simple semigroups, that is $V \cap CS = T$, the trivial variety – equivalently, if it satisfies identities*

(∗) $x^{n+1} = x^n$, $(x^n y^n)^n = (y^n x^n)^n$ *for some positive integer n.*

We note that satisfying the preceding equations is a sufficient condition for any variety of semigroups to be congruence-semimodular. It is however clear that the condition cannot be necessary in general. In fact, no nontrivial group will satisfy $x^{n+1} = x^n$.

Inverse semigroups can be regarded as algebras with a binary and a unary operation (the inverse) satisfying the following identities (see Petrich [1984]):

$$(xy)z = x(yz), \qquad xx^{-1}x = x, \qquad (x^{-1})^{-1} = x, \qquad (xy)^{-1} = y^{-1}x^{-1},$$
$$xx^{-1}yy^{-1} = yy^{-1}xx^{-1}.$$

Hence any variety of inverse semigroups is regular.

Let us recall that in any semigroup S, the set E_S of idempotents of S can be given a partial order by setting $e \leq f \Leftrightarrow ef = fe = e$. Inverse semigroups can be alternatively characterized as regular semigroups in which the idempotents commute. If S is inverse, since idempotents commute, it is easily seen that the above ordering is a semilattice ordering, that is, E_S is a meet semilattice where the meet is the semigroup operation and thus E_S is a regular semigroup.

An inverse semigroup is *combinatorial* if Green's relation H is trivial on this semigroup. For a proof of the following characterization of an inverse semigroup to be combinatorial see Petrich [1984], Lemma 1.3, p. 519.

Lemma 9.2.3   *For a variety* $\vee$ *of inverse semigroups the following conditions are equivalent:*

(i) $\vee$ *is combinatorial;*
(ii) *there is a natural n such that the identity $x^{n+1} = x^n$ holds in* $\vee$;
(iii) *no $S \in \vee$ has nontrivial subgroups.*

Thus, combining Lemma 9.2.3 with the preceding definition of inverse semigroups, we may say that a variety of combinatorially inverse semigroups is any variety of type $\langle 2, 1 \rangle$ that for some n satisfies the equations

$$(xy)z = x(yz), \qquad xx^{-1}x = x, \qquad (x^{-1})^{-1} = x, \qquad (xy)^{-1} = y^{-1}x^{-1},$$

$$xx^{-1}yy^{-1} = yy^{-1}xx^{-1}, \quad \text{and} \quad x^{n+1} = x^n.$$

Jones [1988] proved that any nontrivial variety of combinatorially inverse semigroups is congruence-semimodular. For $n = 1$ the variety defined becomes equivalent to the variety of semilattices. In fact, Jones [1988], Theorem 4.1, proved

Theorem 9.2.4   *A variety* $\vee$ *of inverse semigroups is congruence semimodular if and only if either* (1) $\vee$ *is a variety of groups or* (2) $\vee$ *is a combinatorial variety.*

There have also been a number of investigations of the congruence semimodularity of irregular varieties of semigroups. In order to give some of these results we have to formulate the above-mentioned ideal extension in a more general framework.

Let S, T, K be semigroups such that T has an absorbing element 0 and $T \cap K = \emptyset$. We say that S is an ideal extension of K by T if K is an ideal of S and the Rees factor semigroup S/K is isomorphic with T. If K, C, D are classes of semigroups, we say that K is C-by-D if any $S \in$ K is the ideal extension of a member of C by a member of D.

Chrislock [1969] proved that a semigroup S satisfies an irregular equation if and only if it is completely simple-by-nil, that is, it has a completely simple minimal ideal K such that S/K is nil. Jones [1988] proved that any completely simple-by-nilpotent semigroup has semimodular congruences and thus any completely simple-by-nilpotent variety of semigroups is congruence semimodular. It seems to be an open question whether every irregular variety of semigroups is congruence-semimodular.

Notes

A regular ω-semigroup S is a regular semigroup whose set of idempotents E_S forms an ω-chain $e_0 > e_1 > \cdots > e_n > \cdots$ under the natural partial order defined on E_S by the rule $f \geq e$ if and only if $ef = e = fe$. The congruences on this class of inverse semigroups have been described by several authors (see e.g. Baird [1972 a, b], Petrich [1979]), but this description has only recently been used in the investigation of the congruence lattice of this class of semigroups. Scheiblich [1970] characterized bisimple ω-semigroups whose lattice of congruences satisfies Birkhoff's condition (Bi) (called by some authors the "double covering property"). Bonzini & Cherubini [1990a] give a complete characterization of regular ω-semigroups whose congruence lattices satisfy (Bi). Bonzini & Cherubini [1990b] survey some results concerning regular ω-semigroups whose congruence lattice satisfies certain conditions [such as modularity, M-symmetry, upper semimodularity, and Birkhoff's condition (Bi)].

Eberhart & Williams [1978] give necessary and sufficient conditions for the lattice of congruences on a band, normal band, band of groups, or normal band of groups to satisfy (Bi) (called semimodular here). They also give sufficient conditions for a general lattice to satisfy (Bi) and for a lattice of congruences on any semigroup to satisfy (Bi).

References

Agliano, P. [1990] Congruence semimodularity and identities, Algebra Universalis 27, 600–601.

Agliano, P. [1991] The one-block property in varieties of semigroups, Semigroup Forum 42, 253–264.

Agliano, P. [1992] On combinatorial inverse semigroups, Rend. Sem. Mat. Torino 50, 255–275.

Baird, G. R. [1972a] On a sublattice of the lattice of congruences on a simple regular ω-semigroup, J. Austral. Math. Soc. 13, 461–471.

Baird, G. R. [1972b] Congruences on simple regular ω-semigroups, J. Austral. Math. Soc. 14, 155–167.

Bonzini, C. and A. Cherubini [1990a] Semimodularity of the congruence lattice on regular ω-semigroups, Monatsh. Math. 109, 205–219.

Bonzini, C. and A. Cherubini [1990b] Conditions similar to modularity on the congruence lattice of a regular ω-semigroup, in: *Proc. Internat. Symp. on Semigroup Theory and Related Fields*, ed. by M. Yamada and H. Tominaga, Ritsumeikan University, Kyoto, pp. 41–52.

Chrislock [1969] A certain class of identities on semigroups, Proc. Amer. Math. Soc. 21, 189–190.

Eberhart, C. and W. Williams [1978] Semimodularity in the lattice of congruences, J. Algebra 52, 75–87.

Evans, T. [1971] The lattice of semigroup varieties, Semigroup Forum 2, 1–43.

Freese, R. and J. B. Nation [1973] Congruence lattices of semilattices, Pacific J. Math. 49, 51–58.

Hall, T. E. [1971] On the lattice of congruences on a semilattice, J. Austral. Math. Soc. 12, 456–460.

Howie, J. M . [1976] *An Introduction to Semigroup Theory*, Academic Press, New York.

Jones, P. R. [1983b] On congruence lattices of regular semigroups, J. Algebra 82, 18–39.

Jones, P. R. [1988] Congruence semimodular varieties of semigroups, in: *Lecture Notes in Mathematics 1320*, Springer-Verlag, Berlin, pp. 162–171.

Lallement, G. [1967] Demi-groupes réguliers, Ann. Mat. Pura Appl. 77, 47–129.

Ore, O. [1942] Theory of equivalence relations, Duke Math. J. 9, 573–627.

Petrich, M. [1979] Congruences on simple ω-semigroups, Glasgow Math. J. 20, 87–101.

Petrich, M. [1984] *Inverse Semigroups*, Wiley.

Scheiblich, H. E. [1970] Semimodularity and bisimple ω-semigroups, Proc. Edinburgh Math. Soc. 17, 79–81.

Trueman, D. C. [1983] The lattice of congruences on direct products of cyclic semigroups and certain other semigroups, Proc. Roy. Soc. Edinburgh 95, 203–214.

Whitman, P. M. [1946] Lattices, equivalence relations, and subgroups, Bull. Amer. Math. Soc. 52, 507–522.

9.3 Algebras

Summary. The examples given in the two preceding sections suggest that it is worthwhile to investigate semimodularity as a natural congruence condition for classes of algebras. A systematic investigation of the general theory of congruence semimodular varieties of algebras was initiated by Agliano [1990], Kearnes [1991], and Agliano & Kearnes [1994a,b]. Concerning lower semimodularity, it is known that a variety whose 2-generated free algebra is finite consists of algebras with lower semimodular congruence lattices if and only if it is congruence-modular (Kearnes [1991]). We shall here briefly outline some of these results.

We shall have a look at the following two questions:

(i) How much of the structure involved in congruence-modular varieties exists for congruence-semimodular varieties?

(ii) How much more diversity is permitted ?

Agliano & Kearnes [1994a] is devoted to the first question, and Agliano & Kearnes [1994b] to the second one.

Since the variety of sets is congruence-semimodular (CSM for short) (Ore [1942]), the results of Taylor [1973] yield that congruence semimodularity implies no nontrivial Mal'cev condition. This means that some of the most useful techniques of universal algebra are not applicable to CSM varieties. Correspondingly, there seems to be a scarcity of results concerning CSM varieties that are not congruence-modular.

The results that are known include the following facts, which were mentioned in Section 9.1 and Section 9.2:

besides the variety of sets, the variety of semilattices is CSM (Hall [1971]);

the CSM varieties of regular semigroups have been characterized (Jones [1988]);

the CSM varieties of irregular semigroups have been "almost" characterized (Jones [1988]).

There are a number of lattice-theoretic results (see e.g. the chapters of Crawley & Dilworth [1973] on decomposition theory) that seem to have been proved in particular for algebras with semimodular congruence lattices.

Semimodularity is preserved by subdirect products and by interval sublattices. For congruence lattices of algebras we can say something more (in what follows

0_A and 1_A denote the least and greatest congruence relation, respectively, of the given algebra **A**).

Lemma 9.3.1 *Let* K *be a class of algebras closed under homomorphic images. Then* K *is CSM if and only if for all* **A** $\in$ K, *for all* $\theta \in \text{Con}(\mathbf{A})$ *with* $\theta \succ 0_A$, *we have* $\theta \vee \varphi \succ \varphi$ *for all* $\varphi \in \text{Con}(\mathbf{A})$ *with* $\theta \not\leq \varphi$.

Proof. The necessity is obvious. Let now K be a class of algebras closed under homomorphic images, let **A** $\in$ K, and let $\theta, \varphi \in \text{Con}(\mathbf{A})$ with $\theta \succ \theta \wedge \varphi$. Then $\theta/(\theta \wedge \varphi)$ is an atom of $\text{Con}(\mathbf{A}/(\theta \wedge \varphi))$, and, since K is closed under homomorphic images, $\mathbf{A}/(\theta \wedge \varphi) \in$ K. Hence by the hypothesis we get $(\theta \vee \varphi)/(\theta \wedge \varphi) = (\theta/(\theta \wedge \varphi)) \vee (\varphi/(\theta \wedge \varphi)) \succ \varphi/(\theta \wedge \varphi)$. Therefore $\theta \vee \varphi \succ \varphi$ in $\text{Con}(\mathbf{A})$, that is, $\text{Con}(\mathbf{A})$ is semimodular. ∎

Now if K is a CSM class, then so is $H(\mathsf{K})$. In fact if **B** $\in H(\mathsf{K})$, then **B** $\cong \mathbf{A}/\theta$ for some **A** $\in$ K and $\theta \in \text{Con}(\mathbf{A})$ and, by the second isomorphism theorem (see McKenzie et al. [1987], Theorem 4.10, p. 149), $\text{Con}(\mathbf{B})$ is isomorphic with the interval sublattice $[\theta, 1_A]$ of $\text{Con}(\mathbf{A})$. Hence, by Lemma 9.3.1, $\text{Con}(\mathbf{B})$ is semimodular. Now let V be a variety. It is well known that for any algebra **A** $\in$ V we have **A** $\in \mathbf{F}_V(X)$ for a large enough X, where $\mathbf{F}_V(X)$ denotes the *relatively free algebra on* V *generated by* X. Thus a variety is congruence semimodular if and only if each of its free algebras is.

It follows that a variety is CSM if and only if its free algebras have semimodular congruence lattices. This is a property shared by many other well-known congruence conditions, such as congruence modularity, congruence distributivity, and congruence permutability. For the three congruence conditions just mentioned it is even true that a variety V satisfies the congruence condition if and only if $\mathbf{F}_V(4)$ does if and only if the subvariety generated by $\mathbf{F}_V(2)$ does. We refer to Mal'cev [1954] for congruence permutability, Jónsson [1967] for congruence distributivity, and Day [1969] and Kearnes [1991] for congruence modularity. The results of Kearnes [1991] will be considered below in more detail.

Agliano & Kearnes [1994a], Example 7.1, show that there do not exist finite m and n such that a variety V is CSM if and only if $\mathbf{F}_V(m)$ is, or that V is CSM if and only if the subvariety generated by $\mathbf{F}_V(n)$ is.

As already remarked, a lattice satisfying Birkhoff's condition (Bi) is occasionally called weakly semimodular, since semimodularity implies (Bi) but not conversely (see Section 1.7). We know that there are algebraic lattices that are weakly semimodular but not semimodular. Hence there are algebras that are congruence weakly semimodular [i.e., the corresponding congruence lattices satisfy (Bi)] but not congruence-semimodular. An algebra whose congruence lattice is weakly semimodular is also briefly called a CWSM algebra. For varieties it is still unknown whether CWSM implies CSM (Agliano & Kearnes [1994a], Problem

1): Is there a congruence weakly semimodular variety that is not congruence semimodular?

Agliano & Kearnes [1994 a], Section 7, solved this problem negatively for locally finite varieties with the *congruence extension property* (for which see Grätzer [1978], p. 120). They proved that a variety with the congruence extension property (CEP, for short) is CSM if and only if its finitely generated free algebras are. In fact, they showed that if a variety V has the CEP, then V is congruence-semimodular if and only if $F_V(5)$ is and that V is congruence weakly semimodular if and only if $F_V(4)$ is (cf. Agliano & Kearnes [1994a], Theorem 7.3).

Agliano & Kearnes [1994b] also provided a negative solution to the above problem for regular varieties (cf. Theorem 9.3.4 below).

Chapter 8 of Hobby & McKenzie [1988] contains characterizations of well-known congruence conditions for locally finite varieties in terms of the *type set* of the variety and the structure of the minimal sets. For example, a locally finite variety is congruence-modular if and only if its type set is a subset of $\{2, 3, 4\}$ and all $\langle \alpha, \beta \rangle$-minimal sets have empty tail. No such characterization is known for locally finite CSM varieties. We have the following type-omitting theorem for CSM varieties (cf. Agliano & Kearnes [1994a], Theorem 2.6).

Theorem 9.3.2   *If a variety* V *is CSM, then* $1 \notin \mathrm{typ}\{V\}$.

This is the only type-omitting theorem that holds for CSM varieties, since the varieties of sets, vector spaces, Boolean algebras, lattices, and semilattices are all CSM. However, more can be said about the position of the different type labels in the congruence lattice of a finite algebra whose congruence lattice is upper or lower semimodular (cf. Theorem 9.3.15 below).

In Section 9.2 we noted that Jones [1988] characterized the regular varieties of semigroups that are congruence semimodular, and he partially solved the problem of characterizing the irregular CSM varieties of semigroups.

Jones posed the problem of characterizing regular CSM varieties. In order to formulate the answer given by Agliano & Kearnes, we still need the one-block property, a certain congruence property (common to the congruence lattices of semilattices and sets), which is strong enough to force semimodular congruences. An algebra **A** is said to have the *one-block property* (OBP, for short) if any atom $\theta \in$ Con(**A**) has exactly one nontrivial congruence class. This property was introduced by Agliano [1991]. Agliano & Kearnes [1994b], Theorem 2.2, proved

Theorem 9.3.3   *Let* K *be a class of similar algebras closed under homomorphic images, and consider the following conditions:*

  (i)   K *has the* OBP;
  (ii) *for any algebra* **A** $\in$ K, *any atom* $\alpha \in$ Con(**A**), *and any* $\beta \in$ Con(**A**) *we have*
        $\beta \circ \alpha \circ \beta = \alpha \vee \beta$;

(iii) K *is* CSM.

Then (i) $\Rightarrow$ (ii) $\Rightarrow$ (iii).

Proof. (i) $\Rightarrow$ (ii): Choose $A \in K$, and suppose that $\alpha, \beta \in \text{Con}(A)$ and $\alpha \succ 0_A$. If $(a, b) \in \alpha \vee \beta$, then we can find a chain of elements $a = x_0, \ldots, x_n = b$ where $(x_i, x_{i+1}) \in \alpha \cup \beta$ for each $i < n$. If this chain has minimal length, then the fact that α has only one trivial block implies that at most one nontrivial α-link is involved in this chain. Hence $\alpha \vee \beta = \beta \circ \alpha \circ \beta$.

(ii) $\Rightarrow$ (iii): Suppose that congruence semimodularity fails in K. Then there exists an algebra $A \in K$, an $\alpha \succ 0_A$, $a \ \beta \not\geq \alpha$, and a $\gamma \in \text{Con}(A)$ with $\alpha \vee \beta > \gamma > \beta$. Take $(a, b) \in \gamma - \beta$. By hypothesis we have $(a, b) \in \alpha \vee \beta = \beta \circ \alpha \circ \beta$. Hence there are $u, v \in A$ with $\alpha \beta u \alpha v \beta b$. This means $u \beta a \gamma b \beta v$. But $\beta < \gamma$ and hence $(u, v) \in \alpha \wedge \gamma = 0_A$. Therefore $a \beta u 0_A v \beta b$ and $(a, b) \in \beta$ which is a contradiction. ∎

We leave it as an exercise to show that neither implication in the preceding result can be reversed.

For the next result see Agliano & Kearnes [1994b], Theorem 2.3.

Theorem 9.3.4 *If* V *is a nontrivial variety that has the* OBP, *then* V *is not congruence-modular. In fact, for every nontrivial* $A \in V$, *either* S_7 *or the "kite" (cf. Figure 9.4) occurs as a sublattice of* $\text{Con}(B)$ *for some* $B \leq A^2$.

Proof. Assume that A is a nontrivial member of V, $\theta \in \text{Con}(A)$ is compact, and $\delta \in \text{Con}(A)$ is a lower cover of θ. Let $B \leq A^2$ be the subalgebra whose universe is θ. For $\alpha \in \text{Con}(A)$ let α_0 denote the congruence on B consisting of all $((a, b),(c, d)) \in B^2$ such that $(a, c) \in \alpha$, and let α_1 denote the congruence on B consisting of all $((a, b), (c, d)) \in B^2$ such that $(b, d) \in \alpha$. Let α^* denote $\alpha_0 \wedge \alpha_1$.

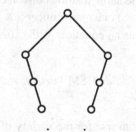

Figure 9.4

From the definition of **B** we have $\theta_0 = \theta_1 = \theta^*$ and $\delta_0, \delta_1 \prec \theta^*$. Now $\delta_0 \prec \theta^*$, but $\delta^* = \delta_0 \wedge \delta_1$ is not a lower cover of $\theta^* \wedge \delta_1 = \delta_1$, since $\delta_1/\delta^* \in \text{Con}(\mathbf{B}/\delta^*)$ has more than one nontrivial equivalence class. This means that $\text{Con}(\mathbf{B})$ is not even lower semimodular, and the first statement is proved.

Since δ^* is not a lower cover of δ_1, we can find δ_1' such that $\delta^* < \delta_1' < \delta_1$. Let $\delta_0' = \{((a, b), (c, d)) \in B^2 : ((b, a), (c, d)) \in \delta_1'\}$. If $\delta_0' \vee \delta_1' < \theta^*$, then $\delta_0' \vee \delta_1', \delta_0$, and δ_1 generate a sublattice of $\text{Con}(\mathbf{B})$ isomorphic to S_7. Otherwise δ_0', δ_1', and δ_0 generate a pentagon in $\text{Con}(\mathbf{B})$. By semimodularity, we cannot have $\delta^* \prec \delta_1'$, and hence we can find δ_1'' such that $\delta^* < \delta_1'' < \delta_1'$. Construct δ_0'' from δ_1'' in the same way that we constructed δ_0' from δ_1'. Either $\delta_0'' \vee \delta_1'', \delta_0$, and δ_1 generate a copy of S_7 or else we can find a δ_1''' and δ_0''' with $\delta^* < \delta_1''' < \delta_1'' < \delta_1'$ and $\delta^* < \delta_0''' < \delta_0'' < \delta_0'$, as in our earlier argument. If this process fails to produce a copy of S_7 as a sublattice of $\text{Con}(\mathbf{B})$, then we end up constructing a copy of the kite. ∎

Agliano & Kearnes [1994b], Theorem 3.3, proved

Theorem 9.3.5   *For a regular variety* V, *the following conditions are equivalent:*

(1) V *is* CSM;
(2) V *is* CWSM;
(3) *Every subdirectly irreducible algebra in* V *has the* OBP *and a zero element for the monolith.*

(We recall that an algebra is subdirectly irreducible if its congruence lattice has exactly one atom, which is called the *monolith*.)

From this theorem one can derive several properties of regular congruence semimodular varieties (see Agliano & Kearnes [1994b], Corollaries 3.4, 3.5, 3.7).

Corollary 9.3.6   *If* V *is a regular* CSM *variety, then* $\text{typ}\{V\} \subseteq \{\mathbf{0}, \mathbf{5}\}$.

We leave it as an exercise to show that the converse of Corollary 9.3.6 fails.

For the next property we need one more concept: A variety of algebras is called *strongly irregular* if it satisfies an equation of the form $t(x, y) = x$, where y is a free variable of t.

Corollary 9.3.7   *If* V *is a regular* CSM *variety, then* V *contains no nontrivial strongly irregular subvarieties.*

Jones [1988] proved the converse for the variety of semigroups: Every regular variety of semigroups that contains no nontrivial strongly irregular subvariety is CSM. In general, however, a regular variety that has no strongly irregular subvarieties need not be CSM. In the next example (due to Nation), let V be the variety of groupoids axiomatized by the following equations: $xy = yx$, $xx = x$, $x(xy) = xy$. Then V is a regular variety with no nontrivial strongly irregular subvarieties. Let now **A** be the algebra whose universe is $\{a, b, c\}$ and whose operation is defined by

	a	b	c
a	a	a	b
b	a	b	b
c	c	b	c

Then $\mathbf{A} \in V$, and **A** is simple of type **3**. Hence V is not CSM, by Corollary 9.3.6.

Corollary 9.3.8 *If V is a regular variety, then V is CSM if and only if for all $\mathbf{A} \in V$ the lattice $\mathrm{Con}(\mathbf{A})$ has a dimension function.*

Let us now turn to the question of congruence semimodularity and (nontrivial) congruence identities. Let V be a variety. If an equation ε in the language of lattices holds in $\mathrm{Con}(\mathbf{A})$ for all $\mathbf{A} \in V$, then we say that ε is a *congruence identity* of V. Given a variety V of algebras, among the interesting questions we can ask about the members of V is the following: Does there exist a lattice identity ε such that for each algebra $\mathbf{A} \in V$, the congruence lattice $\mathrm{Con}(\mathbf{A})$ satisfies ε?

Around 1970 it was conjectured that any nontrivial lattice identity holding as a congruence identity for a variety would imply congruence modularity. This conjecture turned out to be false (see Polin [1976], Day & Freese [1980]), that is, there are nonmodular varieties having nontrivial congruence identities.

On the other hand, there are nonmodular varieties satisfying no nontrivial congruence identity. As already remarked, for the variety Sets (of sets) this was shown by Whitman [1946]. For the variety Semilattices (of semilattices) this is the main result of Freese & Nation [1973]. Agliano [1990] investigated the question whether there are nonmodular CSM varieties having nontrivial congruence identities and showed that the answer is negative in the locally finite case. (We recall that the varieties Sets and Semilattices are locally finite).

In deriving his result Agliano used on the one hand the characterization of modularity within the class of semimodular lattices of finite length: Let **L** be a finite semimodular lattice, and suppose that $S_7 \notin S(\mathbf{L})$. Then **L** is a modular lattice (cf. Theorem 3.1.10). On the other hand, he used the following result due to Hobby & McKenzie [1988] (cf. Remark 9.9, p. 138):

Theorem 9.3.9 *Let V be a locally finite variety, and let ε be a nontrivial lattice identity. If ε is a congruence identity of V, then for any finite $\mathbf{A} \in V$ we have $S_7 \notin S(\mathrm{Con}(\mathbf{A}))$.*

Both results imply

Theorem 9.3.10 *Let V be a locally finite CSM variety. Then either V has modular congruence lattices or it does not satisfy any nontrivial congruence identity.*

Proof.   Suppose that ε is a congruence identity of V for some nontrivial lattice identity ε. Since V is locally finite, the free algebra $F_V(x, y, z, u) = F$ is finite. Hence, by Theorem 9.3.9, $S_7 \notin S(\text{Con}(F))$. But $\text{Con}(F)$ is finite and semimodular, so by Theorem 3.1.10 it is modular. Hence the whole variety has modular congruences.   ∎

Thus we have seen that upper semimodularity arises in a natural way as a congruence condition. Moreover there are many examples of congruence upper semimodular varieties.

What about examples of nonmodular varieties that are congruence lower semimodular? Kearnes [1991] showed that lower semimodularity does not seem to be an interesting congruence condition. We close this section by giving some of Kearnes's results, which rely on tame congruence theory (see Hobby & McKenzie [1988]).

Day [1969] proved that a variety is congruence-modular if and only if it satisfies a certain *Mal'cev condition* in four variables [condition (b) below]. Conditions (c) and (d) below follow from Day's proof.

Theorem 9.3.11   *The following conditions are equivalent for a variety V:*

(a) V *is congruence-modular.*
(b) *There are an n and 4-ary terms $m_0(w, x, y, z), \dots, m_n(w, x, y, z)$ such that V satisfies the identities*

   (i) $m_0(w, x, y, z) = w,\ m_n(w, x, y, z) = z,$
   (ii) $m_i(w, y, y, w) = w,\ i \leq n,$
   (iii) $m_i(w, w, y, y) = m_{i+1}(w, w, y, y)$ *for even $i < n$,*
   (iv) $m_i(w, y, y, z) = m_{i+1}(w, y, y, z)$ *for odd $i < n$.*

(c) $V_4 = HSP(F_V(4))$ *is congruence-modular.*
(d) $\text{Con}(F_V(4))$ *is modular.*

Applying modular commutator theory (for which see Freese & McKenzie [1987]), Gumm [1981] improved the preceding result of Day's: he showed that it is not necessary to work with four-variable terms or with $F_V(4)$ to prove that a variety is congruence-modular.

Theorem 9.3.12   *The following conditions are equivalent for a variety V:*

(a) V *is congruence-modular.*
(b) *There are an n and 3-ary terms $p(x, y, z)$ and $q_0(x, y, z), \dots, q_n(x, y, z)$ such that V satisfies the identities*

   (i) $q_0(x, y, z) = x,$
   (ii) $q_i(x, y, z) = x,\ i \leq n,$
   (iii) $q_i(x, y, y) = q_{i+1}(x, y, y)$ *for even $i < n$,*
   (iv) $q_i(x, x, y) = q_{i+1}(x, x, y)$ *for odd $i < n$,*

(v) $q_n(x, y, y) = p(x, y, y)$,

(vi) $p(x, x, y) = y$.

(c) $V_3 = HSP(\mathbf{F}_V(3))$ is congruence-modular.

Gumm [1981] proved the equivalence of (a) and (b); condition (c) easily follows from his proof. The following example due to Kearnes [1991] shows that the conditions of the preceding result are not equivalent to the condition that $\text{Con}(\mathbf{F}_V(3))$ is modular.

In the variety of sets, $\mathbf{F}_V(3)$ has three elements and $\text{Con}(\mathbf{F}_V(3))$ is the lattice of equivalence relations on this three-element set. Thus $\text{Con}(\mathbf{F}_V(3))$ is isomorphic to the modular lattice M_3. On the other hand, the variety of sets is not congruence-modular (every set of more than three elements has a nonmodular congruence lattice).

Kearnes [1991] also provided an example to the effect that if congruence modularity implies a nontrivial Mal'cev condition, then some of the terms in the Mal'cev condition must have at least three variables. This shows that the number 3 appearing in condition (b) of Gumm's theorem is indeed optimal. However, Kearnes [1991] shows that condition (c) of the preceding theorem can be improved:

Theorem 9.3.13 *The following conditions are equivalent for a variety* V:

(a) V *is congruence-modular;*

(b) $V_2 = HSP(\mathbf{F}_V(2))$ *is congruence-modular;*

(c) *The subalgebra* $\mathbf{S}$ *of* $\mathbf{F}_V(u, v) \times \mathbf{F}_V(u, v)$ *that is generated by* $\{(u, u), (u, v),$ $(v, u), (v, v)\}$ *is congruence-modular.*

The equivalence of conditions (a) and (b) in Theorem 9.3.13 implies that if the free algebra on two generators in V is finite, then V is congruence-modular if and only if a certain finitely generated subvariety is. In what follows this will be used along with results of tame congruence theory (for which we refer to Hobby & McKenzie [1988]) to derive the main result (Theorem 9.3.18 below).

It is not possible to replace $\mathbf{F}_V(2)$ by $\mathbf{F}_V(1)$ in condition (b) of the preceding theorem, since there are nonmodular varieties in which every basic operation is idempotent (e.g. the variety of sets or the variety of semilattices). In such a variety $V_1 = HSP(\mathbf{F}_V(1))$ is a trivial (and hence congruence-modular) variety even though V is not.

Concerning congruence semimodularity let us first make the following observation.

Lemma 9.3.14 *If* $\mathbf{F}_V(4)$ *is finite and* V *is both congruence upper semimodular and congruence lower semimodular, then* V *is congruence-modular.*

The assertion follows from Theorem 9.3.11(d) and from the fact that, in a lattice of finite length, upper and lower semimodularity together imply modularity.

Lemma 9.3.14 remains true if we delete the hypothesis that V is congruence upper semimodular, but is false if we delete the hypothesis that V is congruence lower semimodular. Kearnes [1991], Theorem 3.4, also proved

Theorem 9.3.15   *Let* α, β, *and* γ *be congruences on a finite algebra* **A** *that satisfy* $\alpha \vee \gamma = \beta \vee \gamma$, $\alpha \wedge \gamma = \beta \wedge \gamma$, *and* $\alpha < \beta$.

(a) *If* **A** *is congruence upper semimodular, then* typ$\{\alpha, \beta\} \subseteq \{\mathbf{1}, \mathbf{5}\}$.

(b) *If* **A** *is congruence lower semimodular, then* typ$\{\alpha, \beta\} \subseteq \{\mathbf{1}\}$.

Sketch of proof.   Assume first that **A** is congruence upper semimodular and that typ$\{\alpha, \beta\} \subseteq \{\mathbf{1}, \mathbf{5}\}$ is not true. Let $\theta = \alpha \vee \gamma = \beta \vee \gamma$ and $\delta = \alpha \wedge \gamma = \beta \wedge \gamma$. We may assume that the interval $[\delta, \theta]$ is a minimal one (with respect to inclusion) in Con(**A**) satisfying the hypotheses.

By our assumption we may find $\alpha', \beta' \in$ Con(**A**) such that $\alpha < \alpha' \prec \beta' < \beta$ and typ$(\alpha', \beta') \notin \{\mathbf{1}, \mathbf{5}\}$. Setting $\alpha = \alpha'$ and $\beta = \beta'$, we may assume that $\alpha \prec \beta$. Thus in Con(**A**) we have the lattice of Figure 9.5(a) as a sublattice.

Upper semimodularity of Con(**A**) implies that the pentagon sublattice of Figure 9.5(a) is located in the sublattice shown in Figure 9.5(b) (see Section 3.1). More specifically, it follows from upper semimodularity that there exists a congruence relation γ' such that $\delta \prec \gamma' < \gamma$ and $\beta \prec \gamma' \vee \beta$. Now the minimality assumption on the interval $[\delta, \theta]$ together with Lemma 6.2 of Hobby & McKenzie [1988] can be used to show that $\gamma' \vee \beta = \theta$. This together with $\theta = \alpha \vee \gamma = \beta \vee \gamma$ and $\delta = \alpha \wedge \gamma = \beta \wedge \gamma$ implies that the lattice shown in Figure 9.5(b) (with the coverings indicated by double lines) is a sublattice of Con(**A**). Hence typ$(\gamma' \vee \alpha, \theta) =$ typ$(\alpha, \beta) \in \{\mathbf{1}, \mathbf{5}\}$, which contradicts Lemma 6.3 of Hobby & McKenzie [1988]. It follows that condition (a) holds.

For condition (b), all arguments but Lemma 6.3 of Hobby & McKenzie [1988] can be dualized and we get a lattice dual to the lattice of Figure 9.5(b). Applying Lemma 6.4 of Hobby & McKenzie [1988], we can complete the proof. ∎

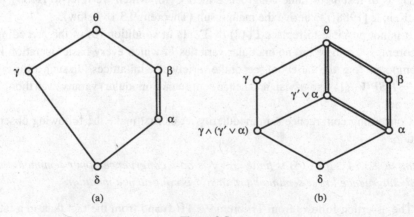

Figure 9.5

Corollary 9.3.16 *Let* **A** *be a finite algebra.*

(a) *If* Con(**A**) *is upper semimodular and* typ{**A**} $\cap$ {**1, 5**} = $\emptyset$, *then* Con(**A**) *is modular.*

(b) *If* Con(**A**) *is lower semimodular and* typ{**A**} $\cap$ {**1**} = $\emptyset$, *then* Con(**A**) *is modular.*

Proof. Condition (a) means that if Con(**A**) is upper semimodular and typ{**A**} $\cap$ {**1, 5**} = $\emptyset$, then $\alpha \vee \gamma = \beta \vee \gamma$, $\alpha \wedge \gamma = \beta \wedge \gamma$, and $\alpha \leq \beta$ together imply $\alpha = \beta$. In other words, Con(**A**) has no sublattice isomorphic to N_5. This is equivalent to the condition that Con(**A**) is modular. Similarly, condition (b) of Theorem 9.3.15 implies condition (b) of the corollary. ∎

Hobby & McKenzie [1988] remark that if a locally finite variety V is congruence n-permutable for some n, or if there is a nontrivial lattice identity satisfied by all the congruence lattices of algebras in V, then typ{V} $\cap$ {**1, 5**} = $\emptyset$. In such a variety, the congruence lattice of a finite algebra is upper semimodular if and only if it is lower semimodular if and only if it is modular.

Corollary 9.3.16 gives rise to the question whether type **1** can occur in a locally finite congruence lower semimodular variety. A negative answer was provided by Kearnes [1991], Theorem 3.7:

Theorem 9.3.17 *Let* V *be a locally finite congruence lower semimodular variety. Then* typ{V} $\cap$ {**1, 5**} = $\emptyset$.

Sketch of proof. It suffices to show that **1** $\notin$ typ{V}. Then $\mathbf{F}_V(4)$ is a finite algebra for which the hypotheses of Corollary 9.3.16 hold. Hence Con($\mathbf{F}_V(4)$) is modular. Theorem 9.3.11 implies that V is congruence-modular. This yields that **5** $\notin$ typ{V}, since Theorem 8.5 of Hobby & McKenzie [1988] shows that typ{V} $\subseteq$ {**2, 3, 4**} for any locally finite congruence-modular variety. ∎

The main result of Kearnes [1991],Theorem 3.8, is

Theorem 9.3.18 *If* V *is a congruence lower semimodular variety and* $\mathbf{F}_V(2)$ *is finite, then* V *is congruence-modular.*

Sketch of proof. By Theorem 9.3.13, V is congruence-modular if and only $V_2 = HSP(\mathbf{F}_V(2))$ is. The variety V_2 is locally finite, and it is a subvariety of V. Hence V_2 is congruence lower semimodular. Theorem 9.3.17 implies typ{V} $\cap$ {**1, 5**} = $\emptyset$. This yields by Corollary 9.3.16 that V_2 is congruence-modular. ∎

In the preceding theorem it is not possible to replace "lower" by "upper": the variety Sets is congruence upper semimodular, it is nonmodular, and typ{Sets} = {**1**}; on the other hand, the variety Semilattices is congruence upper semimodular, it is nonmodular, and typ{Semilattices} = {**5**}.

However, the following result holds (Kearnes [1991], Theorem 3.9 and Corollary 3.10) whose proof is similar to that of Theorem 9.3.18:

Theorem 9.3.19 *If $\vee$ is a congruence upper semimodular variety, $\mathbf{F}_\vee(2)$ is finite, and* $\mathrm{typ}\{HSP(\mathbf{F}_\vee(2))\} \cap \{\mathbf{1}, \mathbf{5}\} = \emptyset$, *then $\vee$ is congruence-modular.*

Corollary 9.3.20 *If $\vee$ is a variety such that $\mathbf{F}_\vee(2)$ is finite, then the following conditions are equivalent:*

(a) $\vee$ *is congruence upper semimodular and congruence join-semidistributive;*
(b) $\vee$ *is congruence lower semimodular and congruence join-semidistributive;*
(c) $\vee$ *is congruence lower semimodular and congruence meet-semidistributive;*
(d) $\vee$ *is congruence-distributive.*

Sketch of proof. It is obvious that condition (d) implies all the other conditions. We have to show that the other conditions imply (d). By Theorem 9.3.18, the statements in (b) imply that $\vee$ is congruence-modular and congruence join-semidistributive. But any modular and join-semidistributive lattice is distributive. Hence $\vee$ is congruence distributive, showing that (b) implies (d). Similarly one can show that (c) implies (d). It remains to show that (a) implies (d). Hobby & McKenzie [1988], Theorem 9.11, yields the following statement: If $V_2 = HSP(\mathbf{F}_\vee(2))$ is congruence join-semidistributive, then $\mathrm{typ}\{V_2\} \cap \{\mathbf{1}, \mathbf{2}, \mathbf{5}\} = \emptyset$. Applying Theorem 9.3.19, we conclude that $\vee$ is congruence-modular. This together with the join semidistributivity of $\vee$ implies that $\vee$ is also congruence-distributive. ∎

The variety Semilattices is locally finite, congruence upper semimodular, and congruence meet-semidistributive, but it is not congruence-distributive. Hence the condition "$\vee$ is congruence upper semimodular and congruence meet-semidistributive" is not equivalent to any one of the conditions in Corollary 9.3.20.

Notes

A variety is called *geometric* if the congruence lattice of any member is a matroid lattice. The variety of sets and any variety of vector spaces are examples of geometric varieties. Are there any others? The answer to this question is yes, although all known examples are built up in a straightforward way from varieties of vector spaces and the variety of sets (Agliano & Kearnes [1994a], Example 8.1).

Berman [1972] characterized those unary algebras whose congruence lattices are semimodular or atomistic. Combining these results, he gives necessary and sufficient conditions for a unary algebra to have a geometric congruence lattice.

Polynomially orderable varieties seem to be the simplest kind of regular CSM varieties. The variety of sets and the variety of semilattices are examples of polynomially orderable varieties. Another example is the variety of directoids introduced by Ježek & Quackenbush [1990]. (See Agliano & Kearnes [1994b], Example 3.14.)

References

Agliano, P. [1990] Congruence semimodularity and identities, Algebra Universalis 27, 600–601.

Agliano, P. [1991] The one-block property in varieties of semigroups, Semigroup Forum 42, 253–264.

Agliano, P. [1992] On combinatorial inverse semigroups, Rend. Sem. Mat. Torino 50, 255–275.

Agliano, P. and K. Kearnes [1994a] Congruence semimodular varieties I: locally finite varieties, Algebra Universalis 32, 224–269.

Agliano, P. and K. Kearnes [1994b] Congruence semimodular varieties II: regular varieties, Algebra Universalis 32, 270–296.

Berman, J. [1972] On the congruence lattice of unary algebra, Proc. Amer. Math. Soc. 36, 34–38.

Crawley, P. and R. P. Dilworth [1973] *Algebraic Theory of Lattices*, Prentice-Hall, Englewood Cliffs, N.J.

Day, A. [1969] A characterization of modularity for congruence lattices of algebras, Canad. Math. Bull. 12, 167–173.

Day, A. and R. Freese [1980] A characterization of identities implying congruence modularity, Canad. J. Math. 32, 1140–1167.

Freese, R. and R. McKenzie [1987] *Commutator Theory for Congruence Modular Varieties*, Cambridge Univ. Press, Cambridge.

Freese, R. and J. B. Nation [1973] Congruence lattices of semilattices, Pacific J. Math. 49, 51–58.

Grätzer, G. [1978] *General Lattice Theory*, Birkhäuser, Basel.

Gumm, H.-P. [1981] Congruence modularity is permutability composed with distributivity, Archiv d. Math. 36, 569–576.

Hall, T. E. [1971] On the lattice of congruences on a semilattice, J. Austral. Math. Soc. 12, 456–460.

Hobby, D. and R. McKenzie [1988] *The Structure of Finite Algebras*, Contemporary Mathematics, Amer. Math. Soc., Providence, R.I.

Ježek, J. and R. Quackenbush [1990] Directoids: algebraic models of up-directed sets, Algebra Universalis 27, 49–69.

Jones, P. R. [1988] Congruence semimodular varieties of semigroups, in: *Lecture Notes in Mathematics 1320*, Springer-Verlag, Berlin, pp. 162–171.

Jónsson, B. [1967] Algebras whose congruence lattices are distributive, Math. Scand. 21, 110–121.

Kearnes, K. K.[1991] Congruence lower semimodularity and 2-finiteness imply congruence modularity, Algebra Universalis 28, 1–11.

Mal'cev, A. I. [1954] On the general theory of algebraic systems, Mat. Sbornik 77, 3–20.

McKenzie, R., G. F. McNulty, and W. Taylor [1987] *Algebras, Lattices, Varieties*, Vol. 1, Wadsworth & Brooks/Cole, Monterey, Calif.

Ore, O. [1942] Theory of equivalence relations, Duke Math. J. 9, 573–627.

Polin, S. V. [1976] The identities of finite algebras, Siberian Math. J. 17, 1356–1366.

Taylor, W. [1973] Characterizing Mal'cev conditions, Algebra Universalis 3, 351–397.

Whitman, P. M. [1946] Lattices, equivalence relations, and subgroups, Bull. Amer. Mat. Soc. 52, 507–522.

Master Reference List

Abels, H. [1991] The geometry of the chamber system of a semimodular lattice, Order 8, 143–158.

Adams, M. E. [1990] Uniquely complemented lattices, in: Bogart et al. [1990], pp. 79–84.

Agliano, P. [1990] Congruence semimodularity and identities, Algebra Universalis 27, 600–601.

Agliano, P. [1991] The one-block property in varieties of semigroups, Semigroup Forum 42, 253–264.

Agliano, P. [1992] On combinatorial inverse semigroups, Rend. Sem. Mat. Torino 50, 255–275.

Agliano, P. and K. Kearnes [1994a] Congruence semimodular varieties I: locally finite varieties, Algebra Universalis 32, 224–269.

Agliano, P. and K. Kearnes [1994b] Congruence semimodular varieties II: regular varieties, Algebra Universalis 32, 270–296.

Aigner, M. [1979] *Combinatorial Theory*, Springer-Verlag, Berlin.

Aigner, M. [1987] Whitney numbers, in: White [1987], pp. 139–160.

Altwegg, M. [1950] Zur Axiomatik der teilweise geordneten Mengen, Comment. Math. Helv. 24, 149–155.

Alvarez, L. R. [1965] Undirected graphs as graphs of modular lattices, Canad. J. Math. 17, 923–932.

Auinger, K. [1990] Atomistic congruence lattices of semilattices, Semigroup Forum 40, 115–117.

Avann, S. P. [1961a] Application of the join-irreducible excess function to semimodular lattices, Math. Annalen 142, 345–354.

Avann, S. P. [1961b] Distributive properties in semimodular lattices, Math. Z. 76, 283–287.

Avann, S. P. [1964a] Dependence of finiteness conditions in distributive lattices, Math. Z. 85, 245–256.

Avann, S. P. [1964b] Increases of the join-excess function in a lattice, Math. Annalen 54, 420–426.

Avann, S. P. [1968] Locally atomic upper locally distributive lattices, Math. Annalen 175, 320–336.

Avann, S. P. [1972] The lattice of natural partial orders, Aeq. Math. 8, 95–102.

Baclawski, K. [1980] Cohen–Macaulay ordered sets, J. Algebra 63, 226–258.

Baclawski, K. [1982] Combinatorics: trends and examples, in: *New Directions in Applied Mathematics* (ed. P. J. Hilton and G. S. Young), Springer-Verlag, pp. 1–10.

Baclawski, K. and A. Björner [1979] Fixed points in partially ordered sets, Advances in Math. 31, 263–287.

Baclawski, K. and A. Björner [1981] Fixed points and complements in finite lattices, J. Combin. Theory Ser. A30, 335–338.

Baer, R. [1942] A unified theory of projective spaces and finite abelian groups, Trans. Amer. Math. Soc. 52, 283–343.

Baird, G. R. [1972a] On a sublattice of the lattice of congruences on a simple regular ω-semigroup, J. Austral Math. Soc. 13, 461–471.

Baird, G. R. [1972b] Congruences òn simple regular ω-semigroups, J. Austral. Math. Soc. 14, 155–167.

Baker, K. [1994] Bjarni Jónsson's contributions in algebra, Algebra Universalis 31, 306–336.

Balbes, R. and Ph. Dwinger [1974] *Distributive Lattices*, Univ. of Missouri Press, Columbia, Mo.

Balbes, R. and A. Horn [1970] Stone lattices, Duke Math. J. 37, 537–546.

Bandelt, H.-J. [1981] Tolerance relations on lattices, Bull. Austral. Math. Soc. 23, 367–381.

Bandelt, H.-J. [1984] Discrete ordered sets whose covering graphs are median, Proc. Amer. Math. Soc. 91, 6–8.

Barbilian, D. [1946] Metrisch-konkave Verbände, Disquisitiones Math. Phys. (Bucureşti), V (1–4), 1–63.

Barbut, M. and B. Monjardet [1970] *Ordre et Classification*, Vols. 1, 2, Hachette, Paris.

Barnabei, M., A. Brini, and G.-C. Rota [1982] Un' introduzione alla teoria delle funzioni di Möbius, in: *Matroid Theory and Its Applications* (ed. A. Barlotti), Liguori Editore, Napoli, pp.7–109.

Barthélemy, J. P., B. Leclerc, and B. Monjardet [1986] On the use of ordered sets in problems of comparison and consensus of classification, J. Classification 3, 187–224.

Basterfield, J. G. and L. M. Kelly [1968] A characterization of sets of n points which determine n hyperplanes, Proc. Cambridge Philos. Soc. 64, 585–588.

Batten, L. M. [1983] A rank-associated notion of independence, in: *Finite Geometries, Proceedings of a Conference in Pullman, Washington, April 1981*, Marcel Dekker, New York, pp. 33–46.

Batten, L. M. [1984] Jordan–Dedekind spaces, Quart. J. Math. Oxford 35, 373–381.

Behrendt, G. [1991] Representation of locally distributive lattices, Portugal. Math. 48, 351–355.

Beltrametti, E. G. and G. Cassinelli [1981] *The Logic of Quantum Mechanics* (Encyclopedia of Mathematics and its Applications, Vol. 15), Addison-Wesley, Reading, Mass.

Bennett, M. K. [1974] On generating affine geometries, Algebra Universalis 4, 207–219.

Bennett, M. K. [1977] Lattices of convex sets, Trans. Amer. Math. Soc. 234, 279–288.

Bennett, M. K. [1987] Biatomic lattices, Algebra Universalis 24, 60–73.

Bennett, M. K. [1989–90] Rectangular products of lattices, Discrete Math. 79, 235–249.

Bennett, M. K. [1987] Biatomic lattices, Algebra Universalis 24, 60–73.

Bennett, M. K. and G. Birkhoff [1983] A Peano axiom for convexity lattices, in: *Calcutta Mathematical Society Diamond Jubilee Commemorative Volume*, pp. 33–43.

Bennett, M. K. and G. Birkhoff [1985] Convexity lattices, Algebra Universalis 20, 1–26.

Beran, L. [1984] *Orthomodular Lattices. Algebraic Approach*, D. Reidel, Dordrecht.

Berman, J. [1972] On the congruence lattice of unary algebra, Proc. Amer. Math. Soc. 36, 34–38.

Birkhoff, G. [1933] On the combination of subalgebras, Proc. Camb. Phil. Soc. 29, 441–464.

Birkhoff, G. [1934] Applications of lattice algebra, Proc. Camb. Phil. Soc. 30, 115–122.

Birkhoff, G. [1935a] Abstract linear dependence and lattices, Amer. J. Math. 57, 800–804.

Birkhoff, G. [1935b] On the structure of abstract algebras, Proc. Camb. Phil. Soc. 31, 433–454.

Birkhoff, G. [1936] On the combination of topologies, Fund. Math. 29, 156–166.

Birkhoff, G. [1940a] *Lattice Theory*, Amer. Math. Soc. Colloquium Publications, Vol. 25, New York.

Birkhoff, G. [1940b] Neutral elements in general lattices, Bull. Amer. Math. Soc. 46, 702–705.

Birkhoff, G. [1944] Subdirect unions in universal algebra, Bull. Amer. Math. Soc. 50, 764–768.

Birkhoff, G. [1948] *Lattice Theory* (2nd edition), Amer. Math. Soc. Colloquium Publications, Vol. 25, New York.

Birkhoff, G. [1958] von Neumann and lattice theory, in: *John von Neumann, 1903–1957*, Bull. Amer. Math. Soc. 64, 50–56.

Birkhoff, G. [1967] *Lattice Theory* (3rd edition), Amer. Math. Soc. Colloquium Publications, Vol. 25, Providence, R.I.; reprinted 1984.

Birkhoff, G. [1982] Some applications of universal algebra, in: *Universal Algebra, Esztergom 1977*, Coll. Math. Soc. J. Bolyai, Vol. 29, North-Holland, Amsterdam, pp. 107–128.

Birkhoff, G. [1987] General remark to Chapter I, in: Rota & Oliveira [1987], pp. 1–8.

Birkhoff, G. and M. K. Bennett [1985] The convexity lattice of a poset, Order 2, 223–242.

Birkhoff, G. and O. Frink [1948] Representation of lattices by sets, Trans. Amer. Math. Soc. 64, 299–316.

Birkhoff, G. and J. von Neumann [1936] The logic of quantum mechanics, Ann. of Math. 37, 823–843.

Birkhoff, G. and M. Ward [1939] A characterization of Boolean algebras, Ann. of Math. 40, 609–610.

Björner, A. [1980] Shellable and Cohen–Macaulay partially ordered sets, Trans. Amer. Math. Soc. 260, 159–183.

Björner, A. [1981] On complements in lattices of finite length, Discrete Math. 36, 325–326.

Björner, A. [1983] On matroids, groups, and exchange languages, in: *Matroid Theory and Its Applications, Proc. Conf. Szeged 1982* (ed. L. Lovász and A. Recski), North-Holland, Amsterdam, pp. 25–60.

Björner, A. [1987] Continuous partition lattice, Proc. Natl. Acad. Sci. U.S.A. 84, 6327–6329.

Björner, A. [1992] Homology and shellability of matroids and geometric lattices, in: White [1992], pp. 226–283.

Björner, A. and L. Lovász [1987] Pseudomodular lattices and continuous matroids, Acta Sci. Math. (Szeged) 51, 295–308.

Björner, A. and I. Rival [1980] A note on fixed points in semimodular lattices, Discrete Math. 29, 245–250.

Björner, A. and G. M. Ziegler [1992] Introduction to greedoids, in: White [1992], pp. 284–357.

Björner, A., A. Garsia, and R. Stanley [1982] An introduction to Cohen–Macaulay partially ordered sets, in: Rival [1982a], pp. 583–615.

Blyth, T. S. and M. F. Janowitz [1972] *Residuation Theory*, Pergamon, Oxford.

Bogart, K. P. [1990] *Introductory Combinatorics* (2nd edition), Harcourt Brace Jovanovich, San Diego (1st edition, Pitman, Marshfield, Mass., 1983).

Bogart, K. P., R. Freese, and J. P. S. Kung (eds.) [1990] *The Dilworth Theorems. Selected Papers of Robert P. Dilworth*, Birkhäuser, Boston.

Bolker, E. [1987] The finite Radon transform, Contemp. Math. 63, 27–50.

Bonzini, C. and A. Cherubini [1990a] Semimodularity of the congruence lattice on regular ω-semigroups, Monatsh. Math. 109, 205–219.

Bonzini, C. and A. Cherubini [1990b] Conditions similar to modularity on the congruence lattice of a regular ω-semigroup, in: *Proc. Internat. Symp. on Semigroup Theory and Related Fields*, ed. by M. Yamada and H. Tominaga, Ritsumeikan University, Kyoto, pp. 41–52.

Boole, G. [1847] *The Mathematical Analysis of Logic. Being an Essay Towards a Calculus of Deductive Reasoning*, Cambridge; reprinted, Oxford, 1951.

Brualdi, R. A. [1987] Transversal matroids, in: White, N. L. [1987], pp. 72–97.

Bruggesser, H. and P. Mani [1971] Shellable decompositions of cells and spheres, Math. Scand. 29, 197–205.

Bruns, G. [1959] Verbandstheoretische Kennzeichnung vollständiger Mengenringe, Arch. d. Math. 10, 109–112.

Brylawski, T. [1975] Modular constructions for combinatorial geometries, Trans. Amer. Math. Soc. 203, 1–44.

Brylawski, T. [1986] Constructions, in: White, N. L. [1986], pp. 127–223.

Budach, L., B. Graw, C. Meinel, and S. Waack [1988] *Algebraic and Topological Properties of Finite Partially Ordered Sets*, Teubner-Texte zur Mathematik, Band 109, Teubner-Verlag, Leipzig.

Buekenhout, F. [1979] Diagrams for geometries and groups, J. Combin. Theory (A) 27, 121–151.

Burosch, G. [1989] Hasse Graphen spezieller Ordnungen, in: *Graphentheorie. Band I: Anwendungen auf Topologie, Gruppentheorie und Verbandstheorie*. (ed. K. Wagner and R. Bodendiek), BI Mannheim, pp. 157–245.

Burris, S. and H. P. Sankappanavar [1975] Lattice-theoretic decision problems in universal algebra, Algebra Universalis 5, 163–177.

Burris, S. and H. P. Sankappanavar [1981] *A Course in Universal Algebra*, Springer-Verlag, New York.

Chajda, I. [1991] *Algebraic Theory of Tolerance Relations*, Univ. Palackeho, Olomous.

Chajda, I. and B. Zelinka [1974] Tolerance relations on lattices, Časop. Pešt. Mat. 99, 394–399.

Chameni-Nembua, C. and B. Monjardet [1992] Les treillis pseudocomplémentés finis, European J. Combin. 13, 89–107.

Chen, C. C. and K. M. Koh [1972] On the lattice of convex sublattices of a finite lattice, Nanta Math. 5, 93–95.

Chrislock [1969] A certain class of identities on semigroups, Proc. Amer. Math. Soc. 21, 189–190.

Cohen, D. W. [1989] *An Introduction to Hilbert Space and Quantum Logic*, Springer-Verlag, New York.

Constantin, J. [1973] Note sur un problème de Maeda, Canad. Math. Bull. 16, 193.

Crapo, H. H. [1965] Single-element extensions of matroids, J. Res. Nat. Bur. Standards Sect. B 69B, 55–65.

Crapo, H. H. [1966] The Möbius function of a lattice, J. Combin. Theory 1, 126–131.

Crapo, H. H. [1967] Structure theory for geometric lattices, Rend. Sem. Mat. Univ. Padova 38, 14–22.

Crapo, H. H. [1968] Möbius inversion in lattices, Arch. d. Math. 19, 595–607.

Crapo, H. H. [1984] Selectors: a theory of formal languages, semimodular lattices, and branching and shelling processes, Advances in Math. 54, 233–277.

Crapo, H. H. [1986] Examples and basic concepts, in: White [1986], Chapter 1.

Crapo, H. H. and G.-C. Rota [1970a] *On the Foundations of Combinatorial Theory: Combinatorial Geometries*, MIT Press, Cambridge, Mass.

Crapo, H. H. and G.-C. Rota [1970b] Geometric lattices, in: *Trends in Lattice Theory* (ed. J. C. Abbott), van Nostrand-Reinhold, New York, pp. 127–172.

Crawley, P. [1959] The isomorphism theorem in compactly generated lattices, Bull. Amer. Math. Soc. 65, 377–379.

Crawley, P. [1961] Decomposition theory for non-semimodular lattices, Trans. Amer. Math. Society 99, 246–254.

Crawley, P. and R. P. Dilworth [1973] *Algebraic Theory of Lattices*, Prentice-Hall, Englewood Cliffs, N.J.

Croisot, R. [1951] Contribution à l'étude des treillis semi-modulaires de longueur infinie, Ann. Sci. Ecole Norm. Sup. (3) 68, 203–265.

Croisot, R. [1952] Quelques applications et propriétés des treillis semi-modulaires de longueur infinie, Ann. Fac. Sci. l'Univ. Toulouse 16, 11–74.

Czédli, G. [1982] Factor lattices by tolerances, Acta Sci. Math. (Szeged) 44, 35–42.

Czédli, G. [1999] *Lattice Theory* (Hungarian), JATEPress, Szeged.

Dauben, J. W. [1986] Review of Mehrtens [1979], Order 3, 89–102.

Davey, B. A. and H. A. Priestley [1990] *Introduction to Lattices and Order*, Cambridge Univ. Press, Cambridge.

Davey, B. A., W. Poguntke, and I. Rival [1975] A characterization of semidistributivity, Algebra Universalis 5, 72–75.

Davis, A. C. [1955] A characterization of complete lattices, Pacific J. Math. 5, 311–319.

Day, A. [1969] A characterization of modularity for congruence lattices of algebras, Canad. Math. Bull. 12, 167–173.

Day, A. and R. Freese [1980] A characterization of identities implying congruence modularity, Canad. J. Math. 32, 1140–1167.

Day, A. and R. Freese [1990] The role of gluing constructions in modular lattice theory, in: Bogart, K. E. et al. [1990], pp. 251–260.

Day, A. and J. Ježek [1983] The amalgamation property for varieties of lattices, Mathematics Report No 1, Lakehead University.

Dean, R. A. and G. Keller [1968] Natural partial orders, Canad. J. Math. 20, 535–554.

Dean, R. A. and R. L. Kruse [1966] A normality relation for lattices, J. of Algebra 3, 277–290.

Dean, R. A. and R. H. Oehmke [1964] Idempotent semigroups with distributive right congruence lattices, Pacific J. Math. 14, 1187–1209.

Dedekind, R. [1872] *Stetigkeit und irrationale Zahlen* (7th edition, 1969), Vieweg, Braunschweig.

Dedekind, R. [1897] Über Zerlegungen von Zahlen durch ihren größten gemeinsamen Teiler, in: *Festschrift der Techn. Univ. Braunschweig*, pp. 1–40; *Gesammelte Werke*, Vol. 2, pp. 103–147.

Dedekind, R. [1900] Über die von drei Moduln erzeugte Dualgruppe, Math. Ann. 53, 371–403; *Gesammelte Werke*, Vol. 2, pp. 236–271.

Diercks, V. [1982] Zerlegungstheorie in vollständigen Verbänden, Diplomarbeit, Universität Hannover.

Dilworth, R. P. [1940] Lattices with unique irreducible decompositions, Ann. of Math. 41, 771–777.

Dilworth, R. P. [1941a] The arithmetical theory of Birkhoff lattices, Duke Math. J. 8, 286–299.

Dilworth, R. P. [1941b] Ideals in Birkhoff lattices, Trans. Amer. Math. Soc. 49, 325–353.

Dilworth, R. P. [1944] Dependence relations in a semimodular lattice, Duke Math. J. 11, 575–587.

Dilworth, R. P. [1945] Lattices with unique complements, Trans. Amer. Math. Soc. 57, 123–154.

Dilworth, R. P. [1946] Note on the Kurosch–Ore theorem, Bull. Amer. Math. Soc. 52, 659–663.

Dilworth, R. P. [1950] Review of Birkhoff [1948], Bull. Amer. Math. Soc. 56, 204–206.

Dilworth, R. P. [1954] Proof of a conjecture on finite modular lattices, Ann. Math. 60, 359–364.

Dilworth, R. P. [1982] The role of order in lattice theory, in: Rival [1982a], pp. 333–353.

Dilworth, R. P. [1984] Aspects of distributivity, Algebra Universalis 18, 4–17.

Dilworth, R. P. [1990a] Background to Chapter 3, in: Bogart et al. [1990], pp. 89–92.

Dilworth, R. P. [1990b] Background to Chapter 4, in: Bogart et al. [1990], pp. 205–209.

Dilworth, R. P. [1990c] Background to Chapter 5, in: Bogart et al. [1990], pp. 265–267.

Dilworth, R. P. and P. Crawley [1960] Decomposition theory for lattices without chain conditions, Trans. Amer. Math. Soc. 96, 1–22.

Dlab, V. [1962] General algebraic dependence relations, Publ. Math. Debrecen 9, 324–355.

Dlab, V. [1966] Algebraic dependence structures, Z. Math. Logik Grundl. Math. 12, 345–377.

Dowling, T. A. [1973] A q-analog of the partition lattice, in: *A Survey of Combinatorial Theory* (ed. J. N. Srivastava), Proc. Internat. Symp. Combin. Math. Appli., Colorado State University, Fort Collins, Colo., September 9–11, 1971, North-Holland, Amsterdam, pp. 101–115.

Dowling, T. A. and R. M. Wilson [1975] Whitney number inequalities for geometric lattices, Proc. Amer. Math. Soc. 47, 504–512.

Dress, A. and L. Lovász [1987] On some combinatorial properties of algebraic matroids, Combinatorica 7, 39–48.

Dress, A., W. Hochstättler, and W. Kern [1994] Modular substructures in pseudomodular lattices, Math. Scand. 74, 9–16.

Dubreil-Jacotin, M. L., L. Lesieur, and R. Croisot [1953] *Leçons sur la théorie des treillis, des structures algébriques ordonnées et des treillis géométriques*, Gauthier-Villards, Paris.

Duffus, D. [1982] Matching in modular lattices, J. Combin. Theory Ser. A 32, 303–314.

Duffus, D. [1985] Matching in modular lattices, Order 1, 411–413.

Duffus, D. and I. Rival [1977] Path length in the covering graph of a lattice, Discrete Math. 19, 139–158.

Duffus, D. and I. Rival [1983] Graphs orientable as distributive lattices, Proc. Amer. Math. Soc. 88, 197–200.

Duffus, D., W. Poguntke, and I. Rival [1980] Retracts and the fixed point problem for finite partially ordered sets Canad. Math. Bull. 23, 231–236.

Dunstan, F. D. J., A. W. Ingleton, and D. J. A. Welsh [1972] Supermatroids, in: *Proc. Conf. Combin. Math.*, Math. Inst., Oxford, pp. 72–122.

Duquenne, V. [1991] The core of finite lattices, Discrete Math. 88, 133–147.

Duquenne, V. and B. Monjardet [1982] Relations binaires entre partitions, Math. Sci. Hum. 80, 5–37.

Duthie, W. D. [1942] Segments of ordered sets, Trans. Amer. Math. Soc. 51, 1–14.

Eberhart, C. and J. Seldon [1972] One-parameter inverse semigroups, Trans. Amer. Math. Soc. 168, 53–66.

Eberhart, C. and W. Williams [1978] Semimodularity in the lattice of congruences, J. Algebra 52, 75–87.

Edelman, P. H. [1980] Meet-distributive lattices and the anti-exchange closure, Algebra Universalis 10, 290–299.

Edelman, P. H. [1986] Abstract convexity and meet-distributive lattices, in: *Contemporary Mathematics*, Vol. 57 (ed. I. Rival), pp. 127–150.

Edelman, P. H. and R. E. Jamison [1985] The theory of convex geometries, Geom. Dedicata 19, 247–270.

Edelman, P. H. and M. E. Saks [1988] Combinatorial representation and convex dimension of convex geometries, Order 5, 23–32.

Erné, M. [1983] On the existence of decompositions in lattices, Algebra Universalis 16, 338–343.

Evans, T. [1971] The lattice of semigroup varieties, Semigroup Forum 2, 1–43.

Faigle, U. [1980a] Geometries on partially ordered sets, J. Combin. Theory Ser. B 28, 26–51.

Faigle, U. [1980b] Über Morphismen halbmodularer Verbände, Aequationes Math. 21, 53–67.

Faigle, U. [1980c] Extensions and duality of finite geometric closure operators, J. Geometry 14, 23–34.

Faigle, U. [1986a] Lattices, in: White [1986], Chapter 3.

Faigle, U. [1986b] Exchange properties of combinatorial closure spaces, Discrete Appl. Math. 15, 249–260.

Faigle, U. [1990] Dilworth's completion, submodular functions, and combinatorial optimization, in: Bogart et al. [1990], pp. 287–294.

Faigle, U. and C. Herrmann [1981] Projective geometry on partially ordered sets, Trans. Amer. Math. Soc. 266, 319–332.

Faigle, U., G. Richter, and M. Stern [1984] Geometric exchange properties in lattices of finite length, Algebra Universalis 19, 355–365.

Felscher, W. [1961] Jordan–Hölder Sätze und modular geordnete Mengen, Math. Z. 75, 83–114.

Finkbeiner, D. T. [1951] A general dependence relation for lattices, Proc. Amer. Math. Soc. 2, 756–759.

Finkbeiner, D. T. [1960] A semimodular imbedding of lattices, Canad. J. Math. 12, 582–591.

Folkman, J. [1966] The homology groups of a lattice, J. Math. Mech. 15, 631–636.

Fort, A. [1974] Una caratterizzazione dei reticoli modulari a catene limitate-finite, Rend. Sem. Mat. Univ. Padova 51, 269–273.

Fraleigh, J. B. [1989] A First Course in Abstract Algebra (4th edition), Addison-Wesley, Reading, Mass.

Freese, R., J. Ježek and J. B. Nation [1995] Free Lattices, Amer. Math. Soc., Providence, R.I.

Freese, R. and R. McKenzie [1987] Commutator Theory for Congruence Modular Varieties, Cambridge Univ. Press, Cambridge.

Freese, R. and J. B. Nation [1973] Congruence lattices of semilattices, Pacific J. Math. 49, 51–58.

Fried, E. and G. Grätzer [1989] Pasting and modular lattices, Proc. Amer. Math. Soc. 106, 885–890.

Fried, E., G. Grätzer, and E. T. Schmidt [1993] Multipasting of lattices, Algebra Universalis 30, 241–261.

Friedman, H. and D. Tamari [1967] Une structure de treillis fini induite par une loi demi-associative, J. Combin. Theory, Ser. A 2, 215–242.

Frink, O. [1941] Representation of Boolean algebras, Bull. Amer. Math. Soc. 47, 755–756.

Fujishige, S. [1991] Submodular Functions and Optimization, Annals of Discrete Mathematics, Vol. 47, Elsevier Science.

Fujiwara, S. [1965] The generalization of Wilcox lattices, Res. Bull. Fac. Liberal Arts Oita Univ. 2, 1–6.

Funayama, N. and T. Nakayama [1942] On the distributivity of a lattice of lattice congruences, Proc. Imp. Acad. Tokyo 18, 553–554.

Ganter, B. and I. Rival [1973] Dilworth's covering theorem for modular lattices: a simple proof, Algebra Universalis 3, 348–350.

Ganter, B. and I. Rival [1975] An arithmetical theorem for modular lattices, Algebra Universalis 5, 395–396.

Ganter, B. and R. Wille [1996] Formale Begriffsanalyse: Mathematische Grundlagen, Springer-Verlag, Heidelberg.

Garsia, A. [1980] Combinatorial methods in the theory of Cohen–Macaulay rings, Adv. in Math. 38, 229–266.

Gaskill, H. S. and I. Rival [1978] An exchange property for modular lattices, Algebra Universalis 8, 354–356.

Gedeonová, E. [1971] Jordan–Hölder theorem for lines, in: *Acta F.R.N. Univ. Comen. – Math., Memorial Volume*, pp. 23–24.

Gedeonová, E. [1972] Jordan–Hölder theorem for lines, Mat. Časopis 22, 177–198.

Gedeonová, E. [1980] Lattices whose covering graphs are S-graphs, Colloq. Math. Soc. J. Bolyai 33, 407–435.

Gedeonová, E. [1981] The orientability of the direct product of graphs, Math. Slovaca 31, 71–78.

Gedeonová, E. [1983] Lattices with centrally symmetric covering graphs, in: *Contributions to General Algebra, Proc. Klagenfurt Conf., June 10–13, 1982*, pp. 107–113.

Gedeonová, E. [1985] Central elements in CS-lattices, in: *Contributions to General Algebra, Proc. Vienna Conf., June 21–24, 1984*, pp. 143–155.

Gedeonová, E. [1990] Constructions of S-lattices, Order 7, 249–266.

Geissinger, L. and W. Graves [1972] The category of complete algebraic lattices, J. Combin. Theory (A) 13, 332–338.

Gragg, K. and J. P. S. Kung [1992] Consistent dually semimodular lattices, J. Combin. Theory Ser. A 60, 246–263.

Grätzer, G. [1959] Standard ideálok (Hungarian), Magyar Tud. Akad. III. Oszt. Közl. 9, 81–97.

Grätzer, G. [1971] *Lattice Theory: First Concepts and Distributive Lattices*, Freeman, San Francisco.

Grätzer, G. [1978] *General Lattice Theory*, Birkhäuser, Basel.

Grätzer, G. [1979] *Universal Algebra* (2nd edition), Springer-Verlag, New York.

Grätzer, G. [1998] *General Lattice Theory*, Birkhäuser, Basel, Second, completely revised edition.

Grätzer, G. and E. Kiss [1986] A construction of semimodular lattices, Order 2, 351–365.

Grätzer, G. and E. T. Schmidt [1957] On the Jordan–Dedekind chain condition, Acta Sci. Math. (Szeged) 7, 52–56.

Grätzer, G. and E. T. Schmidt [1958a] Ideals and congruence relations in lattices, Acta Math. Acad. Sci. Hungar. 9, 137–175.

Grätzer, G. and E. T. Schmidt [1958b] On the lattice of all join-endomorphisms of a lattice, Proc. Amer. Math. Soc. 9, 722–726.

Grätzer, G. and E. T. Schmidt [1961] Standard ideals in lattices, Acta Math. Acad. Sci. Hungar. 12, 17–86.

Grätzer, G. and E. T. Schmidt [1963] Characterizations of congruence lattices of abstract algebras, Acta Sci. Math. (Szeged) 24, 34–59.

Greechie, R. J. and S. Gudder [1973] Quantum logics, in: *Contemporary Research in the Foundations and Philosophy of Quantum Theory* (ed. C. A. Hooker), D. Reidel, Dordrecht, pp. 143–173.

Greene, C. [1970] A rank inequality for finite geometric lattices, J. Combin. Theory 9, 357–364.

Greene, C. [1975] An inequality for the Möbius function of a geometric lattice, Stud. Appl. Math. 54, 71–74.

Greene, C. [1982] The Möbius function of a partially ordered set, in: Rival (ed) [1982a], pp. 555–581.

Greene, C. and G. Markowsky [1974] A combinatorial test for local distributivity, Research Report RC4129, IBM T.J. Watson Research Center, Yorktown Heights, N.Y.

Greferath, M. and S. E. Schmidt [1992] A unified approach to projective lattice geometries, Geom. Dedicata 43, 243–264.

Grillet, P. A. and J. C. Varlet [1967] Complementedness conditions in lattices, Bull. Soc. Roy. Sci. Liège 36, 628–642.

Grötzsch, H. [1958] Ein Dreifarbensatz für dreikreisfreie Netze auf der Kugel, Wiss. Z. Martin-Luther-Univ. Halle, Math. Nat. Reihe 8, 109–119.

Gumm, H.-P. [1981] Congruence modularity is permutability composed with distributivity, Arch. d. Math. 36, 569–576.

Halin, R. [1993] Lattices related to separation graphs, in: *Finite and Infinite Combinatorics in Sets and Logic* (eds. N. W. Sauer et al.), Kluwer Academic, pp. 153–167.

Hall, M. [1959] *The Theory of Groups*, Macmillan, New York.

Hall, M. [1986] *Combinatorial Theory* (2nd edition), Wiley, New York.

Hall, M. and R. P. Dilworth [1944] The imbedding problem for modular lattices, Ann. Math. 45, 450–456.

Hall, P. [1932] A contribution to the theory of groups of prime power order, Proc. London Math. Soc. 36, 39–95.

Hall, P. [1936] The Eulerian functions of a group, Quart. J. Math. (Oxford), 134–151.

Hall, T. E. [1971] On the lattice of congruences on a semilattice, J. Austral. Math. Soc. 12, 456–460.

Halmos, P. R. [1957] *Introduction to Hilbert Space and the Theory of Spectral Multiplicity*, Chelsea, New York.

Halmos, P. R. [1963] *Lectures on Boolean Algebras*, Van Nostrand, Princeton, N.J.

Hamilton, H. B. [1974] Semilattices whose structure lattice is distributive, Semigroup Forum 8, 245–254.

Hardy, G. H. and E. M. Wright [1954] *An Introduction to the Theory of Numbers*, Oxford.

Harper, L. H. and G.-C. Rota [1971] Matching theory, an introduction, in: *Advances in Probability 1* (ed. P. Ney), Marcel Dekker, New York, pp. 169–215.

Haskins, L. and S. Gudder [1972] Heights on posets and graphs, Discrete Math. 2, 357–382.

Hawrylycz, M. and V. Reiner [1993] The lattice of closure relations on a poset, Algebra Universalis 30, 301–310.

Heron, A. P. [1973] A property of the hyperplanes of a matroid and an extension of Dilworth's theorem, J. Math. Anal. Appl. 42, 191–131.

Herrmann, C. [1973] S-verklebte Summen von Verbänden, Math. Z. 130, 255–274.

Hibi, T. [1992] *Algebraic Combinatorics on Convex Polytopes*, Carslaw, Glebe, Australia.

Higgs, D. [1966] Maps of geometries, J. London Math. Soc. 41, 612–618.

Higgs, D. [1968] Strong maps of geometries, J. Combin. Theory. Ser. A 5, 185–191.

Higgs, D. [1985] Interpolation antichains in lattices, in: *Universal Algebra and Lattice Theory, Proc. Conf. Charleston, July 11–14, 1984*, Lecture Notes in Mathematics 1149, Springer-Verlag, Berlin, pp. 142–149.

Hobby, D. and R. McKenzie [1988] *The Structure of Finite Algebras*, Contemporary Mathematics, Amer. Math. Soc., Providence, R.I.

Hochstättler, W. and W. Kern [1989] Matroid matching in pseudomodular lattices, Combinatorica 9, 145–152.

Hoffman, A. J. [1951] On the foundations of inversion geometry, Trans. Amer. Math. Soc. 71, 218–242.

Holland, S. S., Jr. [1970] The current interest in orthomodular lattices, in: *Trends in Lattice Theory* (ed. J. C. Abbott), van Nostrand Reinhold, New York, pp. 41–126.

Howie, J. M. [1976] *An Introduction to Semigroup Theory*, Academic Press, New York.

Huang, S. and D. Tamari [1972] Problems of associativity: a simple proof for the lattice property of systems ordered by a semi-associative law, J. Combin. Theory Ser. A 13, 7–13.

Huhn, A. [1972] Schwach distributive Verbände, Acta Sci. Math. (Szeged) 33, 297–305.

Huntington, E. V. [1904] Sets of independent postulates for the algebra of logic, Trans. Amer. Math. Soc. 5, 288–309.

Igoshin, V. I. [1980] Lattices of intervals and lattices of convex sublattices of lattices (Russian), Uporjadočennyje Množestva i Rešotki 6, 69–76.

Igoshin, V. I. [1988] Semimodularity in interval lattices (Russian), Math. Slovaca 38, 305–308.

Inaba, E. [1948] On primary lattices, J. Fac. Sci. Hokkaido Univ. 11, 39–107.

Ingleton, A. W. and R. A. Main [1975] Non-algebraic matroids exist, Bull. London Math. Soc. 7, 144–146.

Isaacs, I. M. [1994] Helmut Wielandt on subnormality, in: Wielandt [1994], pp. 299–306.

Ito, N. [1951] Note on (LM-)groups of finite order, Kodai Math. Sem. Rep., 1–6.

Iwasawa, K. [1941] Über die endlichen Gruppen und die Verbände ihrer Untergruppen, J. Univ. Tokyo 4, 171–199.

Jakubík, J. [1954a] On the graph isomorphism of lattices (Russian), Czech. Math. J. 4, 131–142.

Jakubík, J. [1954b] On the graph isomorphism of semimodular lattices (Slovak), Matem. Fyz. Časopis 4, 162–177.

Jakubík, J. [1956] Graph-isomorphism of multilattices (Slovak), Acta Fac. Rer. Nat. Univ. Comen. Math. 1, 255–264.

Jakubík, J. [1958] Note on the endomorphisms of a lattice, Čas. Pěst. Mat. 83, 226–229.

Jakubík, J. [1972] Weak product decompositions of partially ordered sets, Colloq. Math. 25, 13–26.

Jakubík, J. [1975a] Modular lattices of locally finite length, Acta Sci. Math. (Szeged) 37, 79–82.

Jakubík, J. [1975b] Sublattices with saturated chains, Czech. Math. J. 25, 442–444.

Jakubík, J. [1975c] Unoriented graphs of modular lattices, Czech. Math. J. 25, 240–246.

Jakubík, J. [1984] On lattices determined up to isomorphisms by their graphs, Czech. Math. J. 34, 305–314.

Jakubík, J. [1985a] On isomorphisms of graphs of lattices, Czech. Math. J. 35, 188–200.

Jakubík, J. [1985b] Graph isomorphisms of semimodular lattices, Math. Slovaca 35, 229–232.

Jakubík, J. [1985c] On weak direct product decompositions of lattices and graphs, Czech. Math. J. 35, 269–277.

Jakubík, J. [1986] Covering graphs and subdirect decompositions of partially ordered sets, Math. Slovaca 36, 151–162.

Jakubík, J. and M. Kolibiar [1954] On some properties of a pair of lattices, Czech. Math. J. 4, 1–27.

Janowitz, M. F. [1965a] A characterisation of standard ideals, Acta Math. Acad. Sci. Hungar. 16, 289–301.

Janowitz, M. F. [1965b] IC-lattices, Portugal. Math. 24, 115–122.

Janowitz, M. F. [1968] Section semicomplemented lattices, Math. Z. 108, 83–76.

Janowitz, M. F. [1970] On the modular relation in atomistic lattices, Fund. Math. 66, 337–346.

Janowitz, M. F. [1975] Examples of statisch and finite-statisch AC-lattices, Fund. Math. 89, 225–227.

Janowitz, M. F. [1976] On the "del" relation in certain atomistic lattices, Acta Math. Acad. Sci. Hungar. 28, 231–240.

Janowitz, M. F. and N. H. Coté [1976] Finite-distributive atomistic lattices, Portugal. Math. 35, 80–91.

Ježek, J. and R. Quackenbush [1990] Directoids: algebraic models of up-directed sets, Algebra Universalis 27, 49–69.

Jipsen, P. and H. Rose [1992] *Varieties of Lattices*, Springer-Verlag, Berlin.

Jipsen, P. and H. Rose [1998] Varieties of Lattices, Appendix F, in: Grätzer [1998], pp. 555–574.

Johnston, K. G. and P. R. Jones [1984] The lattice of full regular subsemigroups of a regular semigroup, Proc. Royal. Soc. Edinburgh 89A, 203–204.

Jones, A. W. [1946] Semi-modular finite groups and the Burnside basis theorem (Abstract), Bull. Amer. Math. Soc. 52, 418.

Jones, J. T. [1972] Pseudocomplemented semilattices, Ph.D. Dissertation, UCLA.

Jones, P. R. [1978] Semimodular inverse semigroups, J. London Math. Soc. 17, 446–456.

Jones, P. R. [1979] A homomorphic image of a semimodular lattice need not be semimodular: an answer to a problem of Birkhoff, Algebra Universalis 9, 127–130.

Jones, P. R. [1983a] Distributive, modular and separating elements in lattices, Rocky Mountain J. Math. 13, 429–436.

Jones, P. R. [1983b] On congruence lattices of regular semigroups, J. of Algebra 82, 18–39.

Jones, P. R. [1988] Congruence semimodular varieties of semigroups, in: *Lecture Notes in Mathematics 1320*, Springer-Verlag, Berlin, pp. 162–171.

Jones, P. R. [1990] Inverse semigroups and their lattices of inverse subsemigroups, in: *Lattices, Semigroups, and Universal Algebra* (ed. J. Almeida et al.), Plenum, New York, pp. 115–127.

Jónsson, B. [1959] Lattice-theoretic approach to projective and affine geometry, in: *The Axiomatic Method* (ed. L. Henkin, P. Suppes, and A. Tarski), Studies in Logic, Amsterdam, pp. 188–203.

Jónsson, B. [1961] Sublattices of a free lattice, Canad. J. Math. 13, 256–264.

Jónsson, B. [1967] Algebras whose congruence lattices are distributive, Math. Scand. 21, 110–121.

Jónsson, B. [1990] Dilworth's work on decompositions in semimodular lattices, in: Bogart et al. [1990], pp. 187–191.

Jónsson, B. and G. Monk [1969] Representation of primary arguesian lattices, Pacific J. Math. 30, 95–139.

Jónsson, B. and I. Rival [1979] Lattice varieties covering the smallest non-modular variety, Pacific J. Math. 82, 463–478.

Kalmbach, G. [1983] *Orthomodular Lattices*, Academic Press, London.

Kalmbach, G. [1986] *Measures and Hilbert Lattices*, World Scientific, Singapore.

Karzel, H. and H.-J. Kroll [1988] *Geschichte der Geometrie seit Hilbert*, Wissenschaftliche Buchgesellschaft Darmstadt.

Kearnes, K. K. [1991] Congruence lower semimodularity and 2-finiteness imply congruence modularity, Algebra Universalis 28, 1–11.

Kelly, D. and I. Rival [1974] Crowns, fences and dismantlable lattices, Canad. J. Math. 26, 1257–1271.

Klein-Barmen, F. [1937] Birkhoffsche und harmonische Verbände, Math. Z. 42, 58–81.

Klein-Barmen, F. [1941] Molekulare Verbände, Math. Z. 47, 373–394.

Knaster, B. [1928] Un théorème sur les fonctions d'ensembles, Ann. Soc. Polon. Math. 6, 133–134.

Kogalovskiĭ, S. R. [1965] On the theorem of Birkhoff (Russian), Uspehi Mat. Nauk 20, 206–207.

Koh, K. M. [1972] On the lattice of convex sublattices of a lattice, Nanta Math. 5, 18–37.

Koh, K. M. [1973] On sublattices of a lattice, Nanta Math. 6, 68–79.

Kolibiar, M. [1965] Linien in Verbänden, Analele Ştiinţifice Univ. Iaşi 11, 89–98.

Kolibiar, M. [1982] Semilattices with isomorphic graphs, in: *Coll. Math. Soc. J Bolyai, Vol. 29: Universal Algebra, Esztergom (Hungary) 1977*, North-Holland, Amsterdam, pp. 473–481.

Kolibiar, M. [1985] Graph isomorphisms of semilattices, in: *Contributions to General Algebra 3, Proc. Vienna Conf., June 21–24, 1984*, pp. 225–235.

Korte, B. and L. Lovász [1984] Shelling structures, convexity and a happy end, in: *Graph Theory and Combinatorics, Proc. Cambridge Combin. Conf. in Honor of Paul Erdős* (ed. B. Bollobás), Academic Press, London, pp. 219–232.

Korte, B., L. Lovász, and R. Schrader [1991] *Greedoids*, Springer-Verlag, Berlin.

Kotzig, A. [1968a] Centrally symmetric graphs (Russian), Czech. Math. J. 18, 606–615.

Kotzig, A. [1968b] Problem, in: *Beiträge zur Graphentheorie* (ed. H. Sachs), Teubner-Verlag, Leipzig, p. 394.

Kung, J. P. S. [1985] Matchings and Radon transforms in lattices I. Consistent lattices, Order 2, 105–112.

Kung, J. P. S. [1986a] *A Source Book in Matroid Theory*, Birkhäuser, Boston.

Kung, J. P. S. [1986b] Radon transforms in combinatorics and lattice theory, in: *Combinatorics and Ordered Sets*, Contemporary Mathematics 57 (ed. I. Rival), pp. 33–74.

Kung, J. P. S. [1986c] Basis-exchange properties, in: White [1986], pp. 62–75.

Kung, J. P. S. [1986d] Strong maps, in: White [1986], pp. 224–253.

Kung, J. P. S. [1987] Matchings and Radon transforms in lattices II. Concordant sets, Math. Proc. Camb. Phil. Soc. 101, 221–231.

Kung, J. P. S. [1990a] Dilworth truncations of geometric lattices, in: Bogart et al. [1990], pp. 295–297.

Kung, J. P. S. [1990b] Dilworth's proof of the embedding theorem, in: Bogart et al. [1990], pp. 458–459.

Kung, J. P. S. and H. Q. Nguyen [1986] Weak maps, in: White [1986], Chapter 9.

Kurinnoi, G. C. [1973] A new proof of Dilworth's theorem (Russian), Vestnik Har'kov. Univ. Mat.-Meh. 38, 11–15.

Kurinnoi, G. C. [1975] Condition for isomorphism of finite modular lattices (Russian), Vestnik Har'kov. Univ. Mat. Meh. 40, 45–47.

Kurosch, A. G. [1935] Durchschnittsdarstellungen mit irreduziblen Komponenten in Ringen und in sogenannten Dualgruppen, Mat. Sbornik 42, 613–616.

Lakser, H. [1973] A note on the lattice of sublattices of a finite lattice, Nanta Math. 6, 55–57.

Lallement, G. [1967] Demi-groupes réguliers, Ann. Mat. Pura Appl. 77, 47–129.

Larson, R. E. and S. J. Andima [1975] The lattice of topologies: a survey, Rocky Mountain J. Math. 5, 177–198.

Lea, J. W. [1974] Sublattices generated by chains in modular topological lattices, Duke Math. J. 41, 241–246.

Leclerc, B. [1990] Medians and majorities in semimodular lattices, SIAM J. Discrete Math. 3, 266–276.

Lee, J. G. [1986] Covering graphs of lattices, Bull. Korean Math. Soc. 23, 39–46.

Lennox, J. C. and S. E. Stonehewer [1987] *Subnormal Subgroups of Groups*, Clarendon Press, Oxford.

Leone, A. and M. Maj [1982] Gruppi finiti minimali non submodulari, Ricerche Mat. 31, 377–388.

Leone, A. and M. Maj [1982–3] Gruppi finiti non submodulari a quozienti propri submodulari, Rend. Accad. Sci. Fis. e Mat. Ser. IV XLX, 185–193.

Levigion, V. and S. E. Schmidt [1995] A geometric approach to generalized matroid lattices, in: *General Algebra and Discrete Mathematics* (ed. K. Denecke and O. Luders), Heldermann-Verlag, Berlin, pp. 181–186.

Libkin, L. O. [1992] Parallel axiom in convexity lattices, Periodica Math. Hungar. 24, 1–12.

Libkin, L. O. [1995] n-distributivity, dimension and Carathéodory's theorem, Algebra Universalis 34, 72–95.

Lindenstrauss, J. [1968] On subspaces of Banach spaces without quasicomplements, Israel J. Math. 6, 36–38.

Lindström, B. [1988a] A generalization of the Ingleton–Main lemma and a class of nonalgebraic matroids, Combinatorica 8, 87–90.

Lindström, B. [1988b] Matroids, algebraic and nonalgebraic, in: *Algebraic, Extremal and Metric Combinatorics, 1986*, London Math. Soc. Lecture Notes Series 131 (ed. M. M. Deza, P. Frankl, and I. G. Rosenberg), pp. 166–174.

Lindström, B. [1990] *p*-Independence implies pseudomodularity, Europ. J. Combin. 11, 489–490.

Lovász, L. and M. Saks [1993] Communication complexity and combinatorial lattice theory, J. Comput. System Sci. 47, 322–349.

Mac Lane, S. [1938] A lattice formulation for transcendence degrees and *p*-bases, Duke Math. J. 4, 455–468.

Mac Lane, S. [1943] A conjecture of Ore on chains in partially ordered sets, Bull. Amer. Math. Soc. 49, 567–568.

Mac Lane, S. [1976] Topology and logic as a source of algebra, Bull. Amer. Math. Soc. 82, 1–40.

MacNeille, H. M. [1937] Partially ordered sets, Trans. Amer. Math. Soc. 42, 416–460.

Maeda, F. [1958] *Kontinuierliche Geometrien*, Springer-Verlag, Berlin.

Maeda, F. and S. Maeda [1970] *Theory of Symmetric Lattices*, Springer-Verlag, Berlin.

Maeda, S. [1967] On atomistic lattice with the covering property, J. Sci. Hiroshima Univ. Ser. A-I 31, 105–121.

Maeda, S. [1974a] Locally modular and locally distributive lattices, Proc. Amer. Math. Soc. 44, 237–243.

Maeda, S. [1974b] Independent complements in lattices, in: *Coll. Math. Soc. J. Bolyai, 14. Lattice Theory*, Szeged, Hungary, pp. 215–226.

Maeda, S. [1977] Standard ideals in Wilcox lattices, Acta Math. Acad. Sci. Hungar. 29, 113–118.

Maeda, S. [1981] On finite-modular atomistic lattices, Algebra Universalis 12, 76–80.

Maeda, S. and Y. Kato [1974] The completion by cuts of an *M*-symmetric lattice, Proc. Japan. Acad. 50, 356–358.

Maeda, S., N. K. Thakare, and M. P. Wasadikar [1985] On the "del" relation in join-semilattices, Algebra Universalis 20, 229–242.

Malliah, C. and S. P. Bhatta [1986] Equivalence of *M*-symmetry and semimodularity in lattices, Bull. London Math. Soc. 18, 338–342.

Mal'cev, A. I. [1954] On the general theory of algebraic systems, Mat. Sbornik 77, 3–20.

Mangione, C. and S. Bozzi [1993] *Storia della Logica da Boole ai Nostri Giorni*, Garzanti Editore, Milano.

Markowsky, G. [1980] The representation of posets and lattices by sets, Algebra Universalis 11, 173–192.

Markowsky, G. [1992] Primes, irreducibles and extremal lattices, Order 9, 265–290.

Mason, J. H. [1977] Matroids as the study of geometrical configurations, in: M. Aigner (ed.) *Higher Combinatorics*, Reidel, Dordrecht, pp. 133–176.

McGrath, J. H. [1991] Quantum theory and the lattice join, Math. Intelligencer 13, 72–79.

McKenzie, R. [1972] Equational bases and nonmodular lattice varieties, Trans. Amer. Math. Soc. 74, 1–43.

McKenzie, R., G. F. McNulty, and W. Taylor [1987] *Algebras, Lattices, Varieties*, Vol. 1. Wadsworth & Brooks/Cole, Monterey, Calif.

McLaughlin, J. E. [1956] Atomic lattices with unique comparable complements, Proc. Amer. Math. Soc. 7, 864–866.

Mehrtens, H. [1979] *Die Entstehung der Verbandstheorie*, Gerstenberg Verlag, Hildesheim.

Menger, K. [1936] New foundations of projective and affine geometry, Ann. of Math. 37, 456–482.

Mihalek, R. J. [1960] Modularity relations in lattices, Proc. Amer. Math. Soc. 11, 9–16.

Mitsch, H. [1983] Semigroups and their lattices of congruences, Semigroup Forum 26, 1–63.

Monjardet, B. [1977] Caractérisation métrique des ensembles ordonnées semi-modulaires, Math. Sci. Hum. 56, 77–87.

Monjardet, B. [1981] Metrics on partially ordered sets – a survey, Discrete Math. 35, 173–184.

Monjardet, B. [1985] A use for frequently rediscovering a concept, Order 1, 415–417.

Monjardet, B. [1990] The consequences of Dilworth's work for lattices with unique irreducible decompositions, in: Bogart et al. [1990], pp. 192–200.

Mosesjan, K. M. [1972a] Strongly basable graphs (Russian), Akad. Nauk Armjan. SSR Dokl. 54, 134–138.

Mosesjan, K. M. [1972b] Certain theorems on strongly basable graphs (Russian), Akad. Nauk. Armjan. SSR Dokl. 54, 241–245.

Mosesjan, K. M. [1972c] Basable and strongly basable graphs (Russian), Akad. Nauk. Armjan. SSR Dokl. 55, 83–86.

Motzkin, T. S. [1951] The lines and planes connecting the points of a finite set, Trans. Amer. Math. Soc. 70, 451–464.

Mycielski, J. [1955] Sur le colorage des graphes, Colloq. Math. 3, 161–162.

Napolitani, F. [1973] Submodularità nei gruppi finiti, Rend. Sem. Mat. Univ. Padova 50, 355–363.

Nation, J. B. [1982] Finite sublattices of a free lattice, Trans. Amer. Math. Soc. 269, 311–337.

Nation, J. B. [1985] Some varieties of semidistributive lattices, in: *Universal Algebra and Lattice Theory Proc. Conf. held at Charleston, July 11–14, 1984*, Lecture Notes in Mathematics 1149 (ed. S. D. Comer), Springer-Verlag, Berlin, pp. 198–223.

Nation, J. B. [1994] Jónsson's contributions to lattice theory, Algebra Universalis 31, 430–445.

Nguyen, H. Q. [1986] Semimodular functions, in: White [1986], pp. 272–279.

Nieminen, J. [1985a] 2-ideals of finite lattices, Tamkang J. Math. 16, 23–37.

Nieminen, J. [1985b] A characterization of the Jordan–Hölder chain condition, Bull. Inst. Math. Acad. Sinica 13, 1–4.

Nieminen, J. [1990] The Jordan–Hölder chain condition and annihilators in finite lattices, Tsukuba J. Math 14, 405–411.

Noether, E. [1921] Idealtheorie in Ringbereichen, Math. Ann. 83, 24–66.

Ogasawara, T. and U. Sasaki [1949] On a theorem in lattice theory, J. Sci. Hiroshima Univ. Ser. A 14, 13.

Ore, O. [1935] On the foundations of abstract algebra I, Ann. of Math. 36, 406–437.

Ore, O. [1936] On the foundations of abstract algebra II, Ann. of Math. 37, 265–292.

Ore, O. [1942] Theory of equivalence relations, Duke Math. J. 9, 573–627.

Ore, O. [1943a] Chains in partially ordered sets, Bull. Amer. Math. Soc. 49, 558–566.

Ore, O. [1943b] Combinations of closure relations, Ann. of Math. 44, 514–533.

Ore, O. [1962] *Theory of Graphs*, Amer. Math. Soc., Providence, R.I.

Oxley, J. G. [1992] Infinite matroids, in: White [1992], pp. 73–90.

Oxley, J. G. [1993] *Matroid Theory*, Oxford University Press.

Padmanabhan, R. [1974] On M-symmetric lattices, Canad. Math. Bull. 17, 85–86.

Papert, D. [1964] Congruence relations in semi-lattices, J. London Math. Soc. 39, 723–729.

Petrich, M. [1979] Congruences on simple ω-semigroups, Glasgow Math. J. 20, 87–101.

Petrich, M. [1984] *Inverse Semigroups*, Wiley, New York.

Petrich, M. [1987] Congruences on completely regular semigroups, Working paper II, University of Vienna, June 1987.

Pezzoli, L. [1981] Sistemi di independenza modulari, Boll. Un. Mat. Ital. 18-B, 575–590.

Pezzoli, L. [1984] On D-complementation, Adv. in Math. 51, 226–239.

Piziak, R. [1990] Lattice theory, quadratic spaces, and quantum proposition systems, Foundations Phys. 20, 651–665.

Piziak, R. [1991] Orthomodular lattices and quadratic spaces: a survey, Rocky Mountain J. Math 21, 951–992.

Polat, N. [1976] Treillis de séparation des graphs, Canad. J. Math. 28, 725–752.

Polin, S. V. [1976] The identities of finite algebras, Siberian Math. J. 17, 1356–1366.

Pretzel, O. [1986] Orientations and reorientations of graphs, in: *Combinatorics and Ordered Sets, Proc. AMS–IMS SIAM Joint Summer Res. Conf., Arcata/Calif. 1985*, Contemporary Mathematics 57, pp. 103–125.

Provan, S. [1977] Decompositions, shellings, and diameters of simplicial complexes and convex polyhedra, Thesis, Cornell Univ., Ithaca, N.Y.

Pták, P. and S. Pulmannová [1991] *Orthomodular Structures as Quantum Logics*, Kluwer Academic, Dordrecht.

Pudlák, P. and J. Tůma [1977] Every finite lattice can be embedded in the lattice of all equivalences over a finite set, Comment. Math. Carolinae 18, 409–414.

Pudlák, P. and J. Tůma [1980] Every finite lattice can be embedded in a finite partition lattice, Algebra Universalis 10, 74–95.

Pym, J. S. and H. Perfect [1970] Submodular functions and independence structures, J. Math. Anal. Appl. 30, 1–31.

Quackenbush, R. W. [1985] Non-modular varieties of semimodular lattices with a spanning M_3, Discrete Math. 53, 193–205.

Quillen, D. [1978] Homotopy properties of the poset of non-trivial p-subgroups of a group, Advances in Math. 28, 101–128.

Race, D. M. [1986] Consistency in lattices, Ph.D. thesis, North Texas State Univ., Denton, Tex.

Ramalho, M. [1994] On upper continuous and semimodular lattices, Algebra Universalis 32, 330–340.

Ratanaprasert, C. and B. Davey [1987] Semimodular lattices with isomorphic graphs, Order 4, 1–13.

Recski, A. [1989] *Matroid Theory and Its Applications in Electric Network Theory and in Statics*, Akadémiai Kiadó, Budapest, and Springer-Verlag, Berlin.

Reeg, S. and W. Weiß [1991] Properties of finite lattices, Diplomarbeit, Darmstadt.

Regonati, F. [1996] Upper semimodularity of finite subgroup lattices, Europ. J. Combinatorics 17, 409–420.

Reuter, K. [1985] Counting formulas for glued lattices, Order 1, 265–276.

Reuter, K. [1987] Matchings for linearly indecomposable modular lattices, Discrete Math. 63, 245–247.

Reuter, K. [1989] The Kurosh–Ore exchange property, Acta Math. Hungar. 53, 119–127.

Richter, G. [1982] The Kuroš–Ore theorem, finite and infinite decompositions, Studia Sci. Math. Hungar. 17, 243–250.

Richter, G. [1983] Applications of some lattice theoretic results on group theory, in: *Proc. Klagenfurt Conf., June 10–13, 1982*, Hölder-Pichler-Tempsky, Wien, pp. 305–317.

Richter, G. [1991] Strongness in J-lattices, Studia Sci. Math. Hungar. 26, 67–80.

Richter, G. and M. Stern [1984] Strongness in (semimodular) lattices of finite length, Wiss. Z. Univ. Halle 39, 73–77.

Rival, I. [1974] Lattices with doubly irreducible elements. Canad. Math. Bull. 17, 91–95.

Rival, I. [1976a] A note on linear extensions of irreducible elements in a finite lattice, Algebra Universalis 6, 99–103.

Rival, I. [1976b] Combinatorial inequalities for semimodular lattices of breadth two, Algebra Universalis 6, 303–311.

Rival, I. [1980] The problem of fixed points in ordered sets, Ann. Discrete Math. 8, 283–292.

Rival, I. (ed.) [1982a] *Ordered Sets. Proc. NATO Adv. Study Inst. Conf. Held at Banff, Canada, Aug. 28–Sep. 12, 1981.*

Rival, I. [1982b] A bibliography (on ordered sets), in: Rival [1982a], pp. 864–966.

Rival, I. [1985a] The diagram, in: *Graphs and Order* (ed. I. Rival), NATO ASI Ser. C: Mathematical and Physical Sciences 147, Reidel, pp. 103–133.

Rival, I. [1985b] The diagram, "Unsolved Problems," Order 2, 101–104.

Rival, I. [1989] Graphical data structures for ordered sets, in: *Algorithms and Order* (ed. I. Rival), NATO ASI Ser. C: Mathematical and Physical Sciences 255, Reidel, pp. 3–31.

Rival, I. [1990] Dilworth's covering theorem for modular lattices, in: Bogart et al. [1990], pp. 261–264.

Rival, I. and M. Stanford [1992] Algebraic aspects of partition lattices, in: White [1992], pp. 106–122.

Robinson, D. J. S. [1982] *A Course in the Theory of Groups*, Springer-Verlag, New York.

Rose, H. [1984] *Nonmodular Lattice Varieties*, Memoirs Amer. Math. Soc. 47, No. 292.

Rose, J. S. [1978] *A Course in Group Theory*, Cambridge Univ. Press, Cambridge.

Rota, G.-C. [1964] On the foundations of combinatorial theory. I. Theory of Möbius functions, Z. Wahrsch. Verw. Gebiete 2, 340–368.

Rota, G.-C. and J. S. Oliveira (eds.) [1987] *Selected Papers on Algebra and Topology by Garrett Birkhoff*, Birkhäuser, Boston.

Rudeanu, S. [1964] Logical dependence of certain chain conditions in lattice theory, Acta Sci. Math. (Szeged) 25, 209–218.

Saarimäki, M. [1982] Counterexamples to the algebraic closed graph theorem, J. London Math. Soc. 26, 421–424.

Saarimäki, M. [1992] Disjointness of lattice elements, Math. Nachr. 159, 169–774.

Saarimäki, M. [1998] Disjointness and complementedness in upper continuous lattices, Report 78, University of Jyväskylä.

Saarimäki, M. and P. Sorjonen [1991] On Banaschewski functions in lattices, Algebra Universalis 28, 103–118.

Sabidussi, G. [1976] Weak separation lattices of graphs, Canad. J. Math. 28, 691–724.

Sachs, D. [1961] Partition and modulated lattices, Pacific J. Math. 11, 325–345.

Sagan, B. E. [1995] A generalization of Rota's NBC theorem, Adv. in Math. 111, 195–207.

Salii, V. N. [1980] Some conditions for distributivity of a lattice with unique complements (Russian), Izv. Vysš. Uceb. Zaved. Mat. 5, 47–49.

Salii, V. N. [1988] *Lattices with Unique Complements*, Translations of Mathematical Monographs 69, Amer. Math. Soc., Providence, R.I. (translation of the 1984 Russian edition, Nauka, Moscow).

Sankappanavar, H. P. [1974] A study of congruence lattices of pseudocomplemented semilattices, Ph.D. Thesis, Univ. of Waterloo.

Sankappanavar, H. P. [1979] Congruence lattices of pseudocomplemented semilattices, Algebra Universalis 9, 304–316.

Sankappanavar, H. P. [1982] Congruence-semimodular and congruence distributive pseudocomplemented semilattices, Algebra Universalis 14, 68–81.

Sankappanavar, H. P. [1985] Congruence-distributivity and join-irreducible congruences on a semilattice, Math. Japonica 30, 495–502.

Sasaki, U. [1952–3] Semi-modularity in relatively atomic upper continuous lattices, J. Sci. Hiroshima Univ. Ser. A 16, 409–416.

Sasaki, U. [1954] Orthocomplemented lattices satisfying the exchange axiom, J. Sci. Hiroshima Univ. A 17, 293–302.

Savel'zon, O. I. [1978] 0-modular lattices (Russian), Uporyad. množestva i reshotky 5, 97–107.

Scheiblich, H. E. [1970] Semimodularity and bisimple ω-semigroups, Proc. Edinburgh Math. Soc. 17, 79–81.

Schmidt, E. T. [1965] Remark on a paper of M. F. Janowitz, Acta Math. Acad. Sci. Hungar. 16, 435.

Schmidt, E. T. [1969] *Kongruenzrelationen algebraischer Strukturen*, VEB Deutscher Verlag der Wissenschaften, Berlin.

Schmidt, E. T. [1990] Pasting and semimodular lattices, Algebra Universalis 27, 595–596.

Schmidt, R. [1994] *Subgroup Lattices of Groups*, de Gruyter, Berlin.

Schmidt, S. E. [1987] Projektive Räume mit geordneter Punktmenge, Mitt. Math. Sem. Gießen 182, 1–77.

Schreiner, E. A. [1966] Modular pairs in orthomodular lattices, Pacific J. Math. 19, 519–528.

Schreiner, E. A. [1969] A note on O-symmetric lattices, Caribbean J. Sci. and Math. 1, 40–50.

Shevrin, L. N. and A. J. Ovsyannikov [1983] Semigroups and their subsemigroup lattices, Semigroup Forum 27, 1–154.

Shevrin, L. N. and A. J. Ovsyannikov [1990] *Semigroups and their Subsemigroup Lattices, Part I: Semigroups with Certain Types of Subsemigroup Lattices and Lattice Characterizations of Semigroup Classes* (Russian), Ural State University Publishers, Sverdlovsk.

Shevrin, L. N. and A. J. Ovsyannikov [1991] *Semigroups and their Subsemigroup Lattices, Part II: Lattice Isomorphisms* (Russian), Ural State University Publishers, Sverdlovsk.

Shevrin, L. N. and A. J. Ovsyannikov [1996] *Semigroups and their Subsemigroup Lattices*. Translated and revised from the 1990/1991 Russian originals by the authors. Kluwer, Dordrecht.

Sikorski, R. [1964] *Boolean Algebras* (2nd edition), Academic Press, New York.

Skornjakov, L. A. [1961] *Complemented Modular Lattices and Regular Rings* (Russian), Gosudarst. Izdat. Fiz.-Mat. Lit., Moscow.

Skorsky, M. [1992] Endliche Verbände. Diagramme und Eigenschaften, Ph.D. Thesis, TH Darmstadt, FB Mathematik.

Soltan, V. P. [1973] The Jordan form of matrices and its connection to lattice theory (Russian), Mat. Issled. (Kishinev), 1(27), 152–170.

Soltan, V. P. [1975] Jordan elements of lattices and subordinate sums (Russian), Mat. Issled. (Kishinev) 10, 230–237.

Stanley, R. [1971a] Modular elements in geometric lattices, Algebra Universalis 1, 214–217.

Stanley, R. [1971b] Supersolvable semimodular lattices, in: *Möbius algebras (Proc. Conf., Univ. of Waterloo, Waterloo, Ont., 1971)*, Univ. of Waterloo, Waterloo, Ont., pp. 80–142.

Stanley, R. [1972] Supersolvable lattices, Algebra Universalis 2, 197–217.

Stanley, R. [1974] Finite lattices and Jordan–Hölder sets, Algebra Universalis 4, 361–371.

Stanley, R. [1983] *Combinatorics and Commutative Algebra*, Progress in Mathematics 41, Birkhäuser, Basel.

Stanley, R. [1986] *Enumerative Combinatorics I*, Wadsworth & Brooks/Cole, Monterey, Calif.

Stern, M. [1978] Generalized matroid lattices, in: *Algebraic Methods in Graph Theory*, Coll. Math. Soc. J. Bolyai 25, pp. 727–748.

Stern, M. [1982] Semimodularity in lattices of finite length, Discrete Math. 41, 287–293.

Stern, M. [1990a] Strongness in semimodular lattices, Discrete Math. 82, 79–88.

Stern, M. [1990b] The impact of Dilworth's work on the Kurosh–Ore theorem, in: Bogart et al. [1990], pp. 203–204.

Stern, M. [1991a] *Semimodular Lattices*, B. G. Teubner, Stuttgart.

Stern, M. [1991b] Complements in certain algebraic lattices, Archiv d. Math. 56, 197–202.

Stern, M. [1991c] Dually atomistic lattices, Discrete Math. 93, 97–100.

Stern, M. [1992a] On complements in lattices with covering properties, Algebra Universalis 29, 33–40.

Stern, M. [1992b] On meet-distributive lattices, Studia Sci. Math. Hungar. 27, 279–286.

Stern, M. [1996a] On the covering graph of balanced lattices, Discrete Math. 156, 311–316.

Stern, M. [1996b] On centrally symmetric graphs, Math. Bohemica 121, 25–28.

Stern, M. [1996c] A converse to the Kurosh–Ore theorem, Acta Math. Hungar. 70, 177–184.

Stone, M. H. [1937] Topological representation of distributive lattices and Brouwerian logics, Časopis. Pešt. Mat. 67, 1–25.

Stückrad, J. and W. Vogel [1986] *Buchsbaum Rings and Applications*, Springer-Verlag, Berlin.

Suzuki, M. [1956] *Structure of a Group and the Structure of Its Lattice of Subgroups*, Ergebnisse, Vol. 10, Springer-Verlag, Berlin.

Szász, G. [1951–2] On the structure of semi-modular lattices of infinite length, Acta Sci. Math. (Szeged) 14, 239–245.

Szász, G. [1953] Generalized complemented and quasicomplemented lattices, Publ. Math. (Debrecen) 3, 9–16.

Szász, G. [1955] Generalization of a theorem of Birkhoff concerning maximal chains of a certain type of lattices, Acta Sci. Math. (Szeged) 16, 89–91.

Szász, G. [1957–8] Semi-complements and complements in semi-modular lattices, Publ. Math. (Debrecen) 5, 217–221.

Szász, G. [1958] On complemented lattices, Acta Sci. Math. 19, 77–81.

Szász, G. [1963] *Introduction to Lattice Theory*, Akadémiai Kiadó, Budapest, and Academic Press, New York.

Szász, G. [1978] On the De Morgan formulae and the antitony of complements in lattices, Czech. Math. J. 28, 400–406.

Tamari, D. [1951] Monoides préordonnés et chaînes de Malcev, Thése, Université de Paris.

Tamaschke, O. [1960] Submodulare Verbände, Math. Z. 74, 186–190.

Tamaschke, O. [1961] Die Kongruenzrelationen im Verband der zugänglichen Subnormalteiler, Math. Z. 75, 115–126.

Tamaschke, O. [1962] Verbandstheoretische Methoden in der Theorie der subnormalen Untergruppen, Archiv d. Math. 13, 313–330.

Tan, T. [1978] On the lattice of sublattices of a modular lattice, Nanta Math. 11, 17–21.

Tarski, A. [1946] A remark on functionally free algebras, Ann. of Math. 47, 163–165.

Tarski, A. [1955] A lattice-theoretical fixpoint theorem and its applications, Pacific J. Math. 5, 285–309.

Taylor, W. [1973] Characterizing Mal'cev conditions, Algebra Universalis 3, 351–397.

Teichmüller, O. [1936] *p*-Algebren, Deutsche Math. 1, 362–388.

Teo, K. L. [1988] Diagrammatic characterizations of semimodular lattices of finite length, Southeast Asian Bull. Math. 12, 135–140.

Tits, J. [1974] Buildings of spherical type and finite BN-pairs, *Lecture Notes in Mathematics 386*, Springer-Verlag, Berlin.

Tomková, M. [1982] On multilattices with isomorphic graphs, Math. Slovaca 32, 63–74.

Topping, D. M. [1967] Asymptoticity and semimodularity in projection lattices, Pacific J. Math. 20, 317–325.

Trueman, D. C. [1983] The lattice of congruences on direct products of cyclic semigroups and certain other semigroups, Proc. Roy. Soc. Edinburgh 95, 203–214.

Varlet, J. C. [1963] Contribution à l'étude des treillis pseudo-complémentés et des treillis de Stone, Mém. Soc. Roy. Sci. Liège 8, 5–71.

Varlet, J. C. [1965] Congruence dans les demi-lattis, Bull. Soc. Roy. Liège 34, 231–240.

Varlet, J. C. [1968] A generalization of the notion of pseudo-complementedness, Bull.
    Soc. Roy. Sci. Liège 36, 149–158.
Varlet, J. C. [1974–5] *Structures Algébriques Ordonnées*, Univ. de Liège.
Veblen, O. and J. W. Young [1916] *Projective Geometry*, Vol. 1, Ginn, New York.
Vernikov, B. M. [1992] Semicomplements in lattices of varieties, Algebra Universalis 29,
    227–231.
Vilhelm, V. [1955] The selfdual kernel of Birkhoff's condition in lattices with finite chains
    (Russian), Czech. Mat. J. 5, 439–450.
Vincze, A. and M. Wachs [1985] A shellable poset that is not lexicographically shellable,
    Combinatorica 5, 257–260.
von Neumann, J. [1936–7] *Lectures on Continuous Geometries*, Inst. for Advanced Study,
    Princeton, N.J.
von Neumann, J. [1960] *Continuous Geometry* (ed. I. Halperin), Princeton Mathematical
    Ser. 25, Princeton Univ. Press, Princeton, N.J.
Walendziak, A. [1990a] Meet-decompositions in complete lattices, Periodica Math.
    Hungar. 21, 219–222.
Walendziak, A. [1990b] The Kurosh–Ore property, replaceable irredundant
    decompositions, Demonstratio Math. 23, 549–556.
Walendziak, A. [1991] Lattices with doubly replaceable decompositions, Ann. Soc. Math.
    Polon. Ser. I: Comment. Math. 30, 465–472.
Walendziak, A. [1994b] On consistent lattices, Acta Sci. Math. (Szeged) 59, 49–52.
Walendziak, A. [1994c] Strongness in lattices, Demonstratio Math. 27, 569–572.
Wanner, T. and G. M. Ziegler [1991] Supersolvable and modularly complemented matroid
    extensions, European J. Combin. 12, 341–360.
Ward, M. [1939] A characterization of Dedekind structures, Bull. Amer. Math. Soc. 45,
    448–451.
Weisner, L. [1935] Abstract theory of inversion of finite series, Trans. Amer. Math. Soc.
    38, 474–484.
Welsh, D. [1976] *Matroid Theory*, Academic Press, London.
White, N. L. (ed.) [1986] *Theory of Matroids*, Encyclopedia of Mathematics and Its
    Applications, Vol. 26, Cambridge Univ. Press, Cambridge.
White, N. L. (ed.) [1987] *Combinatorial Geometries*, Encyclopedia of Mathematics and
    Its Applications, Vol. 29, Cambridge Univ. Press, Cambridge.
White, N. L. (ed.) [1992] *Matroid Applications*, Encyclopedia of Mathematics and Its
    Applications, Vol. 40, Cambridge Univ. Press, Cambridge.
Whitman, P. M. [1941] Free lattices, Ann. of Math. 42, 325–329.
Whitman, P. M. [1946] Lattices, equivalence relations, and subgroups, Bull. Amer. Math.
    Soc. 52, 507–522.
Whitney, H. [1935] On the abstract properties of linear dependence, Amer. J. Math. 57,
    509–533.
Wielandt, H. [1939] Eine Verallgemeinerung der invarianten Untergruppen, Math. Z. 45,
    209–244.
Wielandt, H. [1994] *Mathematical Works, Vol. 1: Group Theory* (ed. B. Huppert and
    H. Schneider), de Gruyter, Berlin.
Wilcox, L. R. [1938] Modularity in the theory of lattices, Bull. Amer. Math. Soc. 44, 50.
Wilcox, L. R. [1939] Modularity in the theory of lattices, Ann. of Math. 40, 490–505.
Wilcox, L. R. [1942] A note on complementation in lattices, Bull. Amer. Math. Soc. 48,
    453–458.
Wilcox, L. R. [1944] Modularity in Birkhoff lattices, Bull. Amer. Math. Soc. 50, 135–138.
Wilcox, L. R. [1955] Modular extensions of semi-modular lattices (Abstract), Bull. Amer.
    Math. Soc. 61, 542.
Wild, M. [1992] Cover-preserving order embeddings into Boolean lattices, Order 9,
    209–232.

Wild, M. [1993] Cover preserving embedding of modular lattices into partition lattices, Discrete Math. 11, 207–244.

Wille, R. [1966] Halbkomplementäre Verbände, Math. Z. 94, 1–31.

Wille, R. [1967] Verbandstheoretische Charakterisierung n-stufiger Geometrien, Archiv Math. 18, 465–468.

Wille, R. [1974] Jeder endlich erzeugte modulare Verband endlicher Weite ist endlich, Mat. Čas. 24, 77–80.

Wille, R. [1976] On the width of sets of irreducibles in finite modular lattices, Algebra Universalis 6, 257–258.

Wille, R. [1982] Restructuring lattice theory: an approach based on hierarchies of concepts, in: Rival [1982a], pp. 445–470.

Wille, R. [1983] Subdirect decomposition of concept lattices, Algebra Universalis 17, 275–287.

Wille, R. [1985] Complete tolerance relations of concept analysis, in: *Contributions to General Algebra 3, Proc. Vienna Conf. June 21–24, 1984*, pp. 397–415.

Woodall, D. R. [1976] The inequality $b \geq v$, in: *Proc. Fifth British Combinatorial Conf. (Univ. of Aberdeen, 1975)*, Congressus Numerantium 15, Utilitas Math., Winnipeg, Manitoba, pp. 661–664.

Zappa, G. [1965] *Fondamenti di Teoria dei Gruppi*, Vol. I, Edizioni Cremonese, Roma.

Zaslavsky, T. [1987] The Möbius function and the characteristic polynomial, in: White [1987], pp. 114–138.

Zassenhaus, H. [1958] *The Theory of Groups*, (2nd edition), New York.

Zelinka, B. [1970] Centrally symmetric Hasse diagrams of finite modular lattices, Czech. Math. J. 20, 81–83.

Žitomirskii, G. I. [1986] Some remarks on properties of lattices which are generated by dual atoms (Russian), Uporyad. Množestva i Reshotki 9, 16–18.

Table of Notation

For most of the symbols that follow, the page number where they first appear is given in parentheses.

Sets and Logic

$\varnothing$	empty set
$\in$	element inclusion
$\subseteq$	set inclusion
$\cup, \bigcup$	set-theoretic union
$\cap, \bigcap$	set-theoretic intersection
$\{x : (P)\}$	set of all elements having property (P)
(a, b)	ordered pair
(a, b, c)	ordered triple
$|A|$	cardinality of the set A
$A - B$	set-theoretic difference
$A - a$	shorthand for $A - \{a\}$
$\times$	Cartesian product
A^2	$A \times A$
$\mathbb{P}$	set of positive integers
$\mathbb{Z}$	set of integers
$\mathbb{Q}$	set of rational numbers
$\mathbb{R}$	set of real numbers
$\Rightarrow$	implication
$\Leftrightarrow$	logical equivalence
$X \rightarrow Y$	used in figures to indicate $X \subset Y$ (proper containment) where X and Y are classes
$X \leftrightarrow Y$ or $X \equiv Y$	used in figures to indicate that the classes X and Y coincide

356

Lattices and Posets

L	lattice
L^*	lattice dual to L
$\leq$	partial ordering relation
$\vee, \bigvee, \sqcup$	join
$\wedge, \bigwedge, \sqcap$	meet
$x \rceil y$	pseudointersection of x and y (260)
$x \prec y$	x is a lower cover of y
$x \succ y$	x is an upper cover of y
$y \downarrow x$	$y \succ x$ or $x = y$ (236)
Θ	congruence relation, tolerance relation
L/Θ	factor lattice
$[a]_\Theta$	block of a tolerance relation (181)
$\Sigma(L)$	skeleton of a finite lattice L (181)
$K + L$	disjoint union of the lattices K and L (27)
$K \times L$	direct product of the lattices K and L
$K * L$	concatenation of disjoint lattices K and L (174)
$L_1 \uparrow L_2 \uparrow \dots \uparrow L_k$	parallel union of pairwise disjoint lattices $L_1, L_2, \dots, L_k$ (174)
$D_k(L)$	the kth Dilworth truncation of L (224)

Special Elements, Pairs, and Triples

0	least element
1	greatest element
x^+	join of all elements covering x (19)
x_+	meet of all elements covered by x (19)
$a \, M \, b$	(a, b) is a modular pair (11)
$a \, \bar{M} \, b$	(a, b) is not a modular pair
$a \, M^* \, b$	(a, b) is a dual modular pair (11)
$a \, \bar{M}^* \, b$	(a, b) is not a dual modular pair
$(a, b, c)D$	(a, b, c) is a distributive triple (76)
$(a, b, c)D^*$	(a, b, c) is a dual-distributive triple (76)
$a \perp b$	a is (semi)orthogonal to b (12, 144)
$a \sim b$	a and b are perspective elements (75)
$a \sim_s b$	a and b are strongly perspective elements (109)
$a < \vert b$	conjunction of $a \wedge b = 0$ and $b \prec a \vee b$ (98)
$a \parallel b$	a and b are parallel elements (conjunction of $a < \vert b$ and $b < \vert a$) (98)
$a \, C \, b$	a commutes with b (106)
$C_r(P)$	$\{(x, y) \in P \times P : x \prec y\}$ (165)

Special Lattices, Semilattices, Posets, and Lattice Properties

2^n	Boolean lattice consisting of 2^n elements
M_3	diamond (9)
N_5	pentagon (9)
S_7	centered hexagon (18, 19)
S_7^*	dual of S_7
T_n	Tamari lattice (27)
$\Pi(X)$	partition lattice on the set X (42)
Π_n	partition lattice on a set of n elements (155)
AC lattice	atomistic lattice with covering property (41)
DAC lattice	AC lattice whose dual is also an AC lattice (125)
$L_c(H)$	lattice of closed subspaces of a Hilbert space H (13)
$L = \Lambda - \underline{S}$	Wilcox lattice (92)
$L(\Delta)$	face lattice of a finite simplicial or polyhedral complex Δ (61)
C_n	chain of n elements (43)
$\text{Int}(C_n)$	lattice of subintervals of C_n (43)
$\text{Sub}(L)$	lattice of all sublattices of a lattice L (49)
$\text{Csub}(L)$	lattice of all convex sublattices of a lattice L (49)
$\text{Co}(P)$	lattice of convex subsets of the poset P (284)
A_p	lattice of partial alphabets (245)
CS lattice	centrally symmetric lattice (205)
S lattice	symmetric lattice (205)
SS lattice	supersolvable lattice (161)
K_{2n}	special CS lattice (206)
PCS	pseudocomplemented (meet)semilattice (315)
E_2	two-element meet-semilattice
Sl_4	$E_2 \times E_2$ (313)
(Js)	join-symmetry (256)
(QS)	quasi upper semimodular (305)
(QS*)	quasi lower semimodular (305)
(W)	Whitman's condition (18)

Join-irreducibles and Meet-irreducibles

$J(L)$	set of all join-irreducible elements ($\neq 0$) of a lattice L (of finite length) (16)
$M(L)$	set of all meet-irreducible elements ($\neq 1$) of a lattice L (of finite length) (17)
$J(1, L) = J(L) \cup \{0\}$	set of all join-irreducible elements of a lattice L (of finite length) (173)

$M(1, L) = M(L) \cup \{1\}$	set of all meet-irreducible elements of a lattice L (of finite length) (173)
$J_c(1, L)$	set of all consistent join-irreducible elements of a lattice L (of finite length) (173)
$D(L)$	set of all doubly irreducible elements of a lattice L (of finite length) (210)
$J(x)$	shorthand for $J([0, x]) = \{j \in J(L) : j \le x\}$ (269)
$J_k(L)$	set of elements having exactly k lower covers (218)
$M_k(L)$	set of elements having exactly k upper covers (218)
$J(k, L)$	set of k-covering elements (set of elements covering at most k elements) (219)
$M(k, L)$	set of k-covered elements (set of elements covered by at most k elements) (219)
j'	uniquely determined lower cover of a join-irreducible $j (\ne 0)$ (16)
m^*	uniquely determined upper cover of a meet-irreducible $m (\ne 1)$ (17)
x'	$\bigvee(j' : j \in J(L), j \le x)$ (derivation of the element x) (179, 182)
x^*	$\bigwedge(m^* : m \in M(L), x \le m)$ (182)
$j \nearrow m$	arrow relation ($j \in J(L), m \in M(L)$) (180)
$j \swarrow m$	arrow relation ($j \in J(L), m \in M(L)$) (180)

Complementation

$\bar{a}$	complement of a
$a^\perp$	orthocomplement of a (11)
$g(a)$	meet-pseudocomplement of a (24)
$a * b$	pseudocomplement of a relative to b (25)
(OC)	orthocomplemented (11)
(RC)	relatively complemented (29)
(UC)	uniquely complemented (29)
(RUC)	relatively uniquely complemented (29)
(SC)	semicomplemented (29)
(SeC)	section complemented (30)
(SeSC)	section semicomplemented (30)

Subsets and Families of Subsets of Posets and Lattices

$[x, y]$	(closed) interval
(x, y)	open interval

$(x]$	principal ideal generated by x
$[x)$	principal dual ideal generated by x
$\mathbf{I}(L)$	lattice of ideals of a lattice L
$\mathbf{D}(L)$	lattice of dual ideals of a lattice L
$\text{ord}(P)$	set of all order ideals of a poset P (16)
$T_P(\Im(P))$	meet-subsemilattice of the trimmed ideals of a poset P (228)
$\bar{L} = L - \{0, 1\}$	proper part of a lattice L of finite length (154)
L^x	sublattice of L generated by the elements covering x (154)
L_x	sublattice of L generated by the elements covered by x (215)
$F(L)$	set of finite elements of a lattice L (95)
$\text{Mod}(L)$	set of (right) modular elements of a lattice L (74)
M-chain	modular chain (161)
L_k	$\{x : r(x) \le k\}$ (subset of lower elements of a finite semimodular lattice L) (216)
U_k	$\{x : r(x) \ge n - k\}$ (subset of upper elements of a finite semimodular lattice L) (216)
R_k	$\{x : \text{reach}(x) \le k\}$ (216)
S_k	$\{x : \text{coreach}(x) \le k\}$ (216)

Functions

$l(P)$	length of a poset P (59)
$r(x)$	natural rank (briefly: rank or height) of x (60)
$d(x)$	dimension function (60)
ζ	zeta function (149)
μ	Möbius function (150)
χ	Euler characteristic (152)
$\text{reach}(x)$	reach of the element x (216)
$\text{coreach}(x)$	coreach of the element x (216)
$\text{g-coreach}(x)$	geometric coreach of the element x (217)
$\tau(x)$	join-order function ($=$ join-rank function) (270)
$\nu(x) = \tau(x) - r(x)$	join-excess function (270)
$\lambda(\alpha)$	length of a word α (244)
$\text{breadth}(L)$	breadth of a lattice L (306)
$\text{dev}(L)$	deviance of a lattice L (231)
$w(L)$	width of a lattice L (308)

Chain Conditions

ACC	ascending chain condition
DCC	descending chain condition
(JD)	Jordan-Dedekind chain condition (60)
(WJD)	weak form of the Jordan-Dedekind chain condition (62)

Distributivity Properties and Modularity Properties

(D)	distributivity (8)
(M)	modularity (10)
(SD$\wedge$)	meet-semidistributivity (18)
(SD$\vee$)	join-semidistributivity (18)
(SD)	semidistributivity (18)
(ULD), (j-d)	upper local distributivity (join-distributivity) (19)
(LLD), (m-d)	lower local distributivity (meet-distributivity) (20)
(ULM)	upper local modularity (44)
(LLM)	lower local modularity (44)
(USP)	upper splitting property (287)
(JID)	join-infinite distributivity identity (25)

Semimodularity and Related Properties

(Sm)	semimodular implication, (upper) semimodularity (2)
(Sm*)	lower semimodularity (3)
(Bi)	Birkhoff's condition (3)
(Bi*)	dual of Birkhoff's condition (38)
(Ms)	M-symmetry (3, 66)
(Ms*)	dual of (Ms) (81)
(Mac)	Mac Lane's condition (112)
(Dil)	Dilworth's condition (122)
(Bi)$_i$	Birkhoff's condition for the lattice of ideals (120)
(Sm)$_i$	(upper) semimodularity for the lattice of ideals (120)
(Ms)$_i$	M-symmetry for the lattice of ideals (120)
(Bi)$_d$	Birkhoff's condition for the lattice of dual ideals (120)
(Sm)$_d$	(upper) semimodularity for the lattice of dual ideals (120)
(Ms)$_d$	M-symmetry for the lattice of dual ideals (120)
(B), (F)	conditions involving maximal chains (117)
(L), (N)	conditions involving maximal chains (125, 126)
(R), (S)	conditions involving maximal chains (125, 126)

Covering Properties and Exchange Properties

(C)	(atomic) covering property (39)
(C*)	dual covering property
(EP)	Steinitz-Mac Lane exchange property (40)
(AEP)	antiexchange property (103)
(GEP)	geometric exchange property (130)
(SEP)	strong exchange property (255)
(E_3)	Menger's exchange axiom (110)

Kurosh-Ore Properties, Crawley's Condition

$\vee$-KORP	Kurosh-Ore replacement property for join-decompositions (34)
$\wedge$-KORP	Kurosh-Ore replacement property for meet-decompositions (34)
$\vee$-KOP	Kurosh-Ore property for join-decompositions (34)
$\wedge$-KOP	Kurosh-Ore property for meet-decompositions (34)
(Cr)	Crawley's condition (174)
(Cr*)	dual of (Cr) (174)

0-Conditions and Disjointness Properties

(Ms_0)	$\perp$-symmetry (M-symmetry at 0) (81)
(Mac_0)	local Mac Lane condition (135)
(D_0)	0-distributivity (133)
(M_0)	0-modularity (133)
(D_0M_0)	logical conjunction of (M_0) and (D_0) (134)
(AD)	atomic disjointness (135)
(GD)	general disjointness (135)
(WD)	weak disjointness (135)

Topology

$\Delta(P)$	order complex of a poset P (152)
$\lvert\Delta(P)\rvert$	the geometric realization of a poset P (157)
$H_n(P, K)$	homology groups of a poset P with coefficients in a field K (158)
CM poset	Cohen-Macaulay poset (165)
EL labeling	edgewise lexicographical labeling (165)
SL-shellable	strongly lexicographically shellable (169)

Covering Graph

$G(P)$	covering graph of a poset P (67)
$K_{2,3}$	covering graph of M_3 (191, 192)
$w(e)$	weight of an edge e (67)
$\delta(a, b)$	distance from a to b (68, 194)
diam	diameter (194)
$V(m, n)$	type of a cell (202)
S graph	symmetric graph (205)
CS graph	centrally symmetric graph (205)

Groups and Lattices of Subgroups

D_4	dihedral group (44)
A_4	alternating group on 4 elements (173)
$\Phi(G)$	Frattini subgroup of a group G (303)
$H \triangleleft G$	H is a normal subgroup of G (300)
$H \triangleleft \triangleleft G$	H is a subnormal subgroup of G (301)
$L(G)$	lattice of all subgroups of a group G (61, 302)
$N(G)$	lattice of normal subgroups of a group G (302)
$W(G)$	lattice of subnormal subgroups of a group G (302)

Universal Algebra

K	class of algebras
$H(K)$	class of all homomorphic images of members of K (13)
$S(K)$	class of all sublattices of members of K (13)
$P(K)$	class of all direct products of members of K (13)
$P_s(K)$	class of all subdirect products of members of K (21)
$P_u(K)$	class of all ultraproducts of members of K (21)
$V(K)$	variety generated by a class K (14)
L	variety of all lattices
D	variety of all distributive lattices
M	variety of all modular lattices
Sets	variety of all sets
Semilattices	variety of all semilattices
Sem	class of all semimodular lattices
Con(A)	congruence lattice of an algebra A
CSM	congruence semimodularity (313)
CWSM	congruence weak semimodularity (325)

$\mathbf{F}_V(X)$ relatively free algebra on V generated by X (325)
CEP congruence extension property (326)
OBP one block property (326)
typ{V} type-set of the variety V (326)

Miscellaneous

$\cong$ isomorphism
$I(M \mid J)$ incidence matrix of J versus M (212)
W_k number of elements of rank k, Whitney numbers (of the second kind) (215)
M_k number of modular elements of rank k, modular Whitney numbers (217)
cl(A) closure of A
(X, cl) closure structure on X
cvx(A) convex hull of A (282)
$F(P)$ Faigle geometry of a poset P (234)
ex(A) set of extreme points of A (283)
(p, A) minimal pair (255)
a/b quotient (286)
$a/b \nearrow c/d$ c/d is perspective upward to a/b (286, 287)
$a/b \searrow c/d$ c/d is perspective downward to a/b (287)
$(C, \nearrow)$ set of prime quotients partially ordered by upward perspectivity (287)
$g(s \prec t)$ generator of the covering pair $s \prec t$ (287)
$a/b \sim c/d$ a/b is perspective to c/d (305)
$a/b \approx c/d$ a/b is projective to c/d (305)

Index